Adaptive Power Quality for Power Management Units using Smart Technologies

This book covers issues associated with smart systems due to the presence of onboard nonlinear components. It discusses the advanced architecture of smart systems for power management units. It explores issues of power management and identifies hazardous signals in the power management units of smart devices. It

- Presents adaptive artificial intelligence and machine learning-based control strategies.
- Discusses advanced simulations and data synthesis for various power management issues.
- Showcases solutions to the uncertainty and reliability issues in power management units.
- Identifies new power quality challenges in smart devices.
- Explains hybrid active power filters, shunt hybrid active power filters, and the industrial internet of things in power quality management.

This book comprehensively discusses advancements of traditional electrical grids, the benefits of smart grids to customers and stakeholders, properties of smart grids, smart grid architecture, smart grid communication, and smart grid security. It further covers the architecture of advance power management units (PMU) of smart devices, and the identification of harmonic distortions with respect to various sensor-based technology. It will serve as an ideal reference text for senior undergraduate and graduate students, and academic researchers in fields including electrical engineering, electronics, communications engineering, and computer engineering.

Future Generation Information Systems
Series editor- Bharat Bhushan

With the evolution of future generation computing systems, it becomes necessary to occasionally take stock, analyze the development of its core theoretical ideas, and adapt to radical innovations. This series will provide a platform to reflect the theoretical progress, and forge emerging theoretical avenues for the future generation information systems. The theoretical progress in the Information Systems field (IS) and the development of associated next generation theories is the need of the hour. This is because Information Technology (IT) has become increasingly infused, interconnected and intelligent in almost all context.

Convergence of Deep Learning and Artificial Intelligence in Internet of Things
Ajay Rana, Arun Rana, Sachin Dhawan, Sharad Sharma and Ahmed A. Elngar

Next Generation Communication Networks for Industrial Internet of Things Systems
Sundresan Perumal, Mujahid Tabassum, Moolchand Sharma and Saju Mohanan

Adaptive Power Quality for Power Management Units using Smart Technologies

Edited by
Arti Vaish
Pankaj Kumar Goswami
Surbhi Bhatia
Mokhtar Shouran

CRC Press
Taylor & Francis Group
Boca Raton London New York

CRC Press is an imprint of the
Taylor & Francis Group, an **informa** business

First edition published 2024
by CRC Press
6000 Broken Sound Parkway NW, Suite 300, Boca Raton, FL 33487-2742

and by CRC Press
4 Park Square, Milton Park, Abingdon, Oxon, OX14 4RN

CRC Press is an imprint of Taylor & Francis Group, LLC

© 2024 selection and editorial matter Gunjan Soni, Om Prakash Yadav, Gaurav Kumar Badhotiya and Mangey Ram; individual chapters, the contributors

Library of Congress Cataloging-in-Publication Data
Names: Vaish, Arti, editor.
Title: Adaptive power quality for power management units using smart technologies / edited by Arti Vaish, Pankaj Kumar Goswami, Surbhi Bhatia, and Mokhtar Shouran.
Description: First edition. | Boca Raton, FL : CRC Press, [2024] | Series: Future generation information systems | Includes bibliographical references and index.
Identifiers: LCCN 2023013361 (print) | LCCN 2023013362 (ebook) | ISBN 9781032392998 (hbk) | ISBN 9781032566306 (pbk) | ISBN 9781003436461 (ebk)
Subjects: LCSH: Smart power grids. | Electric power system stability. | Electromagnetic interference. | Sensor networks.
Classification: LCC TK3105 .A37 2024 (print) | LCC TK3105 (ebook) | DDC 621.319--dc23/eng/20230803
LC record available at https://lccn.loc.gov/2023013361
LC ebook record available at https://lccn.loc.gov/2023013362

ISBN: 978-1-032-39299-8 (hbk)
ISBN: 978-1-032-56630-6 (pbk)
ISBN: 978-1-003-43646-1 (ebk)

DOI: 10.1201/9781003436461

Typeset in Sabon
by SPi Technologies India Pvt Ltd (Straive)

Contents

Preface

Power quality control is a global challenge for all commercial and non-commercial applications. The deviation in power quality is caused by many factors such as heterogeneous load characteristics, solid-state electronic devices, switching modules, variable drives, and many advanced smart sensor networks. With the advent of new technology, smart sensor technologies are playing a vital role in remote sensing and monitoring over a wide application domain, though machine to machine (M2M) communication is leading the world to the next level of technology. This book provides all the possible solutions to overcoming power quality issues and challenges.

Smart technology/new technology helps data transmission, monitoring, and controlling among devices without human intervention. The extensive utility of sensor technology and M2M networking are the future of industries for the production of commercial and non-commercial applications. This book also discusses smart devices which are the universally accepted technology version of the amalgamation of sensors and wide area networking. The soaring applications of smart sensors are continuously enhancing the numbers and types of devices every year. This leads to the conclusion that in the next two decades 80% of devices will be connected to the internet. However, the active involvement of smart devices leads to another issue regarding the insertion of higher-order harmonics in the main power supply. The non-linear current–voltage (I–V) characteristics of smart electronic devices are the main cause of the generation of harmonic current components. The presence of higher-order harmonic current components reduces the power quality of the main supply unit. This results in several severe issues including false reading, malfunctioning of devices, flickers, radio frequency (RF) interference, and low efficiency. This book analyses the effects of harmonic distortion in the power management unit of smart device architectures. The challenges are analysed, and an adaptive neural fuzzy interface system (ANFIS), supervised by an active power filter, is designed for harmonic current compensation. Hysteresis current compensation is the major contribution of the proposed controller. The scheme advanced deals with harmonic distortion caused by the non-linear characteristics of smart devices. Power management is an upcoming area which will help readers in identifying general power quality problems and their solution. It also focuses new technology that will make our power quality management system a smart one. Topics like smart power grids and their benefits to consumers and stakeholders will become of more interest as they familiarise readers with an infrastructure for the monitoring and analysis of smart grids. They will also show readers that we can combine sustainable resources of wind and solar alongside the main grid. This book also focuses on the improved management of distributive energy resources, including micro-grid operations and store management.

The domestic and commercial applications are analyzed for their power quality issues, which are seen in the majority of power converter units due to the significant increase of non-linear components. These issues have been analyzed and are evident in many of the commercial and non-commercial applications. Even household applications, such as televisions, mobile phones, Air Conditioners, refrigerators, personal computers, and microwave ovens, are responsible agents for a reduction in power quality.

They all are equipped with the necessary AC to DC power converter units, which are complex circuits of hybrid components such as diodes, transistors, resistors, and capacitors. Due to the non-linear characteristic of the active components, the higher-order harmonics of a fundamental current are generated and cause interference to the main frequency component. Nowadays, the new generations of devices comprise smart adaptive features related to their monitoring and controlling. The device smartness is the result of embedded systems as an auxiliary system to the main process. Additionally, machine to machine or machine to network communication has included soft computing technologies in the process.

Therefore, IoT devices comprise the soft computing hardware set and smart switching components. This advancement in technologies is a boon to today's world of utility though at the same time the power quality issues are the biggest drawback, and this needs to be taken into consideration. This book mainly focuses on the general classes of power quality problems with smart devices. The novel contribution of this book is to improve power quality in the power management units of smart devices. The ANFIS machine learning approach deals with the uncertainty of load variation on a real-time basis.

Editors

Arti Vaish is a professor in the Department of Computer Engineering, School of Engineering and Technology at O. P. Jindal University, Raigarh, CG, India. She earned her PhD degree from Netaji Subhash Institute of Technology, Delhi University, New Delhi, India (2012). She has registered 20 patents and published and presented more than a hundred research papers in SCI indexed/Scopus indexed international journals and at international and national conferences. She has authored six books. She has over 18 years of rich experience in academia and research. She is a recipient of the 2021 *Who's Who in the World® Lifetime Achievement Award* and was successfully nominated as a candidate for the 2018–2021 edition of *Who's Who in America*. She received an academic excellence award from WeGrow and the Ministry of Micro, Small and Medium Enterprises (MSME) in 2020 and the best researcher award in 2020 from the Institute of Technical and Scientific Research (ITSR) Foundation. She had guided six PhD students. Dr Arti has been a reviewer for several archived journals at the international level: *Progress in Electromagnetic Research*, *Wireless Personal Communication*, the *IETE Journal of Research*, and *WSEAS Transactions*, to name a few of them.

Pankaj Kumar Goswami earned B.Tech., M.Tech, and PhD degrees in the area of Electronic Engineering and has 15 years of teaching experience in reputed organisations. He has more than 30 international publications in peer-reviewed journals, including *IEEE Explore*, which are Scopus and SCI indexed. His main areas of research are machine learning, the internet of things, wideband antenna designs, cognitive radios, and power quality issues in non-linear electronic devices. He has organised several national and international conferences, symposia, and webinars. Dr Goswami has also registered several patents in the fields of antenna design, the internet of things, and machine-learning-based applications under the Intellectual Property Rights of the Government of India. He is an e-educator in the field of machine learning using Python. He is actively associated in NAAC SSR preparations at university interfaces and has visited several times the

SAARC countries of Nepal, Bhutan, the seven sisters (Arunachal Pradesh, Assam, Manipur, Meghalaya, Mizoram, Nagaland, and Tripura), and the "brother" state Sikkim for various academic promotional activities. He is a professional member of several national and international bodies such as, IEEE, IETE and ISTE. Dr Goswami has been involved in various educational outreach activities through online sessions on artificial intelligence, machine learning, and smart antennas.

Surbhi Bhatia holds a doctorate in Computer Science and Engineering from Banasthali University, which she earned in 2018 in the area of machine learning and social media analytics. She earned project management professional certification from the Project Management Institute, USA. She is currently an assistant professor in the Department of Data Science, School of Science, Engineering and Environment, University of Salford, Manchester, United Kindgom. She has more than ten years of teaching experience in different universities in India and Saudi Arabia. She is also a consultant in the Research Lab, India. She is an editorial board member for the *International Journal of Hybrid Intelligence* and for *SN Applied Sciences*. She has registered eight national and international patents in India, Australia, and the USA. She has published many papers in reputed journals and in high indexing databases (SCI, SCIE, Scopus, Web of Science, ESCI). She currently serves as a guest editor of special issues for the journals *SN Applied Sciences, Computer, Materials and Continua*, and *Internet of Things*. She has delivered talks as a keynote speaker at IEEE conferences and in faculty development programmes. She has conducted workshops for All India Council for Technical Education (AICTE) programmers and chaired technical sessions at many conferences. She has authored two books published by Springer and Wiley. She has also edited seven books with CRC Press, Elsevier, and Springer. She was selected for the prestigious "Best Young National Award" by the IRDP group of journals in India. She has had many projects approved by the Ministry of Education, Saudi Arabia and holds a Deanship of Scientific Research in Saudi Arabia and DST, India. Her areas of interest are Data Mining, Machine Learning, Sentiment Analysis, Natural Language Processing, and Data Analytics.

Mokhtar Shouran earned a BSc degree in Control Engineering from the College of Electronic Technology, Bani Walid, Libya, in 2011. He earned an MSc degree with distinction in Electronics and Information Technology from the University of South Wales, Treforest, UK, in 2017. He was also awarded a certificate of excellence. Shouran is currently pursuing a PhD degree in Load Frequency Control in Power Systems at the Wolfson Centre for Magnetics, Cardiff University, Cardiff, UK. His research interests include Optimisation Algorithms, Fuzzy Logic Control, Sliding Mode Control, Traditional Control, and Power System Stability and Control.

Contributors

Titus Ajewole
Department of Electrical and
Electronic Engineering
Osun State University
Osogbo, Nigeria

Daniel Akinyele
Olabisi Onabanjo University
Ago-Iwoye, Nigeria

Funso Ariyo
Obafemi Awolowo University
Ile-Ife, Nigeria

J. Ebenesar Anna Bagyam
Department of Mathematics
Avinashilingam Institute for Home
Science and Higher Education for
Women
Coimbatore, India

Jyoti Dargan
School of Engineering and
Technology
Sushant University
Gurugram, India

Muralikrishnan Dhanasekaran
Harrison College of Pharmacy
Auburn University
Auburn, Alabama

Suryya Farhat
Department of Applied Sciences
and Humanities
ADGITM
Delhi, India

Kumar Gautam
Gwangju Institute of Science and
Technology
Gwangju, Republic of Korea

Garima Goswami
Electrical Engineering Department
Teerthanker Mahaveer University
Moradabad, India

Pankaj Kumar Goswami
Electronics & Communication
Engineering Department
Teerthanker Mahaveer University
Moradabad, India

Neha Gupta
School of Engineering and
Technology
Sushant University
Gurugram, India

Uttam Kumar Gupta
Electrical Engineering
JECRC University
Jaipur, India

D. Jeslin
SBMCH, Bharath Institute of
   Higher Education & Research
Chennai, India

R. Rajesh Kanna
Department of Energy and Power
   Electronics
Vellore Institute of Technology
Vellore, India

Manoj Kumar
Department of Mathematics and
   Statistics
Gurukula Kangri Vishwavidyalaya
Haridwar, India

Ignatius Okakwu
Olabisi Onabanjo University
Ago-Iwoye, Nigeria

Olakunle Olabode
Department of Electrical and
   Electronic Engineering
Osun State University
Osogbo, Nigeria

Ezekiel Babatunde Omoniyi
Department of Industrial
   Engineering
University of Ibadan
Ibadan, Nigeria

P. Panneerselvam
Faculty of Pharmacy, SBMCH
Bharath Institute of Higher
   Education & Research
Chennai, India

Harish Parthasarathy
Netaji Subhash University of
   Technology
New Delhi, India

J. Paruvathavardhini
Jai Shriram Engineering College
Dharapuram Road,
   Avinashipalayam
Tiruppur, India

Merin Susan Philip
National Institute of Technology,
   Rourkela
Rourkela, India

R. Rengaraj
Department of Electrical and
   Electronics Engineering
Sri Sivasubramaniya Nadar College
   of Engineering
Chennai, India

Dinesh Sethi
Electrical Engineering
JECRC University, Jaipur, India

Ravi Sharma
Budapest University of Technology
   and Economics
Budapest, Hungary

D. Shruthi
PSN College of Engineering &
   Technology
Tirunelveli, India

Poonam Singh
National Institute of Technology
Rourkela, India

R. Raja Singh
Department of Energy and Power
   Electronics
Vellore Institute of Technology
Vellore, India

K. Preethi Sowndharya
Department of Mathematics
Avinashilingam Institute for Home
  Science and Higher Education for
  Women
Coimbatore, India

R. Sudarmani
Avinashilingam University
Coimbatore, India

Arti Vaish
School of Engineering
O.P. Jindal University
Raigarh, India

D. Vivesini
Department of Mathematics
Avinashilingam Institute for Home
  Science and Higher Education for
  Women
Coimbatore, India

Chapter 1

Machine learning and 5G-based industrial IoT device positioning for location-aware power quality management

Ravi Sharma

Budapest University of Technology and Economics, Budapest, Hungary

CONTENTS

1.1 INTRODUCTION

Location-aware power quality management has gained traction in industry and academia in recent years as a result of the use of intelligent terminals, cloud-edge computing (CEC), machine learning (ML), and the Industrial Internet of Things (IIoT) [1]. A satellite-based device positioning system in the outdoor environment can provide people with convenient device

DOI: 10.1201/9781003436461-1

positioning services to support applications such as cargo tracking, robotic navigation, and faulty area diagnosis [2]. There are numerous research findings and limitations of non-GPS (or non-satellite) positioning systems available, such as using passive information (passive sound, magnetic field, or radio frequency) or visible light communication, but none of these techniques is suitable for location-aware power quality management [1–3]. Therefore, the focus of this chapter is on the ML-assisted cellular-based IIoT device positioning methods and infrastructure that take advantage of improved 5G/6G network coverage.

Real-time precise IIoT device positioning is widely required by location aware power quality management, which is advantageous for radio resource management in 5G compared to traditional cellular systems. Figure 1.1 shows 5G-based IIoT device positioning for smart power grid infrastructure. Numerous innovative technologies, such as ultra-dense network (UDN), millimetre wave (mmWave), device-to-device (D2D) communication, and massive multiple-input multiple-output (MIMO), are being implemented in 5G technology to improve not only communication performance but also positioning accuracy [4]. 5G networks are expected to have high accuracy in locating power equipment and network utility, such as high throughput, low latency, and seamless coverage, putting additional strain on cellular communication systems. Smart power grid IIoT device positioning systems are becoming more common, as GPS is only available in major cities and requires a clear sky view to function properly. Smart grid IIoT device positioning systems are used for real-time power grid maintenance, power fault source positioning, and automated power supply regulation during peak hours.

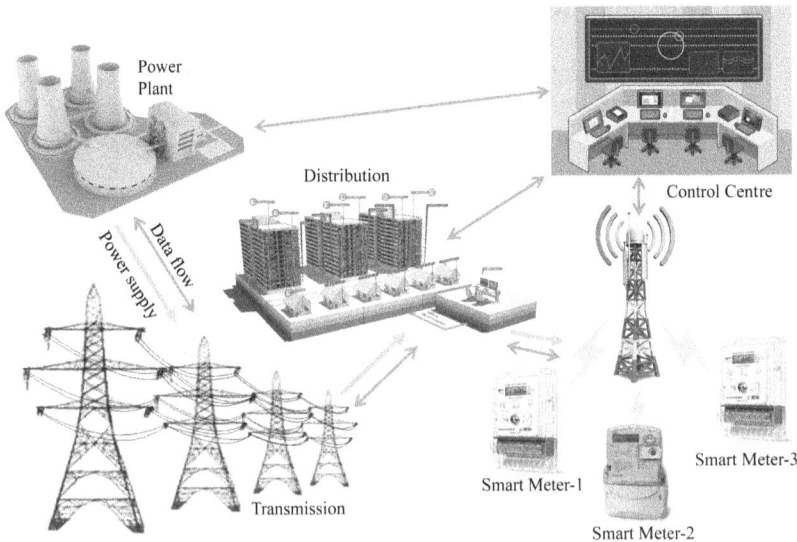

Figure 1.1 An example of a 5G-based IIoT device positioning for smart power grid infrastructure.

The rapid advancement of 5G standardisation and high-bandwidth availability has created a slew of new opportunities for location-aware power quality management and maintenance in eHealth telepresence, augmented reality (AR), and IIoT [5]. Among these, IIoT and AR will benefit the most from 5G network positioning capabilities, because position information can help to optimise and automate the power supply in a variety of vertical sectors, ranging from household appliances to large electric manufacturing machines. Positioning information is extremely beneficial to both sides of the communication, especially in smart grid power quality management: it reduces power distributor maintenance costs and ensures end-user appliance/equipment safety. In addition to new opportunities, improved positioning accuracy will benefit many traditional use cases, such as regulatory needs (emergency power supply regulation), end-user needs (appliance/equipment safety), and network operator needs (electricity distribution companies).

The main contribution of this chapter is to:

1. Demonstrate how emerging 5G communication technologies can be used to position IIoT devices using ML algorithms;
2. Examine different IIoT device positioning techniques for smart power grids based on cellular, WiFi, and 5G systems;
3. Provide an in-depth examination of ML techniques for improving positioning accuracy in 5G for location-aware power quality management;
4. Provide a vision for the future of ML and 5G-based IIoT device positioning in smart power grids.

The remainder of this chapter is structured as follows. Section 1.2 is all about cellular-based IIoT device positioning, and we go over cell identity, angle, range, and fingerprinting-based techniques for smart grid power quality management in great detail. Range and fingerprinting techniques for WiFi-based positioning are covered in Section 1.3. Section 1.4 discusses various algorithms and evaluation methods, as well as the 5G-based IIoT device positioning stack architecture for smart power grids. Section 1.5 discusses various ML techniques for positioning over a 5G network. Section 1.6 summarises and concludes the chapter by discussing potential future improvements.

1.2 CELLULAR-BASED POSITIONING TECHNIQUES

In 5G-based IIoT device positioning, downlink transmission from access points to IIoT devices and uplink transmission from IIoT devices to access points can be used to identify the exact desired location of these IIoT devices. Since, all IIoT devices can receive 5G signals in a smart power grid, it is possible to provide a ubiquitous positioning service that works with virtually any IIoT device and consumes no or little energy in addition to standard routine operations. The basic idea is to measure received signal strength (RSS), also known as a fingerprint, during an offline phase and then compare it to cellular signals heard by an IIoT device in an unknown location during an online phase to determine the best match [6]. Nonetheless, current 5G-based IIoT device positioning techniques either use traditional classifiers like K-nearest neighbour (K-NN) or support vector machine (SVM) to learn the pattern of received signal strength or use probabilistic techniques [7]. Based on the object that computes the position, the 5G-based IIoT device positioning methods are classified into two types: (1) device-based, in which the IIoT device computes its own position; and (2) network-based, in which the position of the IIoT device is computed by the network location server. Because of their centralised nature, most 5G-based IIoT device positioning solutions are network-based, allowing the network operator full control of the IIoT device positioning service as well as support for end systems.

1.2.1 Cell-identity-based techniques

Cell-identity or proximity-based techniques are the simplest to implement because they rely on determining whether or not the IIoT device to be placed is within a specific radio coverage range. The position of the serving base station and the incorporated serving cell must be known in order to estimate the position of the IIoT device; however, due to the large number of base stations required, this method is not suitable for large or sparsely populated areas [8].

There is a need for precise grid maintenance location in location-aware power quality management, since relying solely on cellular network signals is ineffective due to their low accuracies. This is due to the difficulties of grid IIoT device positioning, which requires high accuracy despite cellular signal variability and noise over time, as well as the larger coverage area of cellular towers. The basic idea behind these strategies is to use offline radio signal strength measurements to create a model that can differentiate between different locations within the area of interest. Some smart power grids, used by the K-NN classifier to match incoming signals and estimate the IIoT device position, collect GSM data with special enhanced cellular modems to improve system accuracy. Traditional ML techniques are prone to overfitting and are incapable of dealing with the large noise inherent in the wireless channel; thus, probabilistic techniques are better suited to dealing with the inherent wireless signal noise [9].

1.2.2 Angle-based techniques

Angle-based techniques use either the angle-of-arrival (AoA) or the angle-of-departure (AoD), or both. The AoA is the angle from which the radio

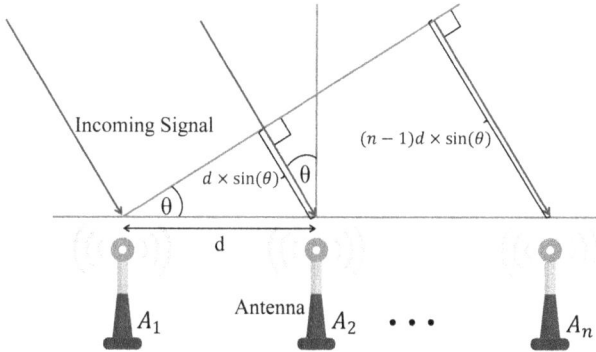

Figure 1.2 The AoA calculation is based on the phase shift of the incoming signal between the antennas.

signal is received, whereas the AoD is the angle from which the signal is transmitted (see Figure 1.2). The AoA positioning method has not been used for short-range positioning. The majority of short-range IIoT device antennas are omnidirectional, with insufficient electrical size to support narrow beam patterns in small ultra high frequency (UHF) band devices. AoA is becoming more popular as a method of incorporating location awareness into communication systems. Angle-based techniques use the angle at which signals arrive at the measuring station to determine IIoT device position. The measuring station and the estimated AoA form a straight line, and the signal source is located on that straight line, which is also known as the bearing line.

A massive MIMO 5G communication system will have a large number of antennas capable of providing a large array aperture and supporting beam-based base station operations [10]. The angle can be easily obtained during the beam search operation in terms of the downlink-AoD and uplink-AoA. While IIoT devices and base stations search for the best beam for the radio signal, downlink-AoD and uplink-AoA reference signal received power (RSRP) are used to determine beam signal quality. The base station can then use pilot signals sent by IIoT devices to analyse uplink channels, allowing it to estimate AoAs with high precision and minimal interdevice interference. As a result of massive MIMO, angle-based positioning approaches in 5G networks, which are not commonly used in 2G-4G networks, will become more prevalent [11].

Unfortunately, the majority of IIoT device positioning requests come from dense multipath power grid environments. In such cases, accurate positioning of IIoT devices necessitates appropriate measurement processing because only line-of-sight (LoS) measured values reflect the actual angular relationship between the IIoT devices and the base station. That's why the two-step initiative is the more commonly used location-aware power quality management approach, in which first a LoS/NLoS recognition is performed, followed by the use of the recognised LoS measurements for positioning [11]. Furthermore, in some cases, a LoS path may not exist in a multipath environment, and all signals are propagated NLoS, resulting in significant positioning errors.

1.2.3 Range-based techniques

In range-based IIoT device positioning techniques, the unknown position of IIoT devices is estimated using distance measurements between transmitters and receivers. Such measurements could be derived from received signal data, such as time-of-arrival (ToA), time difference-of-arrival (TDoA), and received signal strength (RSS) [12]. As shown in Figures 1.3 and 1.4, to esti-mate the IIoT device location using trilateration, the distance between the IIoT devices and at least three base stations must be calculated, or one base station if the base station antennas are sufficiently separated [13].

If the IIoT device receiver clock and the synchronised transmitting clocks are not aligned, a common offset (bias) compromises the measured data, resulting in pseudo-ranges, which are ranged predictions with an unidentified error. This offset must be treated as an unknown quantity for the location of the unidentified IIoT devices. The differences between the pseudo-ranges can be used to estimate or eliminate TDoA. The aforementioned offsets do not occur in non-timing-based systems or fully synchronised ToA (such as RSS) [12, 13].

The new radio system of the Third Generation Partnership Project (3GPP) supports two types of TDoA measurement: (a) observed time difference-of-arrival (OTDoA), also known as downlink-TDoA, which measures the ToA in the downlink; and (b) uplink-TDoA, which is based on sounding reference signals (SRS) [14]. Another advantage of uplink-TDoA over OTDoA is the

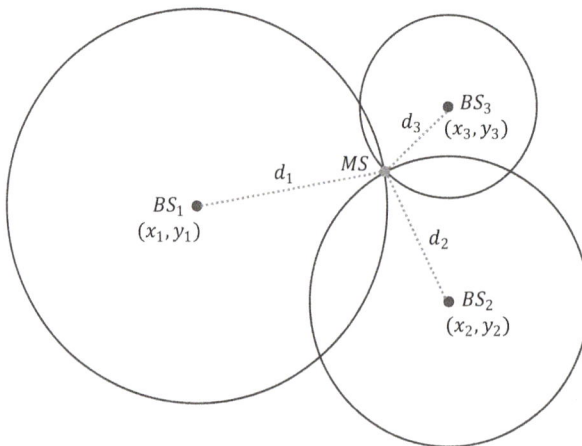

Figure 1.3 IIoT device positioning based on distance estimation using at least three base stations.

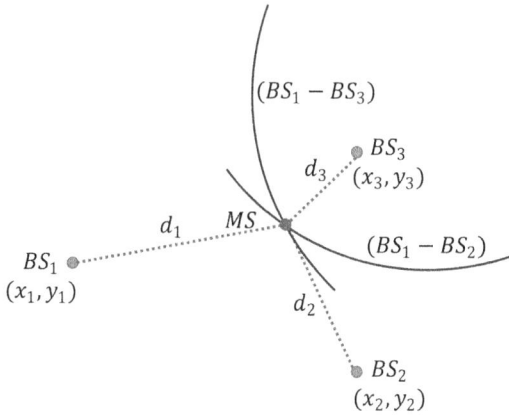

Figure 1.4 IIoT device positioning based on distance difference estimation using at least three base stations.

accuracy of time-stamping, which affects time precision. The accuracy of time stamping is determined by the clock frequency of the hardware and the clock drift. In the case of OTDoA, the ToA is measured at the IIoT devices, which typically produce lower-quality results than measurements at the base station. Because the IIoT devices are measured at different times, time drift between measurements adds another source of error.

In practice, as stated in the angle-based technique discussion, IIoT devices do not have LoS paths to more than one base station, resulting in corrupted ToA measurements. Usually, ToA-based solutions are typically inaccurate in NLoS conditions. The ToA method is notorious in NLoS environments for its complexity and poor performance. The RSS approach has proven useful in smart power grid environments because path loss is easy to predict and is expected to decrease gradually with path length. However, its accuracy in positioning suffers significantly in urban environments.

1.2.4 Fingerprinting-based techniques

The most commonly used fingerprinting-based IIoT device positioning technique has two phases: online (positioning) and offline (training). During the offline or training phase, a data record is created by estimating specific signal or antenna characteristics at known coordinates (i.e., signatures or fingerprints). RSS is one of the signal attributes that IIoT devices in smart power grids can measure. Furthermore, the grid sector, IIoT device orientation, timing advance (TA) or round-trip time (RTT) value (or any other type of timing data), mobile unit type, and so on may be saved. These records of coordinates and associated fingerprints are referred to as the radio map or fingerprint database. This data record helps in the creation of one or more ML algorithms. Positioning is accomplished during the online phase by using the model to determine the IIoT device position based on the features evaluated by the mobile terminal.

The ability of fingerprint-based techniques to remain precise even in highly cluttered multipath environments of location-aware power quality management is their key advantage [11]. It outperforms range-based and angle-based solutions, but at a high cost, because fingerprint collection during the offline phase necessitates extensive human resources. The setup change is the Achilles heel of fingerprinting-based solutions. When the location or number of IIoT devices in the smart power grid environment changes, the fingerprint or radio map must be updated, although a simple adding and removing IIoT device can be modelled by deleting or adding the corresponding values from old fingerprints which requires the collection of new fingerprints. Furthermore, different types of IIoT devices use different types of antennas, which may introduce an additional error component into the measurements. These issues have a negative impact on positioning accuracy.

1.3 WIFI-BASED POSITIONING TECHNIQUES

Location-aware power quality management through IIoT device positioning has piqued the interest of both researchers and industry. The GPS signal strength in the smart power grid area is limited; IIoT device positioning systems typically require adjacent anchor points with known coordinates. Despite the identification of potential needs for smart power grid navigation systems, the best approach in terms of reliability, efficiency, real-time availability, and low cost is still being developed. WiFi-based techniques typically rely on creating a WiFi fingerprint of the overhead WiFi access points during the offline phase to achieve high accuracy for location-aware power quality management. The ideal coordinates are selected during the online phase in order to identify the fingerprint using the received signal strength. This matching may be probabilistic or deterministic in nature.

An unsupervised IIoT device positioning scheme monitors specific locations in the smart power grid environment using recognisable signatures on one or more sensing dimensions and uses WiFi in conjunction with other sensors to combat WiFi channel noise [15]. Other methods [16–18] rely on detailed channel conditions obtained from WiFi rather than radio signal strength readings. Although WiFi fingerprinting techniques are more accurate than cellular-based techniques due to the limited propagation range of WiFi access points [19], they are less widely used than cellular-based techniques because they rely on WiFi coverage and sophisticated IIoT devices.

1.3.1 Range-based techniques

In the range-based location-aware power quality management method, the distances between transceivers are measured using ToA, TDoA, and AoA.

The RSS is first transformed into a range using the data transmission model in these methods. The geometric method is then used to determine the location of the IIoT device within the smart power grid (e.g., trilateral measurement). Implementing ToA, TDoA, and AoA-based positioning techniques takes time and money because they necessitate gateway software or hardware changes. AoA and ToA must be measured using either angular measuring sensors or high-precision timers. These accessories can raise the cost of a smart power grid positioning system. The RSS propagation model, on the other hand, is used to compute the distance between the access point and the IIoT device. The deterioration of each access point signal is determined by the distance travelled as well as various external conditions such as grid infrastructure, walls, and humidity. Therefore, it is impossible to develop a general signal propagation model capable of accurately simulating real-world conditions. In reality, in a complex smart power grid environment, the multipath effect, diffraction, refraction, and reflection all interfere with wireless signal propagation. NLoS propagation positioning has a higher error rate and does not provide satisfactory positioning performance.

1.3.2 Fingerprinting-based techniques

The RADAR positioning module is the primary task in WiFi fingerprint-based IIoT device positioning in a smart power grid. Without assuming LoS, WiFi fingerprinting is a promising approach for signal collection and associated with IIoT device positioning in a smart power grid. In this method, the location characteristic is linked to the mechanism of the signal source, such as the RSS vector emitted by different WiFi access points. Therefore, if the precise location of the IIoT device is unknown, fingerprint identification can provide positioning data without the need for angle or distance measurement. This feature ensures that the multipath effect and non-LoS propagation have no effect on positioning accuracy. Therefore, the WiFi fingerprint-based positioning technique for location-aware power quality management is highly feasible and outperforms other IIoT device positioning techniques.

1.3.2.1 Working principle

The WiFi access points are initially deployed in the smart power grid and are labelled "$A_1, A_2, \cdots A_n$". Monitoring the access point signals with a smart terminal at the point of reference in the smart power grid environment yields the signals emitted by various access points as well as the MAC addresses associated with these access points. The RSS represents the strength of the received signals.

The WiFi signal will suffer path loss during propagation. According to the path loss model, the power density of the signal decreases with increasing distance. The intensity of the WiFi signal emitted by the same access point varies with travel distance. Each location is represented by a distinct RSS vector $(R_{i_1}, R_{i_2}, \cdots, R_{im})$ based on signals emitted by multiple access points. This one-to-one relationship between the RSS vector and the position is analogous to the concept of a human fingerprint, and it can be used to identify location data uniquely. Therefore, the RSS vector evaluated at a given region can be used to estimate the geographical position of that location.

1.3.2.2 Positioning process

The WiFi fingerprint-based positioning procedure has two stages: offline data collection and online comparison [20].

- Step 1: It is difficult to measure the RSS vector at each specific region due to cost and time constraints. Therefore, specified coordinates must be carefully considered in the smart power grid environment. We can construct the RSS vector of the reference point by accumulating the intensity data of the WiFi signal generated by different access points at these points of reference and arranging them in the same order as the access point sequence. These RSS vectors, along with the coordinates of the points of reference, can be saved together in the database. Following that, a smart power grid fingerprint database (radio map) is created.
- Step 2: We can evaluate the position of a target location by making comparisons and matching the RSS vector of the targeted area with the records in the fingerprint database. The fingerprint-based positioning technique does not require access point LoS location information. Therefore, this method is appropriate for use in a sophisticated smart power grid environment.

1.4 5G-BASED POSITIONING TECHNIQUES

As technology advances toward highly assisted IIoT device positioning for smart power grid maintenance and operation, positioning requirements increase: higher accuracy, faster response, and a more robust operation are required even under extreme and special conditions. There are three types of technologies in use today: satellite-based positioning (GNSS), base-station based (2G-4G), and hybrid solutions. The limitations of these technologies are well known: they focus on the control problem or the system situation analysis, but they assume ideal conditions for IIoT device positioning. There are well-defined test scenarios for future challenges: issues in smart grid technologies [21], security and privacy challenges in the smart grid [22], and energy theft detection in the smart grid [23], among others.

Smart power grids are regarded as the most critical structure of many foreign energy approaches in many fast-growing countries around the world.

These smart power grids are based on the assumption that all grid-connected components are meticulously monitored and controlled at all times. A large amount of data must be properly channelled due to the two-way connection between the transmission and distributed sectors. Security concerns are valid, and smart meters and other assets have addressed them. The function of data traffic distribution in a smart power grid network is divided into two parts. The first is the interaction between the utility and its customers through underlying connections such as smart meters and sensors. The second is a direct connection between utilities and power generation. Over the last decade, power line communication has been regarded as the best communication provider between these segments [24].

The 5G-based positioning architecture in smart power grids allows for a diverse set of assets, both transmission and distribution. Multiple domains are involved in the 5G slicing layers at both the user and utility planes. They play an important role in network cost reduction and user-side on-demand deployment. This design reduces capital expenditure and allows network speed to be managed [25]. By utilising all available spectrum resources, cognitive radio networks are a promising technology for providing timely smart power grid wireless communications. The overall outlook for 5G-based IIoT device positioning stack architecture for a smart power grid is shown in Figure 1.5. Wireless communications can advance due to the concurrent connection of wireless transfers with simultaneous power. However, it is a less preferred approach due to the high initial installation cost and limited communication availability during bad weather and natural disasters [26]. The smart power grid is a boom in the new energy internet, where productive and stable communication is in high demand. Efficient data analysis is

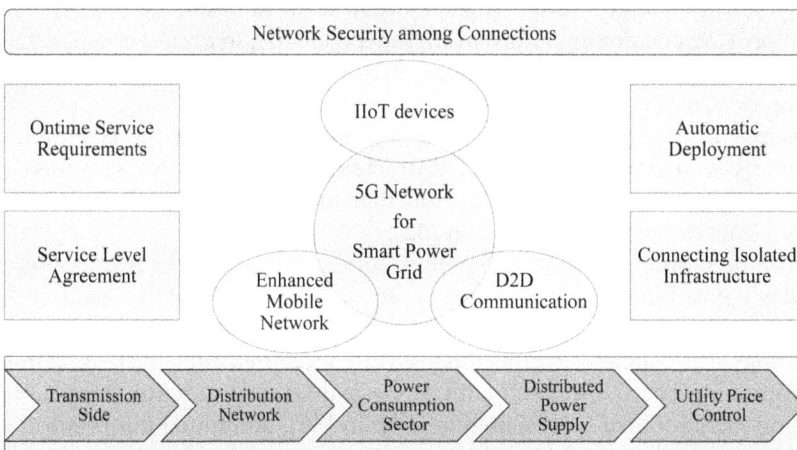

Figure 1.5 5G-based IIoT device positioning stack architecture for a smart power grid.

possible with 5G networks, potentially allowing cities to implement more cost-effective energy plans [27].

1.4.1 5G-based IIoT device positioning algorithms and evaluation methods

In [10], six different positioning technologies are collected and compared. A survey on IIoT device positioning is collected and compared in [28], which gathers and compares the fundamentals of 5G-based positioning. The physical layer of device-to-device relative positioning methods is investigated in [29], and performance bounds for various approaches are evaluated. [30] compares fixed and dynamic thresholds for timing estimators such as correlation receiver and energy detector. New spectral and subspace-based algorithms for improving software-based positioning are presented in [31]. Table 1.1 summarises the technologies to demonstrate the differences and similarities between the various methods and to provide a collection of possible references for the various solutions.

There are several methods for calculating position using 5G. To select one method, the designer must make a number of design decisions. First, the desired accuracy and refreshment rate must be set as a target. Following that, the complexity of measurement and calculation can be tuned by selecting a method from Table 1.1 that meets the hardware and financial constraints.

1.4.2 Hybrid positioning

One approach is to use hybrid technologies to improve IIoT device positioning. Combining multiple methods can result in greater accuracy, faster response time, and a more robust system. In most cases, each component can provide positioning information independently, so even if one part fails, the other can supply position information to the system during reduced operation, making it redundant. In addition to LoS situations, hybrid methods extend the ability of positioning techniques to handle NLoS situations where the signal is penetrated, diffracted, or reflected. An intermediate approach to problem decomposition that improves regular slam methods from start to finish is described in [32].

As the Global Navigation Satellite System (GNSS) is the foundation of today's positioning technology, it is an obvious choice as a resource for hybrid solutions. Figure 1.6 shows how to improve global positioning methodology by combining D2D and GNSS measurements [33]. A cellular information-based approach with a physical-layer abstraction-based simulation methodology is presented in [34]. In [35], additional information on hybrid method sources is used to improve accuracy by way of the network density of 5G deployments and a satellite visibility mask. Table 1.2 collects and compares novel 5G-based hybrid positioning techniques.

Table 1.1 A comparison of 5G positioning technologies

Technologies	Bases of method	Advantages	Disadvantages	Further improvements
Round Trip Time (RTT)	Time-based	Synchronisation error can be ignored	Sensitive to ranging error	Multi-cell-RTT
Received Signal Strength (RSS)	Range-based	Low complexity	Inaccurate (meter-level)	—
Angle of Departure (AoD)	Angle-based	Simple method: least squares (LS)	Error source: Non-Line of Sight (NLoS)	Downlink-AoD
Angle of Arrival (AoA), Direction of Arrival (DoA)	Angle-based	Simple method: least squares (LS)	LoS condition needed	Uplink-AoA, large antenna arrays
Time of Arrival (ToA)	Range-based, spherical	Simple	Clock drift, noise, fading, and Doppler shifts	Energy detector, correlation receiver
Time Difference of Arrival (TDoA)	Time difference (range-based)	No need for synchronisation	Noise-sensitive	Applying equivalent Fisher Information Matrix (FIM)
Observed Time Difference of Arrival (OTDoA)	Reference signal time difference	Commonly deployed	Accurate time-delay estimation needed	Design options are open
Multiple Signal Classification (MUSIC)	Multiple signal classification based on AoA measurements	Extended Kalman Filter (EKF)-based, high accuracy	Complex	Non-line of sight Cancellation Multiple Signal Classification (NCMUSIC)

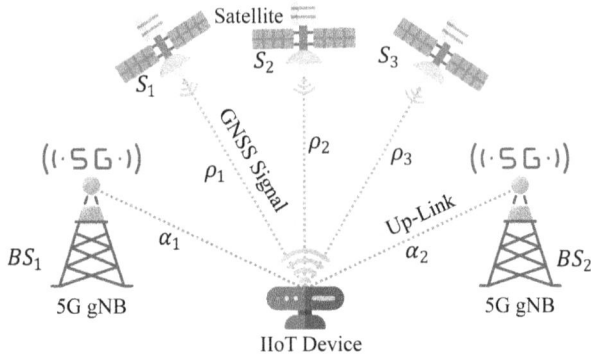

Figure 1.6 Combining D2D and GNSS measurements to improve the global positioning methodology.

As 5G technology is still in its early stages, the results are primarily based on simulation-based analysis. However, it can be seen that hybrid techniques have great potential, with forward-thinking outcomes that can be used as inspiration to improve these technologies further.

1.4.3 Cooperative positioning

In a significant proportion of cases, the IIoT device is surrounded by other IIoT devices in the smart power grid while in use, implying the possibility of cooperation. The main aspects of cooperative positioning are collected in [39], and the paper lays the groundwork for dealing with these methods by providing evaluation and comparison criteria. The two approaches are compared in [40] to a cooperative positioning and mapping method that uses multipath signals generated by the environment as a resource by creating a radio map. A method proposed in [41] for dealing with the challenges of mapping and vehicle state estimation using radio mapping is based on large antenna arrays. The proposed algorithms use a multi-model probability hypothesis density (PHD) filter and a map fusion routine to exploit all available signal paths.

The seamless positioning by utilising mobile terminal collaboration is demonstrated in [42]. D2D communication in dense networks is investigated, using ranges and pseudo-ranges to investigate centralised and decentralised structures. The Cramér–Rao lower bound is used to investigate the performance of cooperative techniques in a typical urban situation and environment. The centralised and distributed GNSS-only and heterogeneous information array processing approaches are compared in [43]. A message-passing algorithm is described in [44] for distributed positioning that has been tested in a variety of network configurations. A DL-based co-operative architecture (DELTA) based on a multi-layered fingerprint roadmap is proposed in [45] that outperforms SVM and K-NN approaches.

Table 1.2 A comparison of hybrid positioning technologies

Technologies	Main approach	Methods	Results	Comments
Inertial Measurement Unit (IMU) [36]	Edge computing	EKF for fusing the estimation results	A simulation environment for scene generation, signal propagation, and position estimation	Sub-meter accuracy, the effect of the number of base stations tested
Simultaneous positioning and mapping (SLAM) [37]	Intermediate approach (does not directly map the wave to the position)	End-to-end processing chain, Poisson Multi-Bernoulli Mixture (PMBM) filter	Problem decomposition with real channel estimation	Mapping and device state estimation handled simultaneously, accurate estimated number, type, and position of landmarks
Global Navigation Satellite System (GNSS) [33]	Newly proposed method: Crossover Multiple-Way Ranging (CO-MWR)	D2D range and angle measurements, integrated algorithms, state dimension reduction	Fewer communication resources needed	Less computation without losing information, accurate and robust method
Global Navigation Satellite System (GNSS) [38]	Physical-layer abstraction-based simulation, Weighted Least Squares (WLS) algorithm	Urban macrocell environments, Gaussian-distributed error model, CDF	Different GNSS constellations examined	5 m horizontal accuracy in 95% of cases in not ideal conditions (AV, UC)
Global Navigation Satellite System (GNSS) [34]	Visibility mask handling cmWave and mmWave signals	Evaluation: CRB, FIM, number of base stations, CDF	Simulation framework to test typical urban scenarios	NLoS signals have to be weighted, sub-meter accuracy achieved

1.5 MACHINE-LEARNING-BASED POSITIONING USING 5G NETWORKS

Deep Learning (DL) is a subset of the ML technique that simulates the data interpretation mechanism of the human brain. The goal is to develop and test a neural network capable of analysing and learning from human brains. DL can be used in a variety of fields, including speech recognition, natural language processing, computer vision, search engines, finance, and online advertising [46]. DL has also been used in wireless-signal-based positioning techniques to improve precision and reduce labour costs. Experimental studies using ML techniques to improve positioning accuracy in 5G technology can be classified into groups based on the previously discussed positioning techniques. No ML experiments have been conducted to investigate cell-identity-based techniques because their only advantage is their simplicity. Although angle-based methods benefit greatly from 5G's massive MIMO, positioning in the LoS scenario does not necessitate an ML approach [11]. Even if angle data cannot be used for precise positioning (as in an NLoS scenario), it can be used to improve the accuracy of other approaches [47]. The AoA/AoD feature is most commonly used as a fingerprint feature, which can reduce positioning error significantly in a LoS scenario. The majority of ML techniques attempt to predict environmental or signal parameters (e.g., LoS/NLoS).

1.5.1 Motivation for using ML techniques for IIoT device positioning in a smart power grid

Many of the limitations of traditional positioning techniques in location-aware power quality management can be effectively addressed by DL and ML algorithms. Traditional methods are often insufficiently scalable, making them unsuitable for large-scale smart power grid environments. Traditional methods of positioning are not very adaptable to rapidly changing environments. Recurrent neural networks could use flight path data to reduce received signal strength indicator (RSSI) variability by sequentially correlating time-varying RSSI measurements [48]. The most difficult problem with IIoT device positioning in smart power grid is RSSI fluctuation, which has a negative impact on location accuracy. The ability of ML algorithms to discover effective information from input data with known or unknown statistics is its most significant advantage.

One of the limitations in the use and accuracy of fingerprinting-based positioning systems is the availability of high-dimensional data. High-dimensional features can be converted to low-dimensional features using both supervised and unsupervised dimension reduction techniques. Reinforcement learning is yet another advantage of the ML technique which can achieve high bandwidth control based on specific learned policies [11]. It is used in the positioning of IIoT devices, allowing them to develop self-adaptive control strategies. Bayesian methods integrate multi-modal IIoT device location and use past records through an iterative tracking process. Unsupervised learning fusion is more practical because it evaluates weights in real-time using online measurements and does not require offline training.

Transfer learning allows ML to quickly learn new things in a new smart power grid environment by comparing them to prior knowledge. For example, in fingerprint-based IIoT device positioning, a transfer learning mechanism can be used to increase system scalability without requiring excessive site surveys or sacrificing accuracy. Aside from transfer learning, DL techniques have shown great promise in improving positioning in complex smart power grid environment scenarios. Non-linearly correlated features are especially difficult to extract and model [49]. Sensor data fusion is used in the smart power grid environment for accurate location estimation, and it is heavily reliant on efficient data fusion techniques. ML algorithms excel at handling multidimensional and multivariate data in dynamic environments. It can be trained to effectively combine results from various positioning sensors, techniques, and methods [47, 49].

1.5.2 Specifics of ML techniques for IIoT device positioning in a smart power grid

Classifier algorithms are primarily used in smart power grid positioning to extract core features of IIoT device signals. Clustering is performed using the fingerprint method based on these extracted features. Feature extraction is also important for identifying and mitigating NLoS. K-NN, SVM, and Decision Tree are three popular classification algorithms [50]. Decision Tree-based IIoT device positioning in a smart power grid outperforms other classification methods such as K-NN and Neural Network in terms of improving IIoT device positioning accuracy. There is a possibility of missing information when dealing with continuous numerical data and categorising it. SVM-based methods become time-consuming and memory-intensive as the number of support vectors increases [50, 51].

In real-world fingerprint recognition situations, a fingerprint map created during the offline phase contains a large amount of data. So, it takes time to compare the information obtained during the online phase with each piece of data from the fingerprint map. Random Forest can be employed to fix the over-fitting problem in this case. Following clustering, a model is built based on the fingerprint information of the points of reference in the group. In fingerprint-positioning techniques, each reference point must differ from other reference points in terms of the extracted features. Some features have been found to be uninformative or to repeat redundant information from other features on a regular basis. Dimensionality reduction is critical in this case for reducing model complexity, shortening training time, and conserving storage space. Principal component analysis (PCA) is useful because it reduces the complexity of multidimensional data while preserving trends and patterns [52, 53].

DL is well known for its distributed computing capabilities and ability to analyse large amounts of unlabelled and uncategorised data. The ability of DL algorithms to extract features from data without the need for manual, feature extraction is their most significant advantage. This removes the need for domain knowledge and hard-core feature extraction. A convolutional neural network (CNN) is a DL algorithm that performs feature extraction and classification. Many IIoT device positioning methods are susceptible to GPS errors and device tampering. RL is the best technique to use in such a challenging circumstance. RL enables the agent to achieve a large goal by interacting with the environment (using a reward and penalty process) and resolving issues caused by signal strength uncertainty [51–53].

1.6 LESSONS LEARNED AND FUTURE IMPROVEMENT POSSIBILITIES

The chapter presented here demonstrates how emerging 5G network communication technologies can be used for 5G-based IIoT device positioning using ML methods for location-aware power quality management. These new technologies increase the volume of information and provide additional positioning features. The majority of these new features, such as AoD and AoA, can be managed using traditional methods such as triangulation, but the integration of these functionalities, combined with the increased volume of data, makes ML techniques more manageable than traditional ones. We can conclude that in smart power grid environments, a mean absolute error (MAE) of less than two metres is achievable under ideal conditions such as low noise, high antenna or BS count, or LoS propagation. The MAE increases dramatically in NLoS and high noise scenarios. 5G positioning faces both technical and non-technical challenges. Non-technical future challenges must be addressed in order to acquire advanced positioning techniques, and the physical layer of 5G must be investigated further in order to gain greater effectiveness from 5G positioning.

ABBREVIATIONS

3GPP	Third Generation Partnership Project
AoA	Angle-of-arrival
AoD	Angle-of-departure
AR	Augmented reality
CDF	Cumulative distribution function
CEC	Cloud-edge computing
CNN	Convolutional neural network
D2D	Device-to-device
DL	Deep learning

EXF	Extended Kalman Filter
FIM	Fisher information matrix
GNSS	Global Navigation Satellite System
GPS	Global Positioning System
GSM	Global System for Mobile Communication
IIoT	Industrial Internet of Things
IMU	Inertial measurement unitK-
NN	K-nearest neighbour
LoS	Line-of-sight
MAC	Media access control
MAE	Mean absolute error
MIMO	Multiple-input multiple-output
ML	Machine learning
MUSIC	Multiple signal classification
NLoS	Non-line-of-sight
OTDoA	Observed time difference-of-arrival
PCA	Principal component analysis
PHD	Probability hypothesis density
RADAR	Radio detection and ranging
RSRP	Reference signal received power
RSS	Received signal strength
RSSI	Received signal strength indicator
RTT	Round trip time
SLAM	Simultaneous localisation and mapping
SRS	Sounding reference signals
SVM	Support vector machine
TA	Timing advance
TDoA	Time difference-of-arrival
ToA	Time-of-arrival
UC	Unified communications
UDN	Ultra-dense network
UHF	Ultra high frequency
WLS	Weighted least squares

REFERENCES

[1] Sigov, A., Ratkin, L., Ivanov, L.A., Xu, L.D.: Emerging enabling technologies for industry 4.0 and beyond. *Information Systems Frontiers*, 1–11 (2022).

[2] Fan, S., Zeng, R., Tian, H.: Mobile feature enhanced high-accuracy positioning based on carrier phase and Bayesian estimation. *IEEE Internet of Things Journal* 9(16), 15312–15322 (2022).

[3] Zhou, T., Ku, J., Lian, B., Zhang, Y.: Indoor positioning algorithm based on improved convolutional neural network. *Neural Computing and Applications* 34(9), 6787–6798 (2022).

[4] Albanese, A., Sciancalepore, V., Banchs, A., Costa-Perez, X.: Loko: Localization-aware roll-out planning for future mobile networks. *IEEE Transactions on Mobile Computing* 1, 1–15 (2022).

[5] Patel, D.: First report on new technological features to be supported by 5g standardization and their implementation impact. Technical report, https://5gsmart.eu/wp-content/uploads/5G-SMARTD5.1.pdf (2020). Accessed: 2022-08-15.

[6] Alhomayani, F., Mahoor, M.H.: Outfin, a multi-device and multi-modal dataset for outdoor localization based on the fingerprinting approach. *Scientific Data* 8(1), 1–14 (2021).

[7] Mohamed, A., Tharwat, M., Magdy, M., Abubakr, T., Nasr, O., Youssef, M.: Deepfeat: Robust large-scale multi-features outdoor localization in lte networks using deep learning. *IEEE Access* 10, 3400–3414 (2022).

[8] Al-Rashdan, W.Y., Tahat, A.: A comparative performance evaluation of machine learning algorithms for fingerprinting-based localization in dmmimo wireless systems relying on big data techniques. *IEEE Access* 8, 109522–109534 (2020).

[9] Rizk, H., Torki, M., Youssef, M.: Cellindeep: Robust and accurate cellularbased indoor localization via deep learning. *IEEE Sensors Journal* 19(6), 2305–2312 (2018).

[10] De Lima, C., Belot, D., Berkvens, R., Bourdoux, A., Dardari, D., Guillaud, M., Isomursu, M., Lohan, E.-S., Miao, Y., Barreto, A.N., et al.: Convergent communication, sensing and localization in 6g systems: An overview of technologies, opportunities and challenges. *IEEE Access* 9, 26902–26925 (2021).

[11] del Peral-Rosado, J.A., Raulefs, R., López-Salcedo, J.A., Seco-Granados, G.: Survey of cellular mobile radio localization methods: From 1g to 5g. *IEEE Communications Surveys & Tutorials* 20(2), 1124–1148 (2017).

[12] Li, Y., Yan, K.: Indoor localization based on radio and sensor measurements. *IEEE Sensors Journal* 21(22), 25090–25097 (2021).

[13] Hinga, S.K., Atayero, A.A.: Deterministic 5g mmwave large-scale 3d path loss model for Lagos Island, Nigeria. *IEEE Access* 9, 134270–134288 (2021).

[14] Lin, X., Li, J., Baldemair, R., Cheng, J.-F.T., Parkvall, S., Larsson, D.C., Koorapaty, H., Frenne, M., Falahati, S., Grovlen, A., et al.: 5g new radio: Unveiling the essentials of the next generation wireless access technology. *IEEE Communications Standards Magazine* 3(3), 30–37 (2019).

[15] Wang, H., Sen, S., Elgohary, A., Farid, M., Youssef, M., Choudhury, R.R.: No need to war-drive: Unsupervised indoor localization. In: *Proceedings of the 10th International Conference on Mobile Systems, Applications, and Services*, pp. 197–210 (2012).

[16] Avola, D., Cascio, M., Cinque, L., Fagioli, A., Petrioli, C.: Person reidentification through wi-fi extracted radio biometric signatures. *IEEE Transactions on Information Forensics and Security* 17, 1145–1158 (2022).

[17] Wu, C., Qiu, T., Zhang, C., Qu, W., Wu, D.O.: Ensemble strategy utilizing a broad learning system for indoor fingerprint localization. *IEEE Internet of Things Journal* 9(4), 3011–3022 (2021).

[18] Liu, Y.-T., Chen, J.-J., Tseng, Y.-C., Li, F.Y.: An auto-encoder multi-task lstm model for boundary localization. *IEEE Sensors Journal* 22(11), 10940–10953 (2022).

[19] Chen, J., Zhou, B., Bao, S., Liu, X., Gu, Z., Li, L., Zhao, Y., Zhu, J., Li, Q.: A data-driven inertial navigation/bluetooth fusion algorithm for indoor localization. *IEEE Sensors Journal* 22(6), 5288–5301 (2021).

[20] Shang, S., Wang, L.: Overview of wifi fingerprinting-based indoor positioning. *IET Communications* 16(7), 725–733 (2022).

[21] Colak, I., Sagiroglu, S., Fulli, G., Yesilbudak, M., Covrig, C.-F.: A survey on the critical issues in smart grid technologies. *Renewable and Sustainable Energy Reviews* 54, 396–405 (2016).

[22] Zhang, H., Liu, B., Wu, H.: Smart grid cyber-physical attack and defense: A review. *IEEE Access* 9, 29641–29659 (2021).

[23] Wen, M., Xie, R., Lu, K., Wang, L., Zhang, K.: Feddetect: A novel privacy preserving federated learning framework for energy theft detection in smart grid. *IEEE Internet of Things Journal* 9(8), 6069–6080 (2021).

[24] Galli, S., Scaglione, A., Wang, Z.: For the grid and through the grid: The role of power line communications in the smart grid. *Proceedings of the IEEE* 99(6), 998–1027 (2011).

[25] Hadjioannou, V., Mavromoustakis, C.X., Mastorakis, G., Batalla, J.M., Kopanakis, I., Perakakis, E., Panagiotakis, S.: Security in smart grids and smart spaces for smooth IoT deployment in 5g. In: *Internet of Things (IoT) in 5G Mobile Technologies*, pp. 371–397. Springer, (2016).

[26] Khan, A.A., Rehmani, M.H., Reisslein, M.: Cognitive radio for smart grids: Survey of architectures, spectrum sensing mechanisms, and networking protocols. *IEEE Communications Surveys & Tutorials* 18(1), 860–898 (2015).

[27] Huang, X., Wang, S.: Aggregation points planning in smart grid communication system. *IEEE Communications Letters* 19(8), 1315–1318 (2015).

[28] Nurmi, J., Lohan, E.-S., Wymeersch, H., Seco-Granados, G., Nykänen, O.: *Multi-technology Positioning*. Springer, (2017).

[29] Zhou, M., Li, Y., Wang, Y., Pu, Q., Yang, X., Nie, W.: Device-to-device cooperative positioning via matrix completion and anchor selection. *IEEE Internet of Things Journal* 9(7), 5461–5473 (2021).

[30] Yang, X., Wen, C.-K., Han, Y., Jin, S., Swindlehurst, A.L.: Soft channel estimation and localization for millimeter wave systems with multiple receivers. *IEEE Transactions on Signal Processing* 70, 4897–4911 (2022).

[31] Leylaz, G., Wang, S., Sun, J.-Q.: Identification of nonlinear dynamical systems with time delay. *International Journal of Dynamics and Control* 10(1), 13–24 (2022).

[32] Ge, Y., Kaltiokallio, O., Kim, H., Jiang, F., Talvitie, J., Valkama, M., Svensson, L., Kim, S., Wymeersch, H.: A computationally efficient ekpmbm filter for bistatic mmwave radio slam. *IEEE Journal on Selected Areas in Communications* 40(7), 2179–2192 (2022).

[33] Yin, L., Ni, Q., Deng, Z.: A GNSS/5g integrated positioning methodology in d2d communication networks. *IEEE Journal on Selected Areas in Communications* 36(2), 351–362 (2018).

[34] Sun, C., Zhao, H., Bai, L., Cheong, J.W., Dempster, A.G., Feng, W.: Gnss-5g hybrid positioning based on toa/aoa measurements. In: *China Satellite Navigation Conference*, pp. 527–537 (2020). Springer.

[35] Rydholm, C., Pommer, W.: *Hybrid Positioning Solution Using 5G and GNSS*. Digitala Vetenskapliga Arkivet 176819 (2021).

[36] Guo, C., Yu, J., Guo, W.-F., Deng, Y., Liu, J.-N.: Intelligent and ubiquitous positioning framework in 5g edge computing scenarios. *IEEE Access* 8, 83276–83289 (2020).

[37] Ge, Y., Kim, H., Wen, F., Svensson, L., Kim, S., Wymeersch, H.: Exploiting diffuse multipath in 5g slam. In: *GLOBECOM 2020-2020 IEEE Global Communications Conference*, pp. 1–6 (2020). IEEE.

[38] del Peral-Rosado, J.A., Renaudin, O., Gentner, C., Raulefs, R., Dominguez-Tijero, E., Fernandez-Cabezas, A., Blazquez-Luengo, F., Cueto-Felgueroso, G., Chassaigne, A., Bartlett, D., et al.: Physical-layer abstraction for hybrid gnss and 5g positioning evaluations. In: *2019 IEEE 90th Vehicular Technology Conference (VTC2019-Fall)*, pp. 1–6 (2019). IEEE.

[39] Zhang, P., Lu, J., Wang, Y., Wang, Q.: Cooperative localization in 5g networks: A survey. *Ict Express* 3(1), 27–32 (2017).

[40] Wymeersch, H., Seco-Granados, G., Destino, G., Dardari, D., Tufvesson, F.: 5g mmwave positioning for vehicular networks. *IEEE Wireless Communications* 24(6), 80–86 (2017).

[41] Kim, H., Granström, K., Gao, L., Battistelli, G., Kim, S., Wymeersch, H.: 5g mmwave cooperative positioning and mapping using multi-model phd filter and map fusion. *IEEE Transactions on Wireless Communications* 19(6), 3782–3795 (2020).

[42] Koivisto, M., Hakkarainen, A., Costa, M., Kela, P., Leppanen, K., Valkama, M.: High-efficiency device positioning and location-aware communications in dense 5g networks. *IEEE Communications Magazine* 55(8), 188–195 (2017).

[43] Bhat, S.J., Venkata, S.K.: An optimization based localization with area minimization for heterogeneous wireless sensor networks in anisotropic fields. *Computer Networks* 179, 107371 (2020).

[44] Wielandner, L., Leitinger, E., Meyer, F., Teague, B., Witrisal, K.: Message passing-based cooperative localization with embedded particle flow. In: *ICASSP 2022-2022 IEEE International Conference on Acoustics, Speech and Signal Processing (ICASSP)*, pp. 5652–5656 (2022). IEEE.

[45] Quraishi, A., Martinoli, A.: Distributed cooperative localization with efficient pairwise range measurements. In: *International Symposium Distributed Autonomous Robotic Systems*, pp. 134–147 (2021). Springer.

[46] Li, Z., Xu, K., Wang, H., Zhao, Y., Wang, X., Shen, M.: Machine-learning-based positioning: A survey and future directions. *IEEE Network* 33(3), 96–101 (2019).

[47] Gante, J., Falcao, G., Sousa, L.: Deep learning architectures for accurate millimeter wave positioning in 5g. *Neural Processing Letters* 51(1), 487–514 (2020).

[48] Hoang, M.T., Yuen, B., Dong, X., Lu, T., Westendorp, R., Reddy, K.: Recurrent neural networks for accurate rssi indoor localization. *IEEE Internet of Things Journal* 6(6), 10639–10651 (2019).

[49] Liu, K., Zhang, H., Ng, J.K.-Y., Xia, Y., Feng, L., Lee, V.C., Son, S.H.: Toward low-overhead fingerprint-based indoor localization via transfer learning: Design, implementation, and evaluation. *IEEE Transactions on Industrial Informatics* 14(3), 898–908 (2017).

[50] Hoang, M.T., Zhu, Y., Yuen, B., Reese, T., Dong, X., Lu, T., Westendorp, R., Xie, M.: A soft range limited k-nearest neighbors algorithm for indoor localization enhancement. *IEEE Sensors Journal* 18(24), 10208–10216 (2018).

[51] Villacrés, J.L.C., Zhao, Z., Braun, T., Li, Z.: A particle filter-based reinforcement learning approach for reliable wireless indoor positioning. *IEEE Journal on Selected Areas in Communications* 37(11), 2457–2473 (2019).

[52] Wang, X., Wang, X., Mao, S.: Deep convolutional neural networks for indoor localization with csi images. *IEEE Transactions on Network Science and Engineering* 7(1), 316–327 (2018).

[53] Nessa, A., Adhikari, B., Hussain, F., Fernando, X.N.: A survey of machine learning for indoor positioning. *IEEE Access* 8, 214945–214965 (2020).

Chapter 2

Overview and comparative application of on-grid and off-grid renewable energy systems in modern-day electrical power technology

Daniel Akinyele
Olabisi Onabanjo University, Ago-Iwoye, Nigeria

Titus Ajewole and Olakunle Olabode
Osun State University, Osogbo, Nigeria

Ignatius Okakwu
Olabisi Onabanjo University, Ago-Iwoye, Nigeria

CONTENTS

DOI: 10.1201/9781003436461-2

2.1 INTRODUCTION

The provision of reliable energy is one important agenda of governments, policy-makers, industrialists, and other relevant stakeholders around the globe. This stems from the fact that energy is not only central to human existence but is also instrumental to any country's social and economic development. The human development index (HDI), which is a yardstick for differentiating between developed and developing countries, is premised on the level of access to modern and reliable energy. Therefore, developed countries are believed to have relatively high HDIs and their citizens experience a high level of human development [1]. Having a high HDI is often translated as having a high quality of life or standard of living in terms of people's economic and well-being, including their access to knowledge or information, all of which are connected to having access to a good energy supply in one way or the other. Several other indicators or yardsticks for measuring the availability of a good energy supply system in a country include GDP and the level of industrialization.

The foregoing is illustrative of the fact that the level of economic advancement and sustainability in developed countries is associated with good energy infrastructure. However, the lack of such infrastructure in several developing countries is among the factors that affect economic prosperity and development. A recent report shows there still exists 0.77 billion people in the world without access to a modern energy supply [2], with the sub-Saharan Africa region, for instance, having the most deficit at over 70% of the total global deficit. Though there has been a drastic reduction in the electricity deficit over the last decade, recent information clearly indicates that the provision of energy has an important role to play in addressing the issue of energy poverty and poor human standards of living in several developing and other energy-impoverished countries (e.g., under-developed states) around the world.

Fossil-fuel-based energy resources form the highest percentage of total global energy [2, 3]. The contribution of fossil fuel in global electrical power generation was 62% in the year 2021, while renewable electricity and nuclear power have shares of 28.3 and 10.0%, respectively. The global renewable energy report in [2] also reveals that the shares of fossil fuels, renewable energy (RE), and nuclear energy in 2011 was 68.0, 20.4, and 12.0%, respectively, implying that the RE proportion in power generation has increased by about 8% over the last decade. Some of the factors responsible for the increased renewable electricity production are the finite nature and cost volatility of fossil fuels and their associated environmental concerns in terms of the anthropogenic activities that are leading to global warming and climate change [3]. The nuclear energy resource, though a low-carbon technology, also presents some concerns in the form of how to store and/or dispose of radioactive materials or wastes and the threat from the increased production of nuclear weapons.

Furthermore, there is a trending global paradigm shift from centralized to distributed generation (DG) against the traditional norm of supplying a changing demand from a centralized electricity generation fueled by fossil and nuclear energy resources [4, 5]. It is projected that in the future the largest fraction of electrical power generation is likely to be sourced from RE resources. With the exception of large-scale hydroelectric and off and on-shore wind power systems, RE-based generating systems are generally smaller than the fossil-fuel and nuclear-based power systems that are currently deployed in different parts of the world for grid-tie and non-grid-tie applications. In addition, small electrical generators such as those that are fueled by renewable energies cannot be integrated with the transmission network as a result of an economic constraint – namely the cost of high voltage (HV) switchgear and transformers [4]. The power transmission network system is usually a long distance away from the small generator which is constrained by the available resource in the location. It is against these backdrops that small generators are connected to the power distribution network rather than the transmission system. Such generating systems are thus regarded as DGs, which are embedded or dispersed generation; and the integration of a storage device is an important consideration to mitigate RE intermittent generation.

The size, configuration, and application of DGs do not only favor the utilization of RE resources but also pave a way for increased deployment of RE technologies and other distributed energy resources (DERs) such as storage and fuel cells systems in future electrical power grids (smart grids) [6]. The integration of RE systems in the radial power grid architecture entails the connection of renewable power generation to the radial power distribution system at the point of common coupling (PCC). The traditional radial distribution system permits a unidirectional flow of power. However, the electrical configuration of the grid when the DG is integrated with it is such that it permits a bi-directional flow of power, in which case power can be allowed to flow to the grid from the DG, and power can also flow from the grid to the DG to a load. This aspect is usually achieved by employing properly designed power electronic converters and a well-coordinated energy management system to ensure stable, quality, and optimal electrical power delivery. The DG in this aspect may be run either in the grid-connected or the islanded mode.

The efforts toward realizing global sustainability and energy transitions are also part of the factors that drive accelerating interests in the utilization of RE resources for electricity generation purposes both for on-grid and off-grid configurations in several parts of the world. For instance, it is one thing to design, plan, and implement a small RE-based generating system, it is another to sustain such a system to achieve long-term viability for the intended use. This is where the use of RE resources is associated with the concept of sustainability that seeks to satisfy the social, technical, environmental, economic, and policy (STEEP) dimensions [7, 8]. Part of the sustainability plans is to ensure global environmental sustainability in which case human activities and energy-generating technologies will be expected to be clean and eco-friendly. This requirement makes the environmentally friendly attribute of RE resources appreciated in terms of mitigating the issue of climate change. Several existing research studies have also suggested that RE systems can help realize a relatively cost-effective energy supply.

Global demand is continuously increasing, which is driven by increased population growth in several regions of the world, accelerated industrial activities and developments, and the use of consumer electronic devices as societies become more affluent [4, 9]. It is also reasonable for the generating facilities to satisfy this demand, which is why the electricity systems around the world are incorporating a wide range of energy sources, some of which are REs. The inability of generation to balance load demand is the reason why loads are being shed (load shedding) – in certain districts curtailing the load to a level that the available generation capacity can handle. Based on this premise, RE resources and other DERs can also serve the purpose of providing some ancillary services (ASs) to support the existing grid [5]. Several viable existing RE-based electricity systems in off-grid communities of some developing countries [10] show that RE resources have a crucial role to play in modern-day electrical power supply technology.

This chapter is concerned in particular with presenting an overview and an exhaustive comparison of the use of RE systems in the context of grid-connected and grid-independent applications in modern-day electrical power technology. It carefully deals with the issues associated with RE system applications using the on-grid and off-grid configurations as the basis of explanation, including some of the fundamental electrical topologies that include power electronics and storage devices and non-RE source(s).

This first section presents an overview of the characteristics of RE resources and systems; the following section will present the reliability of RE technology in terms of energy sources, the combination of these energy resources in different off-grid configurations, and their performances. The third section focuses on the grid-connected configurations of RE systems and applications, while the discussions on ancillary services provided by RE-based energy systems form the basis for the fourth section; this is followed by the conclusions of the chapter.

2.1.1 Characteristics and merits of renewable resources

2.1.1.1 Renewable energy sources

One of the questions that may be in the mind of the reader is what are REs and the characteristics or behavior they exhibit. REs have been defined as energy sources that are obtained from natural resources such as the sun, wind, water, biomass, tides, waves, and geothermal energy [11]. One unique feature of these resources is that they are naturally replenished – the essence of the term "renewable".

It is important to have an understanding of the inherent features and characteristics of RE resources to be able to appreciate how they behave. Failure to understand this will be tantamount to the inability of the planner to optimally harness the resources, which may result in poor engineering design, implementation, and operation, including the associated economic implication. Some of the inherent features of REs are:

1. They are naturally occurring, which is why they are replenished naturally (renewable);
2. They exhibit intermittent or variable characteristics;
3. They are clean and eco-friendly.

2.1.1.2 *Characteristics of renewable energy resources*

RE resources, when compared with fossil-fuel resources, are clean, which implies that they are non-polluting sources of energy [12]. This means that they possess the potential for mitigating the environmental concerns of climate change; this advantage has been an important basis for several research studies conducted or projects implemented on RE-based power generation systems around the world [13–16]. In addition, they are non-depleting and are available in every part of the world, meaning that they are sustainable.

Though during their useful operational life RE systems are believed to generate no emissions, their manufacture is associated with some chemicals and emissions and energy requirements [17]. The time that it will take a given solar photovoltaic module to produce the energy that is equal to the amount of energy used in its manufacture will depend on the location's solar energy resource, that is irradiation, and the method employed to manufacture the solar module [17–20].

2.1.2 Energy conversion in renewable energy systems

RE resources include solar, wind, hydro, biomass, tidal, wave, and geothermal energy. The global total installed renewable power capacity is 3,146 GW; hydropower, solar photovoltaic, wind, bio-power, geothermal, concentrating solar thermal power (CSP), and ocean power account for 1,195, 942, 845, 143, 14.5, 6.0, and 0.5 GW as shown in Table 2.1. Solar energy, for instance, is available as long as there is sunlight; the energy from the sun in this case is referred to as solar insolation or radiation [12]. Also, the wind is obtained from the mechanism of differential solar heating of the surface

Table 2.1 Global total installed renewable power capacity [2]

Resource type	Capacity (GW)	Contribution (%)
Hydropower	1,195	37.98
Solar photovoltaic power	942	29.94
Wind power	845	26.86
Bio-power	143	4.55
Geothermal power	14.5	0.46
CSP	6.0	0.19
Ocean power	0.5	0.016

of the earth. This process occurs as a result of more thermal input being available at the equator with associated water and thermal energy transfer through the mechanism of evaporation and precipitation [12]. Flowing from this, hydro energy is also an indirect form of solar energy obtained through rivers and dams.

Another interesting aspect of the feature and relevance of solar energy is its conversion to biomass through the process of photosynthesis. Plants are developed through the process of photosynthesis and they serve as food for animals; an animal product such as oil is extracted from animal fat and biogas, for instance, that is produced from manure or biomass waste through the anaerobic digestion process, which is also indirectly sourced from solar energy. Apart from this, plant materials such as wood and energy crops can be processed to produce heat energy, which may then be used to develop mechanical/electrical energy. Also, food material/waste may be used to produce bio-diesel through the trans-esterification process. In a nutshell, solar energy serves as a source of wind, hydro, and biomass, but not tidal and geothermal resources [4, 17].

Geothermal energy evolves from the heat from inside the earth [12]. This is brought about by the mechanism of the decay of radioactive particles and the residual thermal energy resulting from the gravitational effect when the earth was being formed. Volcanoes, which are a product of volcanic eruption, are a typical example of geothermal energy appearing on the surface from the interior; the fact that volcanoes are very hot indicates that geothermal energy is a huge thermal resource that may be employed for electricity generation and other applications that require a heat supply. One of the features that distinguish geothermal energy from other sources – solar, wind, and hydro – is the fact that the average thermal flow of the earth is far less than low-density solar irradiation [12].

Given that the geothermal flow is small, there are many sites around the world with an average thermal flow of 300 mW per m² compared to a global average of 60 mW per m² where appreciable co-generation application can be realized [12]. The term "co-generation" is usually referred to as combined heat and power. Another major disparity between geothermal and other resources is the variability of wind and solar energy resources over short time scales and the seasonality of hydro resources. Geothermal resources only decline as thermal energy is extracted, with lifetimes of ten decades or more – because geothermal reservoirs are large and will generate energy for a very long time. However, tidal energy is developed from the gravitational interaction of the moon and the sun. The mechanism responsible for this energy is such that the water level in large oceans on the earth increases and decreases in predictable fashions [17]. These tides occur at around 24 hr and 12 hr 25 minutes during the diurnal and the semidiurnal periods. The difference between the high and low tides (in terms of height) is regarded as the tidal range, and this varies from around half a meter to about 10 meters in certain locations such as those near the continental land masses. Tidal currents are generated by the movement of water, and they can attain speeds of around 5 m/s in coastal and inter-island channels. Though the power obtained from tidal currents, that is, tidal stream power, may be utilized similarly to wind power, its source is the gravitational fields of the moon and the sun, and not solar energy, as is the case with wind energy.

Furthermore, it is important to have a deep understanding of how RE generating systems are sized [18]; in fact, the available resources in a particular location and the users' demand are among the major factors that determine the system capacity for the specified application. Other factors include the type of application, type and nature of the load, and the users to be served.

2.1.2.1 Output power and energy of a solar photovoltaic array

The output energy of a single solar photovoltaic (PV) module per year, in kWh per year, can be calculated by Equation (2.1) [16, 19]:

$$PV_{o/p} = \eta_{pv} \cdot I_S \cdot A_{pv} \cdot PR \tag{2.1}$$

where η_{pv}, I_S, A_{pv} and PR represent the solar module efficiency, incidental solar irradiation in kWh per m² per year, the module's active surface area in m², and the performance ratio. Thus, the PV array's total energy output, that is $PV_{o/p(total)}$, will then be calculated by factoring in the number of modules, that is n, into Equation (2.1), leading to Equation (2.2):

$$PV_{o/p(total)} = n \cdot PV_{o/p} \tag{2.2}$$

The performance ratio, otherwise called the derate ratio or factor, in Equations (2.1) and (2.2), defines the difference between the rated performance of the PV modules (DC) (i.e., the product of the PV efficiency and the solar irradiation) and the actual energy generation (AC) [20].

In other words, PR relates the PV's actual energy output to the theoretical energy output, thus providing insight into quantifying the energy losses, that is, due to heat and conduction losses [21]. PR is also defined in the PVSyst simulation tool as the quantity of energy that is effectively generated, divided by the amount of energy that would have been generated if the solar PV array was continuously operated at the standard test condition (STC) [22]. It is also described in the International Electrotechnical Commission standard 61724.

The fact that the value of PR is independent of the orientation of the solar PV array and the incident solar irradiation, makes it a basis for comparing PV plants in different locations around the world [21]. However, PR depends on the type of electrical installation and considers the solar PV array's performance deviation from the STC. For instance, the operation of the PV modules under ambient temperature values above 25°C results in temperature losses especially for crystalline PV technologies [20].

With the use of the location-specific PR values and the other parameters – solar irradiation, the area of the solar PV module, the conversion efficiency, and the number of solar modules – the total energy output of the PV array will be obtained. However, "default" PRs of 75 and 80% are also recommended in [20, 23, 24] for roof and ground-mounted utility systems, respectively. The default PR values also include the effect of degradation, but in the case where the designer is using the location-specific PR values, it is expected that the losses due to degradation will be added to the performance analysis [20].

Equations (2.1) and (2.2) present the energy output of the PV generating system. It is also important to be able to determine the power capacity of the solar array system. It was established earlier that the location's resource and the load are among the major factors for RE system sizing. The solar power capacity in kW may be calculated by:

$$PV \ Array \ Capacity = \frac{D_{ED}}{Avg.I_S} \cdot S_F \qquad (2.3)$$

where D_{ED}, $Avg.\ I_S$, and S_F represent daily total demand in kWh/d, the location's average solar irradiation (kWh/m²/d), and the solar module safety factor. $Avg.\ I_S$ is also regarded as the peak sun hours (PSH) of the specified location – the average daily solar energy (i.e., solar irradiation) that is received on a particular location or surface, which is equivalent to the hours needed for the solar irradiance to be at the STC value of 1 kW/m² (i.e., peak sun) to realize the total quantity of daily energy that is received [25]. PSH can be expressed using the mathematical relation [25]:

$$PSH\left(hours/d\right) = \frac{Avg.Daily\ Irradiation\left(kWh/m^2/d\right)}{Peak\ Sun\left(1\ kW/m^2\right)} \qquad (2.4)$$

The idea of PSH arises from the fact that solar energy can be described as the total irradiation for a year, that is, $kWh/m^2/yr$, or usually on an average daily basis for a given month or year in $kWh/m^2/d$. However, when solar irradiation is considered on an average day, then the total daily solar energy may be assumed to be the same quantity of solar energy that is received at the irradiance of 1 kW/m^2 for a given number of hours [25].

Furthermore, it is a common practice to overrate the solar PV arrays so as to be able to account for losses. The quantity $\frac{D_{ED}}{Avg.I_S}$ gives the ideal or basic solar PV capacity for the application, assuming there are no losses. However, in a typical solar PV application, there are losses due to temperature, wiring, dust, aging, incomplete absorption of light, and so on [26]. The safety factor, S_F takes care of the additional capacity that can help achieve a practicable or viable solar PV array size for the specified application.

In practice, Equation (2.3) is an easy means of calculating the power capacity of the solar PV array. However, a question is raised as to how to ascertain the appropriate value of the safety factor, S_F, rather than relying just on assumptions. It is against this backdrop that Equation (2.5) is employed to determine the size of a solar PV array, while also providing the opportunity to quantify the temperature losses by using the temperature coefficient of the power, cell temperature, cell temperature at STC, and the other losses by using the derating factor [27].

$$PV\ Array\ Capacity = nP_{RT}D_{FT}\left(\frac{G_L}{G_{STC}}\right)\left[1 + \alpha_{TC}\left(T_{CL} - T_{CL,STC}\right)\right] \qquad (2.5)$$

where P_{RT} is the solar module at STC; D_{FT} represents the derating factor; G_L is the location's irradiance; G_{STC} is the irradiance at STC, i.e., the peak irradiance; α_{TC} is the temperature coefficient of power; T_{CL} is the solar cell temperature; and $T_{CL,STC}$ is the solar cell temperature at STC. At STC, the values of the irradiance, temperature, and air mass spectral distribution are 1 kW/m^2, 25°C, and 1.5, respectively [25]. The value of T_{CL} can be determined by employing Equation (2.6) [27]:

$$T_{CL} = T_{AMB} + \left(NOCT - 20\right).\left(\frac{G_L}{G_{RF}}\right) \qquad (2.6)$$

where T_{AMB}, $NOCT$, and G_{RF} stand for the ambient temperature, nominal operating cell temperature, and irradiance at a reference value of 0.8 kW/m². The V/I characteristics/output of solar PV modules are temperature-dependent, which is the essence and significance of α_{TC} [25].

Using the crystalline silicon PV – a mature and leading technology around the world – as a typical example, an increase in solar PV module cell temperature is associated with a decrease in the module's voltage and power, with just a slight increase in current. This implies that the severity of the effect of operating the PV module at a temperature greater than 25°C is such that it presents a higher reduction in the maximum voltage and power of the module compared to only a slight increase in the current. This is why the impact of temperature is crucial to the performance of the crystalline silicon PVs, hence the need to properly analyze the temperature losses [28]. The values of α_{TC}, P_{RT}, $NOCT$, η_{pv}, and A_{pv} in Equation (2.1) are usually provided by the PV manufacturers and can be found on the module's specification sheet.

However, if the value of $NOCT$ is not known, 48°C may be used as recommended in [29] as a reasonable value, which is not far from the values reported by most widely used solar modules. Also, the values of I_S, G_L, and T_{AMB} are determined from the location's data, while D_{ED} is based on the users' load requirement.

In PV systems, the essence of D_{FT} a scaling factor that is employed in array sizing is to account for the unforeseen reduction in the array's output in real-life operating situations compared to the standard test conditions used to characterize and rate the modules [30]. This scaling factor is employed to account for losses other than temperature losses, such as PV module soiling, shading, aging, and wiring, as is the case with Equation 2.5. This implies that the total losses are equated to those of temperature and those estimated by D_{FT}. Therefore, the safety factor, S_F, in Equation (2.3) is expected to combine the losses, that is, temperature losses plus other losses when expressed as a multiplication factor. The derating factor needs to be decided when modeling the system, but typical values may be 0.8, 0.75, or 0.9 based on the application.

One can conclude that analysis of losses in solar PV arrays is a critical step in ascertaining the appropriate power capacity. Based on the above analyses, it can be said that the ideal solar array capacity without losses can be obtained by rewriting Equation (2.4) as:

$$Ideal\ PV\ Array\ Capacity = nP_{RT}\left(\frac{G_L}{G_{STC}}\right) \qquad (2.7)$$

Equations (2.8) and (2.9) represent the solar array capacity with the consideration of other losses and temperature losses, respectively:

$$PV\ Array\ Capacity\ with\ other\ losses = nP_{RT}D_{FT}\left(\frac{G_L}{G_{STC}}\right) \qquad (2.8)$$

PV array capacity with temperature losses =

$$nP_{RT}\left(\frac{G_L}{G_{STC}}\right)\left[1+\alpha_{TC}\left(T_{CL}-T_{CL,STC}\right)\right] \tag{2.9}$$

Equations (2.10) and (2.11) represent the value of the other losses and the temperature losses, while the total losses are expressed in Equation (2.12):

$$\textit{Other losses} = nP_{RT}\left(\frac{G_L}{G_{STC}}\right)-nP_{RT}D_{FT}\left(\frac{G_L}{G_{STC}}\right) \tag{2.10a}$$

$$= nP_{RT}\left(\frac{G_L}{G_{STC}}\right)\left[1-D_{FT}\right] \tag{2.10b}$$

Temperature losses =

$$nP_{RT}\left(\frac{G_L}{G_{STC}}\right)-nP_{RT}\left(\frac{G_L}{G_{STC}}\right)\left[1+\alpha_{TC}\left(T_{CL}-T_{CL,STC}\right)\right] \tag{2.11a}$$

$$= nP_{RT}\left(\frac{G_L}{G_{STC}}\right)-nP_{RT}\left(\frac{G_L}{G_{STC}}\right)\left[1+\alpha_{TC}\left(T_{CL}-T_{CL,STC}\right)\right] \tag{2.11b}$$

$$= -nP_{RT}\left(\frac{G_L}{G_{STC}}\right)\alpha_{TC}\left(T_{CL}-T_{CL,STC}\right) \tag{2.11c}$$

Total losses =

$$\left(nP_{RT}\left(\frac{G_L}{G_{STC}}\right)\left[1-D_{FT}\right]\right)+\left[-nP_{RT}\left(\frac{G_L}{G_{STC}}\right)\alpha_{TC}\left(T_{CL}-T_{CL,STC}\right)\right] \tag{2.12}$$

Suppose that the ideal or basic capacity of Equation (2.3) is 1000 kW and the total loss is 25%. What this translates to is that S_F is 1.25 according to Equation (2.3). Equations (2.4) to (2.12) may then be employed to quantify temperature losses and other losses.

2.1.2.2 Output of a wind generator

The electrical power available in the wind can be calculated by [4, 11]:

$$P_{Wind} = \frac{1}{2}A\rho v^3 C_p \tag{2.13}$$

where ρ, A, v, and C_p stand for the density of air (i.e., 1.225 kg/m³), the rotor blades' swept area in m², wind speed, and the power coefficient, respectively.

C_p also represents the rotor's aerodynamic efficiency [4]. The maximum value of the wind turbine's aerodynamic efficiency is approximately 59%, which is regarded as the Bert limit [4, 31, 32].

The power in the wind, being proportional to the cube of the wind speed, in Equation (2.13), is an indication that wind generators can produce an appreciable amount of energy. It is possible to generate electricity on a wind farm with a very windy resource at a cost comparable to those of traditional electrical power generators [4].

It is also noteworthy that a location's wind speed varies with height (i.e., the hub height), and this can be expressed as [11, 33]:

$$\frac{v}{v_0} = \left(\frac{h}{h_0}\right)^{\alpha} \tag{2.14}$$

where v, v_0, and α represent the estimated wind speed at a height, h, known wind speed at the height, h_0, and the coefficient of friction, respectively. The values of α are usually dictated by the terrain; typical values are 0.10, 0.15, or 0.20 for terrains with smooth, hard ground and calm water; tall grass on level ground; high crops, hedges, and shrubs, respectively [11, 31, 32].

It is necessary to state that ρ depends on two parameters: the temperature and the barometric pressure; this is responsible for wind electrical power decreases, which occur as the elevation decreases, by about 10% per 1,000 m [12]. Equation (2.14) is essentially a power law employed to define a term referred to as wind shear in a wind power system, which provides an insight into the change in the site's wind speed with elevation; α may also be regarded as the wind shear exponent.

Ascertaining the annual electricity production of a wind turbine (WT) is an important step in system performance assessments. The annual energy generation of the wind turbine, in kWh or MWh per year, can be calculated by [12, 33]:

$$AEG = WT_{RP} \cdot CF \cdot 8760 \tag{2.15}$$

where AEG, WT_{RP}, and CF represent the annual energy generation, the WT's rated power, and capacity factor, respectively, while 8,760 is the number of hours in a year. CF, which is likened to an average efficiency, can be calculated using [12, 34]:

$$CF = \frac{WT_{AVP}}{WT_{RP}} \tag{2.16}$$

where WT_{AVP} and WT_{RP} are the average power and the WT's rated power. WT_{AVP} is usually estimated by dividing the energy generation by the hours in that period under consideration, for example, a year or a month. For instance, suppose that AEG is 3×10^6 kWh per year and WT_{RP} is 1,500 kW, then the value of WT_{AVP} will be $\left(\dfrac{3 \times 10^6}{1500}\right) = 342.47$ kW. The value of CF will then be $\left(\dfrac{342.47}{1500}\right) = 0.2283 = 22.83\%$. Also, it is important to note that the value of CF depends on WT_{RP} versus the WT rotor area; this is because WT models can have different generator sizes for the same rotor size or the same generator sizes for different rotor sizes for better operational performance in different wind speed regimes [12].

WTs have CFs of 0.2–0.4 depending on how windy the location under consideration is, and there is no practical electrical power-generating system that can achieve a CF of 1.0 [4, 12]. This is a result of unforeseen unavailability due to system breakdowns or faults, the scheduled maintenance exercises in the year, and so on. However, a CF of 0.85–0.90 is achieved by base load thermal generating plants when they are newly commissioned, though this value depreciates over the years until the power plants are finally decommissioned at the end of their operational lives [4].

2.1.2.3 Output of hydro-generating systems

The power that can be obtained from a water flow rate, Q (cumecs), that is falling through a particular net or effective head (m), is given by [4]:

$$P_{hydro} = \rho g H_{eff} Q \tag{2.17}$$

where ρ, g, H_{eff}, and Q represent the density of water (i.e., 1,000 kg/m³), the acceleration due to gravity (with a value of 9.8 m/s²), the effective head (in meters), and the water flow rate in cumecs, respectively. The term "cumecs" stands for m³ per second. The effective head is given by:

$$H_{eff} = H_{gross} - H_{loss} \tag{2.18}$$

where H_{gross} is the gross head, which is the actual height difference, while H_{losses} is the head loss in the piping. Going forward, Equation (2.19) can be employed to show the theoretical maximum electrical power that a pipeline delivers when the value of H_{loss} is $\left(\dfrac{H_{gross}}{3}\right)$ [27].

The energy output of the hydro system, E_{hydro}, at a time frame, Δt (hr), is given by [35]:

$$E_{hydro} = P_{hydro}\Delta t \tag{2.19}$$

The hydraulic efficiency is the ratio of the power obtained by the runner of the hydro turbine to the electrical power supplied at the inlet of the turbine system [35]; this is mathematically expressed by:

$$\eta_H = \frac{P_{runner}}{P_w} \tag{2.20}$$

where P_{runner} and P_w represent the runner power and water power, respectively. There is also mechanical efficiency, which is defined as the ratio of the available electrical power at the shaft of the hydro system (i.e., P_{shaft}) to the runner power; this is given by [35]:

$$\eta_M = \frac{P_{shaft}}{P_{runner}} \tag{2.21}$$

Now, suppose that the total efficiency of the turbo-generator is represented by η, the hydraulic power and the corresponding energy output of the hydroelectric power system can then be represented by Equations (2.22) and (2.23), where η is also defined by Equation (2.24) [35]:

$$P_{hydro} = \eta \rho g H_{eff} Q \tag{2.22}$$

$$E_{hydro} = \eta P_{hydro} \Delta t \tag{2.23}$$

$$\eta = \eta_H \eta_M \tag{2.24}$$

The essence of the application of η in Equations (2.22) to (2.24) in a real-life situation is that there is no practical hydroelectric system that can attain the efficiency of 100%. The quantification of losses is also an important consideration for sizing the hydropower system. Modern turbines achieve an energy conversion of 0.9, which is in the range of 0.6 to 0.8 in micro-hydro (μ-hydro) schemes [35].

There are large and small-scale hydroelectric schemes. The large-scale hydropower scheme is a major contributor to electric power production around the globe, and it is sometimes described as a power plant that has a capacity of more than 30 MW [12, 27]. This is because the electrical technology is mature and well-developed, and its operation is desirable in terms of system availability and flexibility of electricity supply [4]. Going by the recent report in [12], shown in Table 2.1, hydropower capacity accounts for about 38% of the total global renewable power capacity. However, one of the challenges of large hydro schemes is that it is associated with huge upfront capital costs with profits accumulated over a long time into the future; their implementation is also associated with adverse environmental impacts such as flooding of large land masses and displacement of people – the Three Gorges scheme in China is an example.

In the case of small hydropower (SHP) systems, there is little variation in the output on a minute-to-minute basis; however, there can be substantial changes on hourly or daily cycles because of sudden rainfall [4]. This way, it is important to note that the output of the scheme is highly sensitive to seasons in the year – dry and rainy seasons, for instance [36]. When a large number of SHP systems are integrated into an existing electrical power network, they will have a negligible influence on the minute-to-minute operation of the network as a result of the aggregation of uneven short-term variations brought about by statistical smoothening [4, 5]. Furthermore, the term "SHP" is commonly described as one having a capacity of less than 5 MW [4]. However, SHP has also been considered as that having a generating capacity between 0.1 and 30 MW, while those smaller than 0.1 kW are regarded as μ-hydro schemes [27].

A simple example of the micro-hydro schemes is the run-of-the-river system arrangement, which does not require a dam. The scheme is found attractive as the energy resource is often situated in off-grid rural locations, in which case the size and the load demand requirements of the communities in such areas are consistent with the possible energy supply. The above sizing methodologies assume a typical SHP or μ-hydropower system.

2.1.2.4 Output of tidal power systems

It is established that the natural rise and fall of coastal tidal waters are a result of the gravitational fields of the moon and the sun [4]. The moon is closer to the earth, though less massive, and exerts a dominant effect on tidal resources – the influence of the moon is 2.2 times greater than the influence of the sun, which is why the tidal energy resource can be considered a kind of lunar energy [4]. The tidal range – the vertical rise and fall of tides – is small; however, there is a form of enhancement of the range in certain locations. The tidal range may be enhanced by funneling, resonance, and the Coriolis effect.

It is possible to trap seawater at a high tide in an estuary basin of area, A, behind a barrier or dam to develop a tidal range of power. Suppose that the water is allowed to run out at low tides through the turbine system; then the average electrical power generated is [17]:

$$P_{tidal} = \frac{\rho A R^2 g}{2\tau} \tag{2.25}$$

where ρ, A, R, g, and τ represent the seawater density, area of the estuarine basin, the range, the acceleration due to gravity, and the period, respectively. For example, if $A = 12$ km², $R = 3$ m, $\tau = 12$ h 25 min, $g = 9.81$ m/s, and 1,025 kg/m³ is used for ρ, then the value of P_{tidal} is 12.15 MW.

However, in case the maximum and minimum tidal range values exist or are recorded, then the value of R will then be the mean of the square of the maximum and the square of the minimum values. Therefore, Equation (2.25) can be translated to [17]:

$$P_{tidal} = \frac{\rho A g}{2\tau}\left[\frac{R_{max}^2 + R_{min}^2}{2}\right] = \frac{\rho A g\left(R_{max}^2 + R_{min}^2\right)}{4\tau} \tag{2.26}$$

2.1.2.5 Output of biogas systems

Using a typical illustration as presented in [28], biogas is referred to as gasified biomass. The idea is that biomass feedstock such as energy crops (e.g., giant reed, miscanthus, wood waste, agricultural residue) can be taken through a process of gasification either by a thermo-chemical or biological method. The end product, which is a gaseous fuel, includes synthesis gas, syngas, producer gas, or wood gas. Although biogas has a low heating value (LHV) compared to a fossil fuel alternative (e.g., natural gas), it constitutes a large quantity of nitrogen (that is non-combustible), and its advantages over solid biomass materials include higher efficiency, relatively clean combustion, and better control.

The annual electricity production of a biomass gasifier can be estimated by [37]:

$$E_{bg} = P_{sr} \cdot 8760 \cdot C_{UF} \tag{2.27}$$

where P_{sr}, C_{UF}, and 8760 are the biomass gasifier system rating, the capacity utilization factor, and the total hours in the year.

The operation of a co-fired generator is one application that is of interest in biogas systems, which runs on a mixture of biogas and fossil fuel resources [29]. The estimation of the output of this generator presented in this section is based on the method and assumptions discussed in the Hybrid Optimization of Multiple Energy Resources (HOMER Pro Index). HOMER Pro Index is software used for designing microgrid systems. It is possible to calculate the output and the mass flow rates of biogas and fossil fuel resources at each time step. The assumptions are [29]:

1. The substitution ratio, Zg, of the biogas resource represents a constant parameter, which is independent of the electrical power output of the fuel mixture of the engine;
2. The energy system arrangement is such that it always seeks to maximize the utilization of biogas resources while minimizing the fossil fuel resource utilization;
3. The system has been configured in such a way that the fossil fuel percentage cannot be less than a certain pre-determined minimum level;

4. The generating system can deliver up to 100% generation provided the fossil fuel percentage is high enough, even if the derating factor introduced to the dual-fuel mode of operation is below 100%.

Furthermore, the fuel curve of a co-fired generator is described in terms of its fuel consumption in the pure fossil-fuel mode, which is given by:

$$m'_o = \rho_{ffl} \cdot \left[F_o Y_{gen} + F_1 P_{gen} \right] \tag{2.28}$$

where m'_o, ρ_{ffl}, F_o, Y_{gen}, F_1, and P_{gen} represent the fossil fuel flow rate in pure fossil mode (kg/hr), the density of fossil fuel (kg/L), the generator's fuel curve intercept coefficient (L/hr/kW), its rated capacity (kW), its fuel curve slope (L/hr/kW), and its power output, respectively. F_o is defined as the ratio of the generator's no-load fuel consumption to its rated capacity, while F_1 is the generator's marginal fuel consumption [29]. However, Equation (2.29) results from the first assumption:

$$m'_o = m'_{ffl} + \frac{m'_g}{Z_g} \tag{2.29}$$

where m'_{ffl}, m'_g, and Z_g represent the fossil fuel flow rate in dual-fuel mode (kg/hr), the biogas flow rate in dual-fuel mode (kg/hr), and the biogas substitution ratio, respectively; m'_g can be estimated by:

$$m'_g = Z_g \cdot \left(m'_o - m'_{ffl} \right) \tag{2.30}$$

where Z_g is the proportion with which the biogas resource replaces fossil fuel in a co-fired generator? The fossil fuel proportion can be defined as the fossil fuel utilized by the generator when in dual-fuel mode divided by the amount of fossil fuel that can produce the same output power when in pure fossil mode [29]. This can be described using the mathematical relation:

$$x_{ffl} = \frac{m'_{ffl}}{m'_o} \tag{2.31}$$

Putting Equation (2.31) into Equation (2.29), we get:

$$m'_g = Z_g \cdot \left(m'_o - x_{ffl} \cdot m'_o \right) \tag{2.32}$$

Therefore, the biogas flow rate in dual-fuel mode (kg/hr) is then given by:

$$m'_g = Z_g m'_o \cdot \left(1 - x_{ffl} \right) \tag{2.33}$$

The goal of the second assumption is to maximize m'_g, which also translates to minimizing x_{ffl}. The constraint represented by Equation (2.34) is based on the third assumption:

$$x^*_{ffl} \leq x_{ffl} \leq 1 \tag{2.34}$$

where x^*_{ffl} is the minimum fossil fuel proportion that is needed for ignition. Diesel engine systems, for example, need a certain minimum quantity of diesel fuel to achieve proper ignition. This may not be the case with spark-ignition engine systems, which might be capable of operating on a pure biogas resource [28]. Employing Equation (2.32), we obtain the target value of biogas flow rate, m''_g, represented by:

$$m''_g = Z_g m'_o \cdot \left(1 - x^*_{ffl}\right) \tag{2.35}$$

However, there are essentially two independent upper limits of the actual quantity of m'_g. The generator's output at x^*_{ffl} is limited to the maximum output given in:

$$Y^*_{gen} = \tau \cdot Y_{gen} \tag{2.36}$$

where τ is the derating factor – the maximum output of the co-fired generator running at x^*_{ffl} – and is less than unity. Putting Equation (2.28) into Equation (2.33), we get:

$$m'^*_g = Z_g \rho_{ffl} \cdot \left[F_o Y_{gen} + F_1 Y^*_{gen} \right] \cdot \left(1 - x^*_{ffl}\right) \tag{2.37}$$

However, the available biomass resource, a_g, also constitutes an upper limit on m'_g, therefore the actual quantity of m'_g as stated in Equation (2.38) implies the minimum of m''_g, m'^*_g, and a_g:

$$m'_g = MIN\left(m''_g, m'^*_g, a_g\right) \tag{2.38}$$

If the value of the quantity m'_g is known, then x_{ffl} can be determined. Therefore, from Equation (2.33):

$$x_{ffl} = 1 - \frac{m'_g}{Z_g m'_o} \tag{2.39}$$

And from Equation (2.31):

$$m'_{ffl} = x_{ffl} \cdot m'_o \tag{2.40}$$

Therefore, at any time step, given certain values of P_{bio} and a_g, the values of biogas and fossil fuel flow rates can be determined by Equations (2.38) and (2.40), respectively.

Equations (2.28) to (2.40) are based on the approaches presented in the HOMER Pro simulation tool for co-firing. These can guide the simulation and analysis of a co-fired generating system.

2.1.3 Limitations of Renewable Energy Resources

The issues of intermittency and the relatively low density of the sources are among the major limitations of renewable energies, which invariably causes these resources to have higher initial capital costs [12]. A solar PV system is used in this section to explain the mechanism or significance of the low energy density of RE resources. The average sun intensity outside the atmosphere – that is, the solar constant – is around 1.353 kW per m² [4]. However, because of the effect of attenuation brought about by the atmosphere, the peak solar intensity of about 1 kW per m² is obtained at sea level, resulting in a 24-hour yearly average of 0.2 kW per m² on average over the surface of the earth. This relatively small total energy density is the reason why large surface areas are required for significant energy production from the solar PV system.

However, other perceived issues which are sometimes considered as disadvantages of different kinds of RE resources other than limitations include foul odors from biomass materials and processing, visual pollution, displacement and death of birds by wind turbine blades, displacement of people in the case of a large hydropower plant, and brine from geothermal energy. The problem of intermittency leads to the integration of storage devices with RE systems, one of which is the battery energy system. Some battery cells are made of materials that have environmental concerns such as hazardous cadmium wastes in the case of nickel-cadmium battery technology [5]. Apart from this, a future concern is the question of how to handle the environmental impact of the renewable technologies that are currently within their operational life. For instance, the lifespan of a solar PV module is usually between 25 and 30 years; however, there is the possibility of serious environmental issues after these systems are decommissioned if there are no plans for component recycling and end-of-life management.

To this end, this chapter presents an overview of the inherent features, characteristics, and limitations of RE resources to aid understanding of how they may be harnessed for energy supply or grid support purposes. Different configurations of RE system applications are examined to highlight whether or not the energy systems considered have an interaction with the existing grid and how the intended users are served. Some electrical topologies of RE applications including autonomous, backup, grid-tie, and grid-parallel power applications are presented and compared. The chapter is meant to provide useful insights into the use of RE resources in producing and managing electric power.

2.2 RENEWABLE ENERGY RESOURCES: AUTONOMOUS SUPPLY AND RELIABILITY

2.2.1 Intermittency of renewable energy resources

The intermittent characteristics of RE resources are usually reflected in their system output, which makes them *variable*, unlike traditional energy resources – diesel, natural gas, gasoline, etc. This is one of the fundamental technical challenges of renewable energies, which has over the years attracted attention, interest, and solutions both in research and real-life engineering situations.

It is reasonable to use the case of wind energy resources as an example in this section. Because of its RE nature, the wind speed at a particular site is continuously varying. These include: annual changes that take place in the yearly mean wind speed; seasonal changes that occur over the seasons; synoptic changes which take place with passing weather systems; diurnal changes; and turbulence that occurs on a second-to-second basis [4]. These resource characteristics, as they occur on different timescales, present a difficulty in predicting the total wind energy that may be captured from a potential location on an annual and seasonal basis. This development, therefore, requires sound technical efforts to ensure that the variability of renewable electricity generation is being managed for appropriate utilization.

Furthermore, the variable nature of RE resources has led to a growing development and application of energy storage technologies such as batteries, compressed-air, superconducting magnetic, ultra-capacitors, cryogenic and aquiferous, pumped heat, and flywheel energy storage systems, including hydrogen storage and fuel cells that are also currently being considered for managing/balancing fluctuating energy generation [5, 38]. In light of this, storage technologies have been instrumental in the increased penetration of RE resources in the grid system. It is also important to state that the size/scale and type of RE, the duration of operation of the storage devices, the type of storage technology, and the cost are among the factors that determine whether the storage device is to be deployed for a particular application.

Scholarly works in the direction of managing the variability of RE resources are well established in the literature, such as the study on realizing long-duration energy storage and flexible electricity production technologies with high-variable renewable energy networks [39]. A survey of hybrid energy storage systems emphasized their application for intermittent renewable energies [40], and there was a discussion of the issues about grid-scale storage applications for renewable power integration in [41]. Providing an answer to the question of how much energy storage is required to integrate large intermittent renewable energy resources is the aim of the work reported in [42]. An investigation of the sizing of a voltage source for a battery bank was discussed in [43]. The work was conducted for microgrid systems with renewable energy resource utilization. These studies are mentioned in this section to showcase a few contributions in the literature that focused on addressing the concerns of RE intermittency to generate or supply stable energy. Several others exist that considered both storage devices and power conditioning and control systems.

2.2.2 Standalone single-source RE configurations

Standalone single-source renewable energy system configurations are illustrated in Figure 2.1a–e. These systems have been created using the Hybrid Optimization of Multiple Energy Resources (HOMER) Pro Microgrid simulation tool, and are presented in this chapter for illustration, each with different renewable energy resources. These systems also have DC and AC couplings with a bi-directional power converter that connects the battery bank with the AC bus, although Figure 2.1 is based on the AC system. The true picture of the bi-directional converter is that it is a dual converter because it does the work of an inverter to the left side (i.e., DC to AC) and it functions as a rectifier to the right (i.e., AC to DC) of Figure 2.1a–d.

The hydrokinetic component or model in the HOMER tool can be employed to represent different kinds of low-head hydro schemes; this includes the run-of-the-river, tidal, and wave energy systems [29]. It is essentially described as a turbine, but the model can also represent a wave energy converter. However, it is assumed to be a tidal system in this chapter. The renewable energy systems are assumed to satisfy a commercial facility having a load demand of 2,424 kWh/yr with the peak load requirement being 348 kW.

The commonality of the RE systems application shown in Figure 2.1 is due to the electricity supply being from a single-source system such as solar, wind, hydro, hydrokinetic (tidal current), and biomass (biogas). It is important to understand that these energy configurations depend on the available RE resources in the location where the energy supply is required. The size of the load and the type of application are other critical factors that determine the techno-economic feasibility of single-source RE systems (i.e., 100% RE systems). While it may be techno-economically viable to meet relatively low energy demands such as those in domestic/residential premises, it requires a huge capital investment to handle a load size of 2,424 kWh per day as presented in Figure 2.1. Therefore, a single-source RE system is most suitable for low load demand per day; also, though it has a relatively high initial capital cost per unit of power, it has a relatively low operating cost [44].

Single-source energy configurations do not have any interaction with the existing grid, and neither do they have any additional power source(s) that can support the load in the case where RE generation is low or not available. This development makes achieving high system reliability a major issue with systems being run by a single-source RE supply – because the energy supply

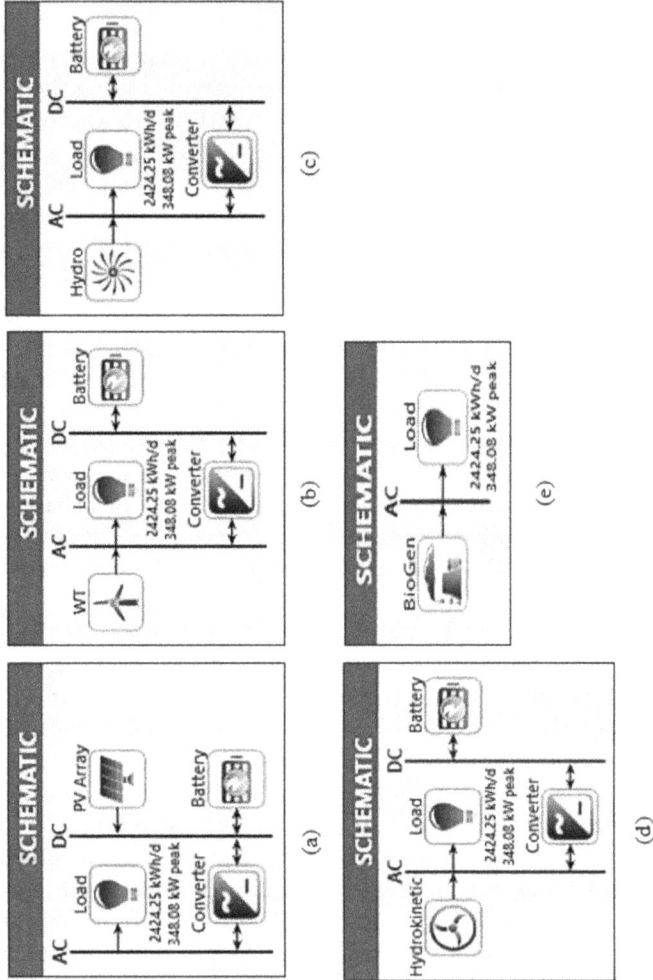

Figure 2.1 Standalone single-source RE configurations: (a) PV/battery/converter and the load; (b) WT/battery/converter and the load; (c) hydro/battery/converter and the load; (d) hydrokinetic/battery/converter and the load; (e) biogas and the load.

is highly dependent on the natural cycle of the RE sources. Therefore, a loss of load (LOL) will be experienced whenever there is no supply from the source or when the load exceeds generation.

LOL is usually expressed in terms of the unmet energy demand. Suppose the energy demand met by the systems in Figure 2.1a–e are assumed to be 1,832, 1,500, 1,644, 1,934, and 2,034 kWh per year. With the assumption that generation is meant to equate to demand, then the unmet demand will be 592.25, 924.25, 780.25, 490.25, and 390.25 kWh/yr. A standard measure of this unmet demand is the loss of load probability (LOLP) which expresses the obtained values as a fraction or percentage. The LOLP is thus 24.43, 38.13, 32.19, 20.22, and 16.10%, respectively, and can be mathematically represented as [45]:

$$LOLP(\%) = \frac{\sum_{i=1}^{8760} (ED_{unmet})_i}{\sum_{i=1}^{8760} (E_{totd})_i} \times 100 \tag{2.41}$$

where ED_{unmet} and E_{totd} stand for the unmet energy demand and total energy demand, respectively.

Furthermore, LOLP may also be translated to the number of hours or days in the year that the users' energy demand is unmet. Since the total hours in the year is 8,760, then the number of hours in the year that the unmet energy demand is experienced is the product of LOLP and 8,760 hours. Therefore, for the obtained LOLP, the users' energy demand will not be met for about 2,140, 3,340, 2,820, 1,771, and 1,410 hours in the year. Similarly, the number of days of unmet energy demand will then be determined by dividing these values by the days in the year (i.e., 365), thus translating to 89.2, 139.2, 117.5, 73.8, and 58.8 days.

The availability of the energy system is another important reliability parameter, which may be considered as the direct opposite of the LOLP parameter. This usually depicts the percentage or fraction of time that an energy-generating system can satisfy the load requirements [46]. Based on the mathematical relation of Equation (2.1), the system availability can then be defined by:

$$Availability(\%) = \left(1 - \frac{\sum_{i=1}^{8760} (ED_{unmet})_i}{\sum_{i=1}^{8760} (E_{totd})_i}\right) \times 100 \tag{2.42}$$

An electrical system that is designed or configured to operate at an availability of 96%, for instance, will be expected to satisfy the user's load requirement 96% of the time. In standalone power arrangements, critical loads usually require an availability of about 99%, while non-critical loads are designed with a system availability of around 95% [46]. The availability of the above examples is 75.57, 61.87, 67.81, 79.78, and 83.9%, respectively, according to the mathematical relation of Equation (2.42). It has also been established that when compared with fossil fuel-based generators, the single-source renewable energy generators, examples of which are shown in Figure 2.1, present low environmental concerns.

2.2.3 Standalone hybrid RE configurations

The essence of a hybrid electricity system generally is that it integrates two or more energy resources, a storage system, and the necessary controls for providing electricity to remote loads [41, 47, 48]. Standalone hybrid renewable energy system configurations are illustrated in Figure 2.2.

The peculiarity of these energy configurations is that they have more than one source of RE unlike those shown in Figure 2.1. One merit of the hybrid RE systems over the single-source systems is that they can realize relatively high reliability as a result of the complementary characteristics of the participating energy sources [44, 49]. Thus, they can be designed or configured to handle a higher load demand requirement than the single-source energy configurations because of the contributions from different power sources.

RE resources do not only behave differently; their availability also differs. Solar energy, for instance, is only available during the day. In addition, there is high solar energy generated during the summer or the dry season compared to the winter or rainy season and cloudy days when there is low solar energy production. However, it is possible to experience windy weather during the day, night, on cloudy days, in winter, or in the rainy season. The combination of these energy sources will ensure that one energy resource complements the other. Thus, the reliability of the energy configuration of Figure 2.2a compared to Figure 2.1a and b is improved.

Furthermore, there are abundant water resources during the rainy season compared to the dry season. It is also common knowledge that drought is usually experienced in a particular year where there is low rainfall (i.e., a dry year). These resource characteristics are opposite to the solar energy resource, which is abundant during the dry season during a dry year. The combination of these two resources can achieve better reliability compared to a single-source electricity configuration. This reveals that RE applications illustrated in Figure 2.2b, c, g, and h are expected to realize improved reliability performance compared to the system in Figure 2.1c that is powered only by a hydro source.

Other RE resources such as biomass, tide, wave, and geothermal also have their availability and seasonal characteristics. This makes a hybrid configuration of these resources a better option compared to an energy supply system that depends only on one source. However, there can still be some periods in the year when the loss of load will be experienced when the RE

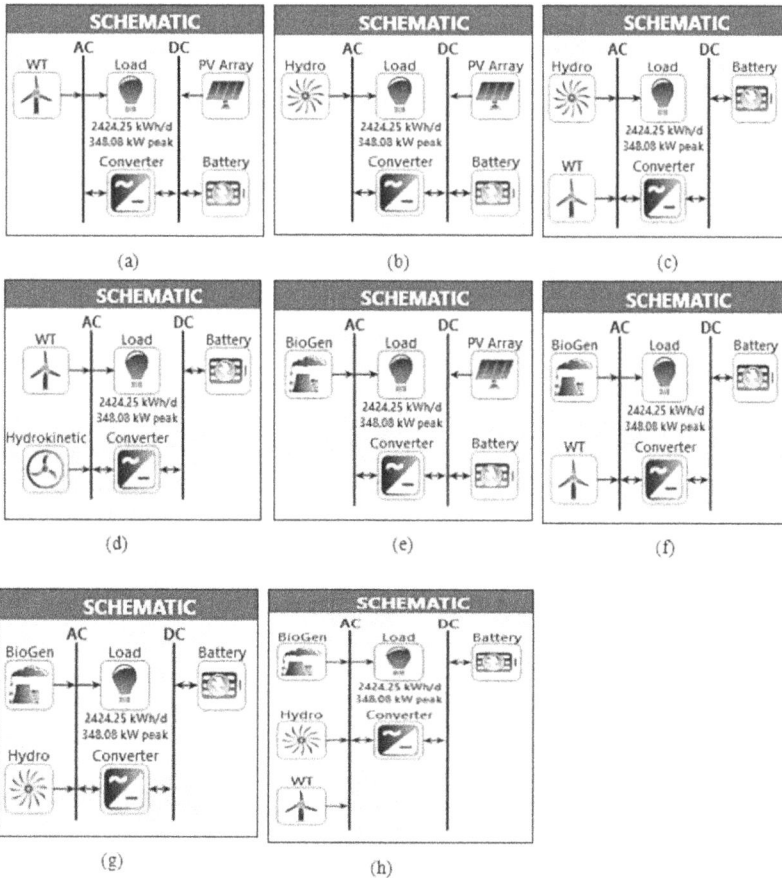

Figure 2.2 Standalone hybrid RE configurations: (a) PV/WT/battery/converter and load; (b) PV/hydro/battery/converter and load; (c) WT/hydro/battery/converter and load; (d) WT/hydrokinetic/battery/converter and load; (e) biogas/PV/battery/comverter and load; (f) biogas/WT/battery/converter and load; (g) biogas/hydro/battery/converter and load; (h) biogas/hydro/WT/battery/converter and load.

resources may be unavailable or the supply is not enough to satisfy the demand requirements. This development is still due to the variable and temporal characteristics of RE resources, which affect system reliability in terms of the LOLP. Hence, the energy supply is highly dependent on the natural cycles of REs.

It is also important to note that the technical and economic feasibility of hybrid renewable electricity systems is also determined by the size of the load, the type of application, and the available RE resources. Such systems usually have a relatively high initial capital cost per unit of power and relatively low operating costs [44].

2.2.4 Standalone hybrid RE and non-RE source configurations

Hybrid renewable energy and non-renewable energy system configurations are presented in Figure 2.3a–e.

The conventional generator in this case is referred to as a fossil-fueled generator such as those run on diesel, gasoline, or natural gas, depending on the design or the choice of fuel made by the user. They are also considered as dispatchable generating systems and are most suitable for consistently high user load demand per day in several electrical installations; although when such generating systems are run as the only source of power, they have a relatively low initial capital cost per unit power and are associated with relatively high operating costs owing to fuel consumption [44]. However, the initial cost of procurement is relatively high.

These standalone power systems benefit from two different energy resources – renewable and non-renewable – which is why such electrical systems are typically configured to include at least one conventional power generating set (i.e., a dispatchable generator) and one renewable energy-based generator as illustrated in Figure 2.3 [44]. A common example of this is shown in Figure 2.3a which may integrate a solar PV array and a diesel generator with the addition of battery storage and a power converter – the control. The addition of a battery bank to the energy configuration helps to achieve more efficient operation of the diesel generator and also provides support for the load when RE generation is low or not available. The hybrid renewable and the non-renewable energy source configurations are suitable for variable, mid-range low, and high load demand, and are associated with moderate to high initial capital cost per unit of power and operating cost [44].

In a practical sense, the electrical systems illustrated in Figure 2.3 are systems usually designed or meant to meet the load demand of multiple customers in a particular locality through electrical wire connections [10]. However, to realize a relatively clean or eco-friendly supply, it is desirable to have a high RE contribution in the total energy mix (i.e., RE and non-RE resources). The RE fraction or contribution may simply be expressed by:

$$RE\ Fraction\,(\%) = \frac{RE\ generation}{Total\ generation} \tag{2.43}$$

where the total generation refers to the sum of the contributions from all the generating systems, that is, the RE and non-RE systems. However, RE penetration can be described by [4, 29]:

$$RE\ Penetration\,(\%) = \frac{Total\ RE\ generation}{Total\ Load\ Served} \tag{2.44}$$

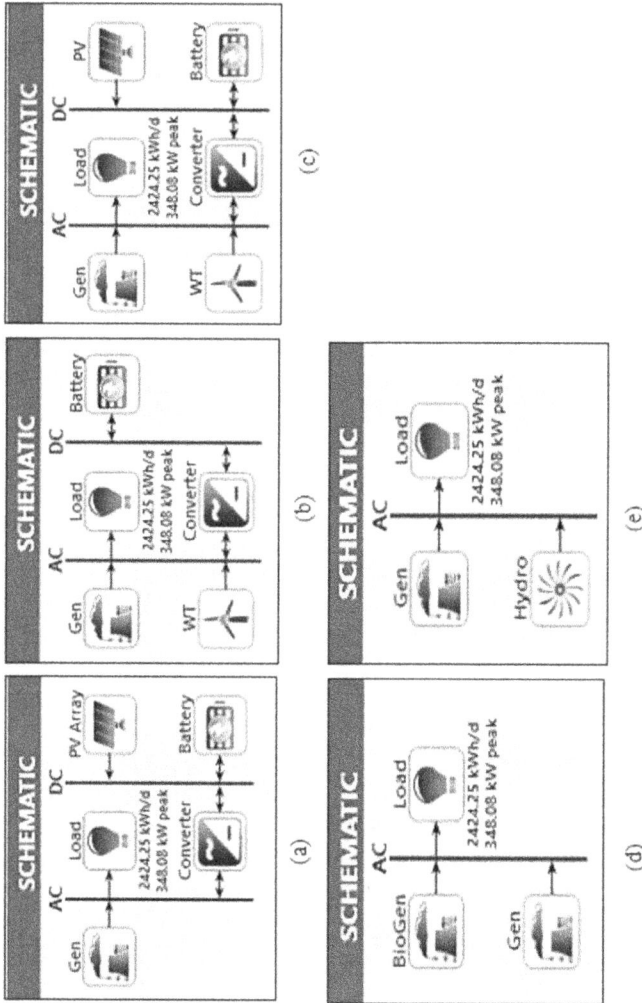

Figure 2.3 Standalone hybrid RE and non-RE resource configurations: (a) PV/Gen/battery/converter and load; (b) WT/Gen/battery/converter with load; (c) WT/hydro/battery/converter and load; (d) biogas/Gen and load; (e) hydro/Gen and load.

Furthermore, another practical aspect of the hybrid RE and non-RE system is that since a high fraction of renewable electricity is desired and thus used as the main source of power, then non-renewable electricity generation may be employed as a backup. The backup in this case is specifically used to cover the periods in the year when there is a loss of load, hence helping to achieve high reliability of energy supply – that is, high energy system availability. The aspect of backup application will be further illustrated in a subsequent section of the chapter.

There is also the possibility of load growth – this has been described as climbing the energy ladder [50]. Another benefit of the non-RE generator shown in Figure 2.3 is that the dispatchable generator, being the additional power source, can be employed to meet the additional load, depending on the level of increase. This is possible because the non-RE generating system is independent of the natural cycles as is the case with RE generators.

There are different types of hybrid RE and non-RE systems. It is apparent that the systems presented in Figures 2.1–2.3 do not have any interaction with the grid, thus they must have the capability of load following, especially those that include a battery bank. In addition, there are three classifications of standalone hybrid RE and non-RE systems: series, switched, and parallel hybrid power systems [10, 51]. Figure 2.3a–e are good examples of the parallel hybrid configuration.

2.2.4.1 Series hybrid systems

Figure 2.4 is an example of the series hybrid configuration based on existing studies [10, 51], where all the energy-generating sources present a DC supply to the battery bank to supply the DC and AC loads. The WT is DC operated. The series hybrid configuration may be run in manual or automatic mode, including the integration of a battery voltage monitoring mechanism and an ON/OFF control for the conventional generator. A unique feature of such an arrangement is that each component is equipped with a charge controller – i.e., controllers 1 and 2 – as is the case with wind and PV sources shown in Figure 2.4, while the conventional generator is fitted with a rectifier (a battery charger). The two charge controllers will help to prevent overcharging of the battery cells. Furthermore, this application has to be configured in such a way that the capacities of the DC-AC converter and the conventional generator must satisfy the peak load requirements for reliable system operation. This is because it is not possible to wire and operate the DC-AC power converter and the conventional one in parallel.

However, a situation whereby a large percentage of the energy generation is fed through the battery results in increased charge/discharge cycles of the battery bank, thus leading to a decrease in the efficiency of the system – through a reduction of the battery's cycle life and the inability of the generator to directly supply the AC load. In addition, conversion losses are also

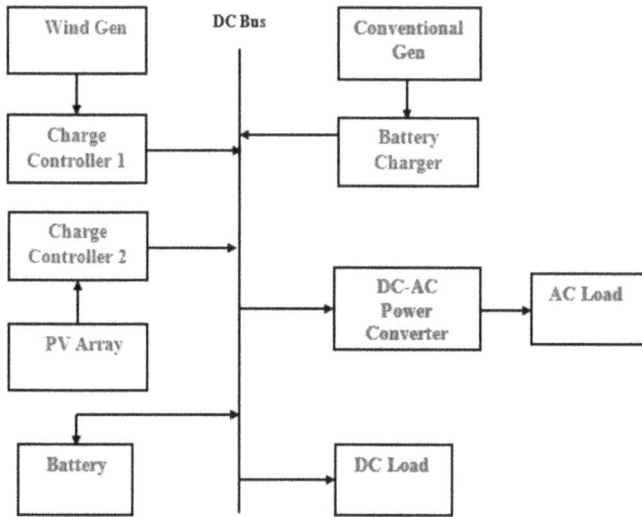

Figure 2.4 Series hybrid application of Wind/PV system.

incurred when the AC supply from the generator is first converted to DC by the battery charger before the generation is then converted back to an AC supply through the converter. The failure of the DC-AC power converter implies that the AC loads cannot be serviced, except in such an emergency where the rectifier in Figure 2.4 is by-passed and the generator is then wired to directly supply the AC loads.

2.2.4.2 Hybrid systems with switching

The switched hybrid power configuration is shown in Figure 2.5 [10, 51]. This system allows AC supply, either from the DC-AC power converter or the generator to the AC, loads through switching – e.g., a change-over switch without running the participating generating sources in parallel. The use of a switch makes this hybrid electrical configuration system differ from the parallel configuration shown in Figure 2.3, and the direct use of the AC bus makes it different from the series hybrid system configuration.

In a practical sense, the generator's capacity will be greater than the peak load [29]. The surplus can then be employed to charge the battery bank. It is also possible to switch off the generator when the load demand requirements are satisfied by the PV system. The switched hybrid system, just like the series hybrid system, can be run in manual mode, but it can also be operated in automatic mode with the use of a dedicated automatic regulator that can monitor the battery voltage and the automatic transfer switch (ATS) for the generator. The efficiency of the hybrid system with switching is expected to improve as the AC load is directly supplied by the generator.

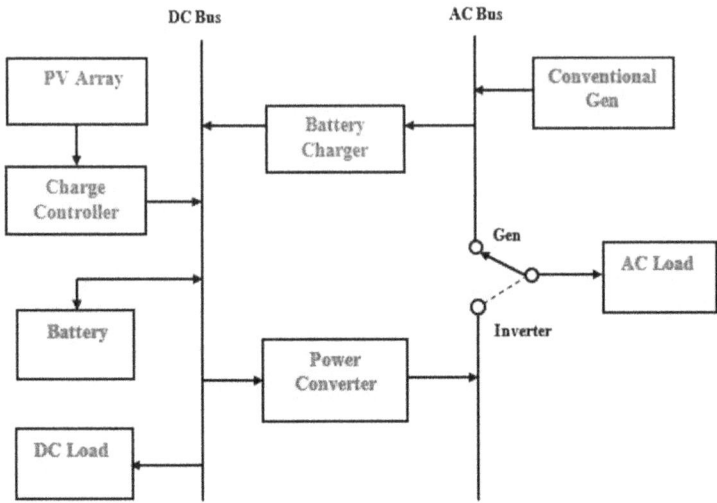

Figure 2.5 Switch hybrid application of PV/generator system.

2.3 GRID-CONNECTED RENEWABLE ENERGY SYSTEMS

Connection of RE systems to the grid takes two different forms. An RE system can be either a grid-tied bi-directional configuration for both DC and AC load requirements, or a grid-paralleled configuration that permits dual-path energy flow.

2.3.1 Grid-tied configuration

Figure 2.6 is an example of a grid-tied RE system for a household application. The figure represents grid-connected solar and SHP systems. The electrical arrangement of Figure 2.6a consists of a bi-directional power electronic converter (a dual converter in this case), a solar PV array, a battery bank, a meter, the DC and AC buses which serve the purpose of satisfying the DC and AC load requirements, and the grid. The arrangement in Figure 2.6b consists of the three-phase synchronous generator, a bi-directional power converter, a meter, a rectifier, switches S1 and S2, and DC and AC loads. In this type of renewable energy application, it is possible to allow an energy supply along three different paths [38]:

1. From the power grid to the loads;
2. From the PV system to the loads;
3. From the PV system to the power grid.

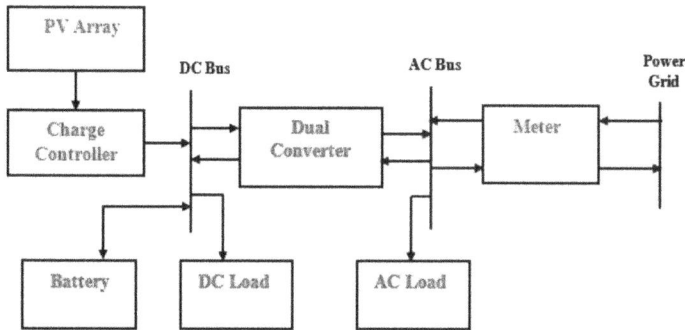

(a)

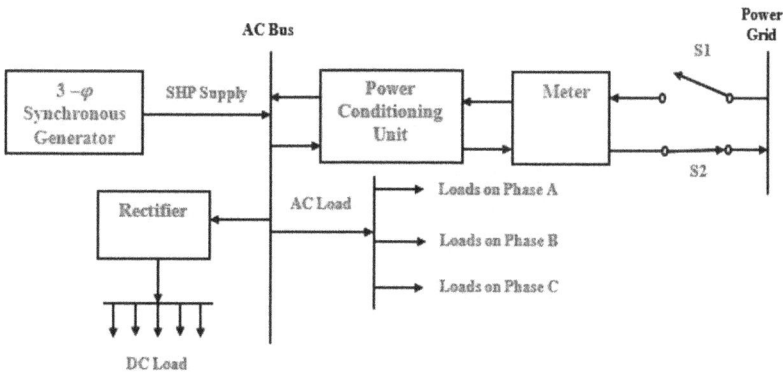

(b)

Figure 2.6 (a) Grid-connected PV system with a battery supplying domestic load; (b) grid-connected SHP application supplying domestic load.

There are two kinds of systems that interact with the existing grid: the one that has no battery backup and the one that includes a battery backup [52]. The system that interacts with the grid with the battery storage system allows the load(s) to be serviced when the grid supply is unavailable as a result of scheduled maintenance, faults, and so on. In the event of an outage, the electrical configuration disconnects from the power grid and then satisfies some part of the intended load demand. The bi-directional link to the battery component in Figure 2.6a allows the battery to be charged through the charge controller on the DC bus in one direction, while in the second direction, the battery is discharged to support the load when the grid is unavailable, say at night. However, in the grid-interactive PV system presented in Figure 2.6b, there will not be any supply to the load at night when the power grid supply is unavailable – a serious reliability issue. The battery bank is not part of Figure 2.6b.

The essence of the bi-directional power converter is that it serves as an interface between the solar PV array system or the SHP and the power grid. This power converter is also referred to as a dual converter, as in the case of Figure 2.6a, while it may just be referred to as a power conditioning unit (PCU) in Figure 2.6b. A distinction between the two power converters is the one shown in Figure 2.6a which is doing the work of inverter and rectifier, while the one in Figure 2.6b does not involve these. Nonetheless, in cases where the electricity supply is greater than the load demand of the PV or the SHP system, there is an excess generation that can then be exported to the electrical power grid. However, when the energy supply from the PV or the SHP is less than the load demand, there is an energy deficit. This energy shortage is addressed by importing energy from the power grid.

Apart from this aspect, another important consideration is the protection of the system, as there is the need to ensure that the grid-connected systems (also referred to as grid or utility-interactive systems), such as the PV and the SHP, are not allowed to supply energy to the power grid when it is not available as a result of a fault or scheduled maintenance [53, 54]. This prevention is considered one of the functions of the PCU – that is the bi-directional power converters shown in Figure 2.6a and b. The utility-interactive system, such as the one discussed in this section, requires standard codes: the IEEE 1547 standards for integrating distributed generation systems is one of the popular examples [55].

The bi-directional energy meter is expected to be an intelligent meter (i.e., smart meter), configured with the capability of quantifying both the energy imported from and exported to the power grid by the prosumer (someone who is connected to the grid but still able to generate their energy), thus providing the net energy (i.e., the difference) employed by the utility for billing purposes. The economic benefit of selling back excess energy to the power grid is that it can help achieve a reduction in the electricity bill [56]. The prosumers desire to see a reduced monthly bill from the exercise.

The DC load in Figure 2.6a is assumed to be matched with the voltage obtained from the battery. However, a DC-DC converter (Chopper) may be used to realize a voltage level that is higher or lower than that supplied by the battery. Switch S1 in Figure 2.6b is open while switch S2 is closed; this is interpreted as the SHP being able to meet the DC and AC load requirements, while the excess generation is being exported to the power grid. This implies that energy generation is greater than the load demand. However, if switch S1 was closed and S2 opened, this means that energy is imported from the grid to support the load – which may be the case during the dry season or years when there is a low hydro resource. In the case of a hydro system in which energy cannot be exported to the power grid, this means that an ordinary meter is required and not a smart meter.

The idea of Figure 2.6b may also be extended to the grid-connected wind power system, where an induction generator may be employed to replace the synchronous generator with appropriately designed power electronic converters serving as an interface between the wind power and the grid. The rectifier, the DC and AC loads, the meter, and the switches will also be part of the system architecture.

2.3.2 Grid-Paralleled configuration

The electrical system arrangement presented in Figure 2.7 is the kind of energy system shown in Figure 2.6 that allows energy supply flow along two different paths only:

1. From the PV system to the loads;
2. From the power grid to the loads.

Such RE generation application only purchases energy from the power grid when the need arises – that is, when the PV energy generation is less than the users' load demand. This may be the case at night when the sun is not available or during cloudy days when the solar PV system has low generation. However, it does not allow energy to be sold back to the power grid, implying that an interconnection standard is not required; also, the meter in Figure 2.7 is thus a normal energy meter [57]. In this case, the solar PV system can be designed to satisfy a high percentage of the load demand, while the power grid is also employed as a backup for reliability purposes during times when there is increased demand or cloudy/rainy days. The renewable energy application in Figure 2.7 does not require batteries [57]. However, it may be practically necessary to include batteries to support the load when the power grid is unavailable and/or if the loads to be serviced are critical.

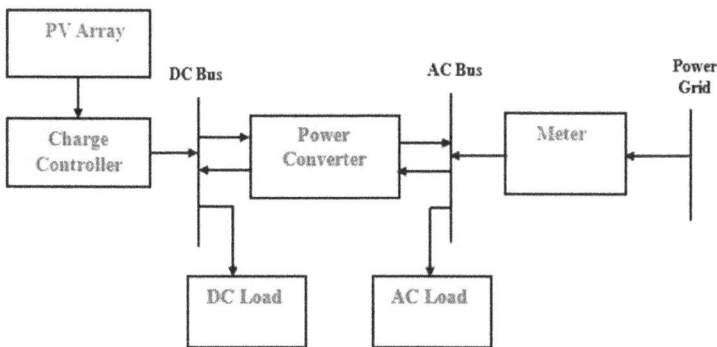

Figure 2.7 Grid-connected PV system with a battery supplying domestic load without the capability to export excess to the grid.

2.4 ANCILLARY SERVICES BY RE-BASED DGs

2.4.1 Background on ancillary services

REs alongside other DERs such as energy storage devices, fuel cells, and micro-turbines are employed by grid operators to support the grid in maintaining the quality and reliability of supply. ASs are defined as services provided to ensure that the electrical power grid operation is kept within an acceptable window or limits for achieving security of supply [58].

Traditionally, it is important to keep the frequency and voltage within specific safe limits and then make efforts to restore these two quantities to the normal window or range after an imbalance has occurred. These ASs can be regarded as frequency response/regulation, voltage regulation/control (popularly realized through the mechanism of reactive power compensation or support), and the capability for black-starting. A black start is the process of restoring an electric power station or a part of an electric grid to operation without relying on the external electric power transmission network to recover from a total or partial shutdown and which can be classified as [58–60]:

1. Frequency ASs that are mainly concerned with balancing (i.e., frequency support);
2. ASs that focus on the management of congestion;
3. Non-frequency ASs which are mainly for controlling voltage, restoring the power grid, etc.

Transmission system operators (TSOs) employ large electrical power-generating plants in transmission systems to provide ASs; however, it is possible nowadays to provide ASs through controllable distributed energy systems by distribution system operators (DSOs) in active distribution networks [61]. It is, therefore, important to state at this point that this section is concerned with how distributed renewable energy generation (DREG) can support the distribution power grid system. The classification of the categories of ASs may also be summarized as frequency support, voltage support, and restoration services [60], with some other forms of ASs such as load following, peak shaving, islanded operations, spinning, and non-spinning reserves.

The paradigm shift from a centralized to distributed power generation system is one of the factors that favor the aggregation and penetration of DREG systems in the distribution system. Such a development has allowed the active participation of several DREG system owners in the operation of the electrical power network; the decentralized approach to the operation of power systems with significant penetration of DREGs presents an avenue for realizing appreciable security of energy supply [58].

In light of this, several studies have also concentrated on ensuring appropriate and/or optimum allocation of distributed generators in the grid, as improper sizing and location of these generators may lead to power quality issues, increase in power losses, reduction of system reliability, increase in voltage at the end of a feeder, and so on [62–68]. This is because some of these technical issues that in the form of voltage control, power quality, protection, fault level, and grid losses are associated with the high penetration of level of DREGs [61].

Building new distribution networks to accommodate DREGs to address the mentioned grid-integration issues is not only trivial but also attracts a very huge, if not an unrealistic, economic implication. This, in itself, is one factor that justifies the need for a greater level of flexibility by the DSO to achieve [61]:

1. Enhancement of the hosting capacity of the distribution network – through increased RE penetration;
2. Addressing the issue of congestion keeps the normal system operation while also maintaining security boundaries;
3. Optimizing network planning;
4. Maintaining the voltage levels within acceptable limits.

The foregoing necessitates the introduction of grid-interfacing/interactive power electronic converters and control mechanisms and some other critical components that can help achieve the T and D deferment. Even though new T and D plants are not built, the conventional one-way power flow configuration of the existing distribution network has been modified by allowing a bi-directional power flow between DREGs or DERs and the grid at the PCC. Realizing grid expansion, therefore, is a means of increasing the penetration of DREGs by optimizing the architecture of the power grid in a way that minimizes power losses and enhances fault detection and/or correction; thus increasing the hosting capacity of renewable energy in the grid by reducing the grid impedance [69, 70].

Recent studies also show the relevance of the aspect of provision of ancillary services by distributed renewable energy resources to the power grid. The authors in [71] presented an integrated market solution to enable an active distribution network (ADN) by providing reactive power ASs using the mechanism of transmission-distribution coordination. Discussed in [72] is a multi-period fast robust optimization strategy for partial DGs for ASs purposes, while [73] discussed a teaching-learning based optimization technique for determining the allocation of DGs for minimizing real power loss on the Nigerian power grid.

2.4.2 Categories of DERs

The classification of DERs is shown in Figure 2.8, namely RE, non-RE, storage and fuel cells, and hydrogen technologies [5, 38, 69, 74]:

1. RE technologies: solar PV, solar thermal, SHP, μ-hydro, wind, biomass, tidal, wave, and geothermal.

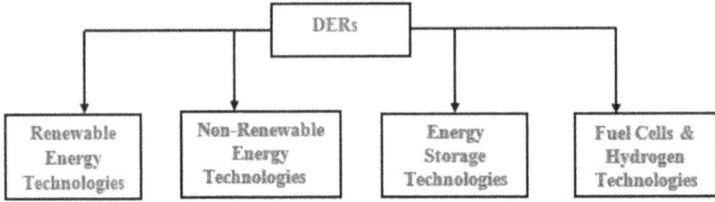

Figure 2.8 Categories of distributed energy resources.

2. Non-RE technologies: gas turbines, μ-gas turbines, reciprocating engines, combustion systems.
3. Energy storage technologies including: mechanical systems (flywheel, compressed-air, pumped hydro storage); electrical systems (ultracapacitors, superconducting magnetic energy storage); chemical systems (battery storage such as lead-acid, nickel-cadmium, vanadium redox, ZEBRA, Li-ion technologies); thermal systems (pumped heat, cryogenic, aquiferous heat systems).
4. Fuel cells and hydro technologies: solid oxide, alkaline, molten carbonate, direct methanol, phosphoric acid, solid polymers, PEM systems, and hydrogen technology, which are harnessed through appropriate energy conversion systems.

2.5 CONCLUSION

This chapter has brought to the fore interesting aspects of applications of RE resources in electrical power generation parlance. It first presented the inherent features, characteristics, and limitations of RE systems, and how this knowledge can help to aid the understanding of the utilization of RE generation for off-grid and grid support purposes. Particular attention was paid to the issue of the intermittency of RE sources, and the application of hybrid energy sources for reliability purposes. This aspect also involved the analysis of the loss of load probability and availability indices. Furthermore, different energy configurations of RE systems were considered, such as those on-grid and off grid with a clear understanding of whether or not the energy systems being considered are connected to the existing power grid in light of how the intended users' load demand requirements are satisfied. The chapter also paid special attention to the electrical topologies of RE in addition to other important components such as storage devices, power electronic converters, and conventional energy resources, including the sizing methodologies of the energy-generating systems. Electrical systems such as standalone, backup, grid-tie, and grid-parallel power applications were also dealt with in the chapter with examples. It is expected that the chapter will help to provide useful insights into the utilization and application of RE resources for generating and managing power.

REFERENCES

[1] Energy education: Human development index. Available: www.energyeducation. ca/encyclopaedia/Human_development_index Accessed: 02/09/2022.

[2] Renewables 2022 Global Status Report (REN21). Available: https://www. ren21.net/wp-content/uploads/2019/05/GSR2022_Full_Report.pdf Accessed: 02/09/2022.

[3] Energy indicators for sustainable development: Guidelines and methodologies. International Atomic Energy Agency (IAEA), Vienna, 2005, pp. 1–171.

[4] Freris L, Infield D. *Renewable energy in power systems*. 1st Ed., John Wiley & Sons, West Sussex, UK, 2008, pp. 1–302.

[5] Akinyele DO, Rayudu RK. Review of energy storage technologies for sustainable power networks. *Sustainable Energy Technologies and Assessments* 2014; 8: 74–91.

[6] Monyei C, Viriri S, Adewumi A, Davidson I, Akinyele D. A smart grid framework for optimally integrating supply-side, demand-side and transmission line management systems. *Energies* 2018; 11: 1038.

[7] Feron S. Sustainability of off-grid photovoltaic systems for rural electrification in developing countries: A review. *Sustainability* 2016; 8: 1326.

[8] Akinyele D, Belikov J, Levron Y. Challenges of microgrids in remote communities: A STEEP model application. *Energies* 2018; 11: 432.

[9] Michaelides (Stathis) EE. *Alternative energy sources*. 1st Ed., Springer-Verlag Berlin Heidelberg, 2012, pp. 1–467.

[10] Louie H. *Off-grid electrical systems in developing countries*. 1st Ed., Springer International Publishing, Switzerland, 2018, pp. 1–492.

[11] Luo FL, Ye H. *Renewable energy systems: Advanced conversion technologies and applications*. CRC Press – Taylor and Francis Group, Boca Raton, FL, 2013, pp. 1–864.

[12] Nelson VC. *Introduction to renewable energy (Energy and environment)*. 1st Ed., CRC Press, Taylor & Francis Group, Boca Raton, FL, 2011, pp. 1–376.

[13] Yuan H, Ye H, Chen Y, Deng W. Research on the optimal configuration of photovoltaic and energy storage in rural microgrid. *Energy Reports* 2022; 8: 1285–1293.

[14] Roy D, Hassan R, Das BK. A hybrid renewable-based solution to electricity and freshwater problems in the off-grid Sundarbans region of India: Optimum sizing and socio-economic evaluation. *Thermal Science and Engineering Progress* 2022; 101450.

[15] Kamal Md. M, Mohammed A, Ashraf I, Fernandez E. Rural electrification using renewable energy resources and its environmental impact assessment. *Environmental Science and Pollution Research* 2022; 1–18.

[16] Akinyele DO, Rayudu RK, Nair NKC. Grid-independent renewable energy solutions for residential use: The case of an off-grid house in wellington, New Zealand. *IEEE PES Asia-Pacific Power and Energy Engineering Conference (APPEEC)*, 2015, pp. 1–5.

[17] Twidell J, Weir T. *Renewable energy resources*. 2nd Ed., Taylor & Francis Group, Oxon, UK, pp. 1–625.

[18] Rajanna S, Saini RP. Development of optimal integrated renewable energy model with battery storage for a remote Indian area. *Energy* 2016; 111: 803–817.

[19] Laleman R, Albrecht J, Dewulf J. Comparing various indicators for the LCA of residential photovoltaic systems. In: Singh A, Pant D, Olsen S. (eds) *Life*

cycle assessment of renewable energy sources. Green Energy and Technology. Springer, London 2013, pp. 211–239.

[20] Fthenakis V, Frischknecht R, Raugei M, Kim HC, Alsema E, Held M, de Wild-Scholten M. Methodology guidelines on life cycle assessment of photovoltaic electricity. 2nd Ed., *International Energy Agency Photovoltaic Power Systems (IEA-PVPS TASK 12)*, 2011.

[21] Performance ratio – Quality factor for the PV plant. Technical Information by SMA. Available: https://files.sma.de/downloads/Perfratio-TI-en-11.pdf Accessed: 19/09/2022.

[22] Performance ratio. PVSyst software. Available: https://www.pvsyst.com/help/performance_ratio.htm Accessed: 19/09/2022.

[23] Mason JM, Fthenakis VM, Hansen T, Kim HC. Energy pay-back and life cycle CO_2 emissions of the BOS in an optimized 3.5 MW PV installation. *Progress in Photovoltaic Research Application* 2006; 14: 179–190.

[24] Fthenakis VM, Kim HC, Alasema E. Emissions from photovoltaic life cycles. *Environmental Science and Technology* 2008; 42: 2168–2174.

[25] Brooks W, Dunlop J. Photovoltaic (PV) installer resource guide. North American Board of Certified Energy Practitioners (NABCEP), March 2012, v. 5.3.

[26] Okakwu IK, Alayande AS, Akinyele DO, Olabode OE, Akinyemi JO. Effects of total system head and solar radiation on the techno-economics of PV groundwater pumping irrigation system for sustainable agricultural production. *Scientific African* 2022; 16: e01118.

[27] Masters GM. *Renewable and efficient electric power systems.* 1st Ed., John Wiley & Sons, Hoboken, NJ, 2004, pp. 211–259.

[28] Ajewole TO, Olabode OE, Alawode OK, Lawal MO. Small-scale electricity generation through thermal harvesting in rooftop photovoltaic picogrid using passively cooled heat conversion devices. *Environmental Quality Management* 2020; 29: 95–102.

[29] HOMER Pro Index. https://www.homerenergy.com/products/pro/index.html Accessed: 26/12/2022.

[30] Markvart T, Castener L. *Practical handbook of photovoltaics: Fundamentals and applications.* 1st Ed., Elsevier Kidlington, Oxford, UK, 2003, pp. 1–1015.

[31] Ajewole, TO, Lawal MO, Omoigui MO. Development of a lab-demo facility for wind energy conversion systems. *International Journal on Energy Conversion* 2016; 4(1): 1–6.

[32] Ajewole TO, Alawode KO, Omoigui MO, Oyekanmi WA. Design validation of a laboratory-scale wind turbine emulator. *Cogent Engineering* 2017; 4(1280888): 1–13.

[33] Okakwu IK, Olabode OE, Alayande AS, Somefun TE, Ajewole TO. Techno-economic assessment of wind turbines in Nigeria. *International Journal of Energy Economics and Policy* 2021; 11(2): 240–246.

[34] Okakwu IK, Alayande AS, Olabode OE. Performance and economic analysis of Kainji hydropower plant in Nigeria. *Arid Zone Journal of Engineering Technology and Environment* 2019; 15(2): 461–469.

[35] Nazari-Heris M, Mohammadi-Ivatloo B. Design of small hydro generation systems. In: Gharehpetian GB, Agah SMM, *Distribution generation systems*, Elsevier, 2017, pp. 301–332.

[36] Ajewole TO, Ajewole OT, Fagbamiye MO, Omoigui MO. A study on the hydro-electric potential of Olumirin waterfall at Erin-Ijesa in Osun State of Nigeria. *UNIOSUN Journal of Science* 2018; 3(1): 64–72.

[37] Singh S, Singh M, Kaushik SC. Feasibility study for an islanded microgrid in rural area consisting of PV, wind, biomass and energy storage system. *Energy Conversion and Management* 2016, 128: 178–190.

[38] Akinyele D, Olabode E, Amole A. Review of fuel cell technologies and applications for sustainable microgrid systems. *Inventions* 2020; 5(3): 42.

[39] Hunter CA, Penev MM, Reznicek EP, Eichman J, Rustagi N, Baldwin SF. Techno-economic analysis of long-duration energy storage and flexible power generation technologies to support high-variable renewable energy grids. *Joule* 2021; 5(8): 2077–2101.

[40] Mamen A, Supatti U. A survey of hybrid energy storage systems applied for intermittent renewable energy systems. *14th International Conference on Electrical Engineering/Electronics, Computer, Telecommunications and Information Technology (ECTI-CON)*, 2017, pp. 1–5.

[41] Castillo A, Gayme DF. Grid-scale energy storage applications in renewable energy integration: A survey. *Energy Conversion and Management* 2014; 87: 885–894.

[42] Solomon AA, Child M, Caldera U, Breyer C. How much energy storage is needed to incorporate very large intermittent renewables? *Energy Procedia* 2017; 135: 283–293.

[43] Ganesan S, Subramaniam U, Ghodke AA, Elavarasan RM, Raju K, Bhaskar MS. Investigation on sizing of voltage source for a battery energy storage system in microgrid with renewable energy sources. *IEEE Access* 2020; 8: 188861–188874.

[44] IEEE Guide for optimizing the performance and life of lead-acid batteries in remote hybrid power systems. IEEE Standard 1561, 2007, pp. 1–35.

[45] Posadillo R, Luque RL. A sizing method for standalone PV installations with variable demand, *Renewable Energy* 2008; 33: 1049–1055.

[46] Wenham SR, Green MA, Watt ME, Corkish R. *Applied photovoltaics.* 2nd Ed., James & James Earthscan, UK and USA, 2007; pp. 1–336.

[47] Olabode OE, Ajewole TO, Okakwu IK, Alayande AS, Akinyele DO. Hybrid power systems for offgrid locations: A comprehensive review of design technologies, applications and future trends. *Scientific African* 2021; 13: e00884.

[48] Ajewole TO, Oyekanmi WA, Babalola AA, Omoigui MO. RTDS Modeling of a Hybrid-Source Autonomous Electric Microgrid. *International Journal of Emerging Electric Power Systems* 2017; 18(2):1–11.

[49] Ajewole TO, Craven PMR, Kayode O, Babalola OS. Simulation of Load-Sharing in Standalone Distributed Generation System. *In the Proceedings of the 7th International Conference on Clean and Green Energy 2018*, Paris, France, pp. 108–115.

[50] Louie H, Dauenhauer P, Wilson M, Zomers A, Mutale J. Eternal light. *IEEE Power and Energy Magazine* 2014; 70–78.

[51] Nayar CV, Islam SM, Dehbonei H, Tan K, Sharma H. Power electronics for renewable energy sources. *In Alternative Energy in Power Electronics* 2011, 1–79.

[52] El Chaar L. Photovoltaic system conversion. In: *Alternative energy in power electronics*, Elsevier 2011, pp. 155–1765.

[53] Types of PV systems, Available: http://energyresearch.ucf.edu/en/consumer/solar-technologies/solar-electricity-basics/types-of-pv-systems/ Accessed: 28/09/2022.

[54] Adefarati T, Bansal RC. Energizing renewable energy systems and distribution generation. In: *Pathways to a smarter power system*, Elsevier, 2019, pp. 29–65.

[55] Basso TS, DeBlasio R. IEEE 1547 series of standards: Interconnection issues. *IEEE Transactions Power Electronics* 2004; 19: 1159–1162.

[56] Emmanuel M, Akinyele D, Rayudu R. Techno-economic analysis of a 10 kWp utility interactive photovoltaic system at Maungaraki school, Wellington, New Zealand. *Energy* 2017; 120: 573–583.

[57] Barbir F. Fuel cell applications. In *PEM fuel cells*, Elsevier 2013, pp. 373–434.

[58] Demoulias CS, Malamaki KND, Gkavanoudis S, Mauricio JM, Kryonidis GC, Oureilidis S, Kontis EO., Ramos JLM. Ancillary services offered by distributed renewable energy services at the distribution grid level: An attempt at proper definition and quantification. *Applied Sciences* 2020; 10: 7106.

[59] ENTSO-E; E.DSO; Eurelectric; GEODE & CEDEC. TSO–DSO REPORT: An integrated approach to active system management with the focus on TSO–DSO coordination in congestion management and balancing, Technical Report. 2019. Available: https://docstore.entsoe.eu/Documents/Publications/Position%20papers%20and%20reports/TSO-DSO_ASM_2019_190416.pdf Accessed: 29/09/2022.

[60] Holttinen H, Cutululis NA, Gubina A, Keane A, Van Hulle F. Ancillary services: Technical specifications, system needs and costs. Deliverable D2.2, REserviceS FP7 Project. 2012. Available: http://orbit.dtu.dk/files/72251308/Ancillary_Services.pdf Accessed: 29/09/2022.

[61] Lamberti F. Ancillary services in smart grids to support distribution networks in the integration of renewable energy resources. Testi Di Docttorato, Universita degli Studi di Salerno, 2015.

[62] Rueda-Medina AC, Padilha-Feltrin A. Optimal allocation of distributed generators providing reactive power support ancillary service. *IEEE PowerTech Conference*, 2011, pp. 1–6.

[63] Rousis AO, Tzelepis D, Pipelzadeh Y, Strbac G, Booth CD, Green TC. Provision of voltage ancillary services through enhanced TSO-DSO interaction and aggregated distributed energy resources. *IEEE Transactions on Sustainable Energy* 2021; 12(2): 897–908.

[64] Rathna A, Loganathan PS. Adaptive voltage control ancillary service for renewable energy in distribution network. *International Journal of Engineering Research and Technology* 2015; 3(16): 1–7.

[65] Kandpal R, Kumar A. Effect of change in number and power factor of DG on optimal allocation for minimal actual power loss in RDS. *International Journal of Engineering, Science and Technology* 2022; 14(3): 56–63.

[66] Samimi A, Kazemi A. A new approach to optimal allocation of reactive power ancillary service in distribution systems in the presence of distributed energy resources. *Applied Sciences* 2015; 5(4): 1284–1309.

[67] Gopiya NS, Khatod DK, Sharma MP. Optimal allocation of distributed generation in distribution system for loss reduction. *IPCSIT* 2012; 28: 42–46.

[68] Okelola MO, Olabode OE, Ojo KE. Application of artificial bee colony for optimal siting and sizing of single distributed generation for real power loss minimization: A case study of Nigerian 11 kV feeder. *Proceedings of 8th International*

Conference on Mobile e-Services and Workshop on Bioinformatics, October 2019, Ladoke Akintola University of Technology, Ogbomoso, Nigeria, pp. 105–117.

[69] Iweh CD, Gyamfi S, Tanyi E, Effah-Donyima E. Distributed generation and renewable energy integration into the grid: Prerequisites, push factors, practical options, issues and merits. *Energies* 2021; 14: 5375.

[70] Bayer B, Mariam A. Innovative measures for integrating renewable energy in the German medium-voltage grids. *Energy Reports* 2020; 6: 336–342.

[71] Chen H, Li H, Lin C, Jin X, Zhang R, Li X. An integrated market solution to enable active distribution network to provide reactive power ancillary service using transmission-distribution coordination. *IET Energy Systems Integration* 2021; 4(1): 98–115.

[72] Zhang J, Cui M, He Y. Multi-period fast robust optimization for partial distributed generators (DGs) providing ancillary services. *Energies* 2021; 14(16): 1–21.

[73] Okelola MO, Olabode OE, Ajewole TO. Teaching-learning based optimization approach for determining size and location of distributed generation for real power loss reduction on Nigerian grid. *UNIOSUN Journal of Engineering and Environmental Sciences* 2019; 1(1): 1–10.

[74] Distributed energy resources, Australian Energy Market Commission (AEMC). Available: http://www.aemc.gov.au/energy-system/electricity/electricity-system/distributed-energy-resources. Accessed: 29/09/2022.

Chapter 3

Conceptual perspective of renewable energy resources

A paradigm shift in combating world climate change

Titus Ajewole and Olakunle Olabode
Osun State University, Osogbo, Nigeria

Funso Ariyo
Obafemi Awolowo University, Ile-Ife, Nigeria

Daniel Akinyele and Ignatius Okakwu
Olabisi Onabanjo University, Ago-Iwoye, Nigeria

Ezekiel Babatunde Omoniyi
University of Ibadan, Ibadan, Nigeria

CONTENTS

DOI: 10.1201/9781003436461-3

3.1 INTRODUCTION

Over time, humanity has witnessed a dramatic change in all facets of life such as cultural practices, religious practices, civic activities, municipal transportation systems, and public health practices due to the advent of conventional power systems [1, 2]. The invention of electricity can be viewed as one of the monumental achievements of humanity because its effect is far-reaching, multi-faceted, and has resulted in a dramatic turnaround in the way daily activities and engagements are carried out [3]. With the arrival of conventional power systems, humanity has gained massive control over a good number of his activities such as traffic management systems, tax collection systems, electronic voting, improved working environment both for private and public ventures, and enhanced educational systems.

Conventional power systems are anchored in sources such as hydro, coal, and natural gases [4–6] and, except for hydro sources, all other sources are fossil-based with massive emissions of gases capable of depleting the ozone layers and thereby causing climate change and global warming. The ozone layer is housed in the stratosphere and lies between 15 and 30 km above sea level [7]. It is needless to re-emphasize that the recent global warming, ozone layer depletion, and climate change have been a major talking point across the globe. Quite a number of climate-change-based research centers (e.g., the Grantham Research Institute, London; the Centre for Energy and Environmental Market, Australia; the Climate Change Science Institute, USA) have been constantly looking for ways of mitigating the challenges associated with the depletion of the ozone layer, global warming, and the attendant climate change scenarios, [8] which are interlinked as environmental issues.

However, conceptual differences exist. For instance, climate change deals predominantly with the mechanism by which carbon-related gases, such as methane (CH_4), carbon dioxide (CO_2), and other greenhouse gases (GHGs), modify the global climate system. Ozone layer depletion is predominantly occasioned by the presence and attendant damaging effect of chlorine or bromine on the stratosphere, which is known to be the protective covering for the earth. The depletion of the ozone layer is heavily occasioned by the presence of chlorofluorocarbon and hydrochlorofluorocarbon. These two chemicals are known as GHGs and, usually, they are common by-products of heavy chemical allied companies. Global warming on the other hand is occasioned by burning fossil fuels, resulting in the protracted rise of the planet's overall temperature.

A critical appraisal of some conventional power plants has shown that their end product contributes in no small measure to the above-mentioned environmental issues. For instance, CO_2 among other gases released in thermal power plants can be viewed as one of the largest contributors to the increased level of CO_2 globally [9]. It is a known fact that thermal power plants use fossil fuels or nuclear materials to generate electrical energy.

The utilization of thermal power plants has a high percentage in the energy generation mix of any country around the globe. In the USA, about 83% of electrical energy comes from thermal technologies, predominantly natural gas, coal, and nuclear [9]. Also, the heat produced in these plants is massive. Despite the fact that several approaches (water cooling and dry cooling) have been developed over time, at high ambient temperatures, however, their contribution to global warming can be considered significant since the above approaches can be less than efficient [9, 10]. The research findings of the authors in [10] showed that 87.3 million metric tonnes/annum of CO_2 emissions occurred from the thermal power stations in Nigeria alone.

The by-product of coal-fired plants has been predominantly ash, which often is channeled into an adjacent river for waste disposal. If the technology surrounding the coal-fired plant is well managed, the potential for carbon emissions is expected to be minimal. However, in the event of low calorific values, high ash content, and inefficient combustion technologies, there is the high risk of increasing GHGs and other pollutants [11]. From a global perspective, the consumption of coal is responsible for about 75% of CO_2 emissions from fossil-based fuels [12]. The authors in [13] show that coal-fired power plants are one of the principal contributors of GHG emissions in China, and it was estimated to be 80% of the nation's electricity sector emissions. The work of the authors in [14, 15] show that China has been ranked as the largest emitter of CO_2 on the globe, as far back as 2006, and its emissions are seen to be more than twice that of the USA, which was ranked second highest in 2018.

Also, diesel remains one of the most used forms of liquid fuel in the generation of electricity. A good number of heavy industries rely on diesel-generation to meet their energy demands. The combustion of diesel results in the heavy release of CO_2 as a by-product. In recent times, most especially in developing countries where an erratic power supply from the national grid has been a major challenge, households, churches, mosques, and private and commercial businesses owners have preferred to use diesel engines either as the main supply of electricity or as a supplement to meet their energy needs. For instance, a good number of rural areas in third-world countries around the globe use diesel generation sets (gensets) for lighting, water pumping for irrigation, and cottage firms. There are about four million small diesel engines used specifically for water pumping meant for the irrigation of farmlands in India [16]. Quite a lot of countries meet their energy demands with the use of diesel power plants, such as Nigeria, India, and Pakistan [17, 18]. The authors in [17] show that diesel-fired power plants are heavily responsible for the massive emission of CO_2 compared to other fossil-fuel-based power stations.

It is imperative to bring to mind that black carbon released in the combustion of diesel can trigger climate change. It does this by absorbing sunlight; consequently, heat is generated in the atmosphere, which is capable of warming the air, thereby affecting both the precipitation pattern and regional cloud formation [19]. It is, therefore, a thing of concern that the amount of carbon and nitrogen oxide related gases emitted could trigger ozone layer depletion and climate change.

Based on the foregoing, it is not an understatement that traditional power systems are one of the major contributors to climate change, ozone layer depletion, and global warming scenarios. The contribution of the energy sector to emissions of GHGs has been estimated to be 74% of total global GHG emissions, which is comparatively higher than any other sector. Furthermore, in 2016 the energy sector alone released about 33 gigatonnes of CO_2, and by 2021 that had grown to 36.3 gigatonnes [20]. Therefore, advocacy for green (clean) energy adoption is increasing globally, and this is unconnected to the fact that green energy resources are known to have zero carbon footprints. With this attribute, green energy promises to combat world climate change. The tides are also turning in favor of green buildings because they aid the preservation of priceless natural resources. They also enhance the quality of modern life, promote waste reduction measures by way of re-use and recycling, and encourage the use of non-toxic materials within their enclosures [20]. Green energy entails the utilization of renewable energy resources (RES) such as wind, solar, tidal and wave energy, geothermal, hydropower, and biomass. These resources are constantly being replenished by nature and are free and abundant. Also, the technologies surrounding their harvesting and utilization is gradually maturing. Globally, nature supports and promotes the utilization of green energy resources. In recent times, there has been a strong desire for transition to the adoption of RES. One good thing about these cleaner forms of energy is their ability to flexibly blend with conventional power systems or to be used in either standalone off-grid or on-grid modes.

In this chapter, the factors capable of triggering climate change globally are discussed. A conceptual review of the available RES, with emphasis on the carbon footprint, is presented to foster a thorough understanding of RES suitability and scalability for household or community-based energy generation and utilization. Some existing work on the utilization of RES for households or communities are appraised, so as to evaluate their suitability and sustainability. The potential of biomass energy and small hydropower systems for sustainable household and community-based energy generation are specifically promoted in this chapter. Likewise, the chapter gives special cognizance to powering municipal transportation by RES, a concept that is examined and recognized as a potential means of curtailing GHG emissions from automobiles.

The organization of the chapter is as follows. In Section 3.2 the available RES options are articulated with zero carbon footprints as the specific emphasis. In Section 3.3, the concept and techniques of carbon-capturing are examined, together with their possible utilization and storage approaches. Section 3.4 discusses the indispensability of fossil fuels while Section 3.5 provides a conclusion to the chapter with some recommendations.

3.2 CLIMATE AND CLIMATE CHANGE

To have a full grasp of the term "climate change", and the factors responsible for it, in addition to the utilization of RES as a way of combating world climate change, it is imperative to examine scholarly input from various researchers who have invested a great deal of effort in defining the term "climate". The founding father who did major pioneering work regarding the definition of "climate" can be traced to A. J. Herbertson in 1907 [21]. Other prominent scholars who modified this pioneering definition include Mark Twai [22], W. Köppen in 1918 [23], and H. H. Lamb in 1972 [24]. The early years of the 20th century witnessed an amazing input from the US National Weather Service [25], the Climate Prediction Center of the National Weather Service [26], and the American Meteorological Society [27]. The contributions from these institutes cannot be underestimated in understanding what climate entails. The common point of view of the above-mentioned scholars and institutes is the fact that climate is the "average of weather over at least a period of 30 years". Climate gives us the weather we see; more explicitly we could say without mincing words that "weather is what we obtained instantly while climate can be viewed as what we get if we keep expectant for a long time" [28].

The seasons of the year can be known and predicted even before the actual season begins with a good understanding of the prevailing climatic conditions of that area over a long period. Only a few parts of the globe are known to have four seasons, and that is mostly those countries within the temperate and polar regions. The rest of the globe in the southern and northern hemispheres has only two seasons in the year. For instance, in Nigeria and many sub-Saharan African countries, it has been established that November to March is usually the dry season. This commences in the early part of November and reaches its peak in March, while the rainy season begins in April and runs through to October. The rainy season is lighter from April to May and heavier in the early part of June to September. In the case of places like the USA and Canada, four major seasons are identifiable: Spring (March to June), Summer (June to September), Fall (September to December), and Winter (December to March). It should be noted that each season of the year has its peculiar attributes which have influenced the coping strategies adopted by people to sustain their continual existence and meet their needs in the areas of agriculture, transportation, occupation, and shelter, among others.

However, in the last few decades the tides have turned against this stable climate, and it is increasingly difficult to predict the next season of the year due to unexpected changes in climatic conditions. Climate change can be regarded as the direct consequence of a change in weather and time. The appearance of the term "climate change" surfaced in the literature in the early 1970s [28]. It is not an overstatement that the widespread song heralded by indigenous and foreign newspapers, local conferences, and global conferences focuses on the need to develop coping strategies to combat the devastating effects of climate change [29, 30]. Several researchers believe that "climate change" refers to alterations of a global nature with severe effects on the life-supporting activities of humanity, such as the health-care system, the design and construction of modern-day buildings, weather forecasting, the practice of agriculture, electrical energy requirements, transportation systems, and which dictate the kinds of clothes we put on [30, 31]. It is no gainsaying that the attendant consequence of these changes in climate patterns has a telling effect on diverse areas of human activity.

In other words, climate change can be perceived as a protracted shifting in the pattern of the world temperature and other elements of weather and climate. Explicitly, climate change appears in the form of a severe rise in global temperature [32], protracted drought [33], incessant flooding [34], harsh weather conditions [35], massive melting glaciers promoting a rise in global sea levels [36, 37], global warming [38], increased levels of precipitation [39], changes in the wind regimes of different locations across the globe [40], and life-threatening weather events [35]. Shifts in the usual patterns of global events of nature influence humanity's perception and invokes consciousness of the evolving world. This consciousness drives humanity towards developing appropriate coping strategies in all their forms to stay afloat in the face of the adverse effects of the changing climate.

3.2.1 Factors capable of triggering climate change globally

With a good understanding of the factors that can trigger climate change, we can see how the adoption of RES can be beneficial in combating/coping with world climate change. From the foregoing, one of the factors that can trigger climate change is the high concentration of CO_2 gas and the heavy concentration of GHGs in the stratosphere caused by the burning of fossil-fuel-based materials. It is also a known fact that CO_2 and GHGs stir up a heat-trapping mechanism that is capable of increasing the earth's global temperature rise. The authors in [41] present multiple illustrative plots that show an increased rise in global earth temperature occasioned by humanity and its activities. Documentation has it that the combustion of fossil material has contributed more to climate change than any other activities of humanity in recent times. A rise in global temperature can increase the rate of water evaporation and significant melting of glaciers, resulting in a rise in the general sea level.

Similarly, a good number of works in the literature support the fact that there is an appreciable rise in global temperatures occasioned by humanity and the year 2022 was the sixth warmest year since global records began in 1880 at 0.86°C (1.55°F) above the 20th-century average of 13.9°C (57.0°F) [41]. Combined forces occasion the rise in the global temperature; some of these forces were classified as human activities (the release of GHGs, depletion of the ozone layer, depletion of land vegetation/cover, and the presence of aerosols in

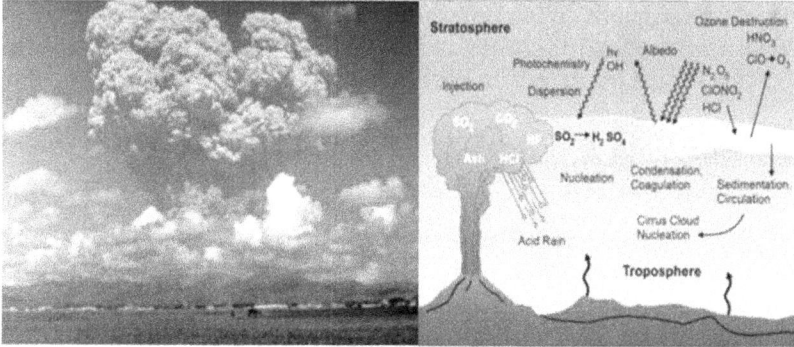

Figure 3.1 Volcanic eruption and potential reaction of release gases [42].

the atmosphere), others were nature-driven factors such as volcanic eruptions, the solar system, and the orbital system [41]. The contribution of a continuous volcanic eruption to an increased level of CO_2 and oxides of sulfur in the atmosphere cannot be over-emphasized. Figure 3.1 shows a volcanic eruption and the process of the reaction of released gases into the atmosphere [42].

Another prominent cause of climate change is deforestation, resulting in desert encroachment. With the desert extending to areas where there was no desert, the attendant consequence of this is the disappearance of natural vegetation, the extinction of wild animals, and alteration of rain patterns. There is documentation to support the fact that approximately 10% of global warming crises originated from CO_2 released from tropical deforestation [43]. This is unconnected to the fact that when forests, whether light or thick, are cleared, or when subjected to burning after being cultivated, the stored CO_2 in them is released into the atmosphere. Figure 3.2 illustrates deforestation in a location named Adekunle Village in Okeho, Oyo State, Nigeria.

Figure 3.2 Deforestation in Adekunle Village, Kajola Local Government Area, Okeho, Oyo State, Nigeria.

It should be pointed out that after indiscriminate cutting down of the thick forest for agricultural purposes, the remains are usually set on fire, which in the process triggered a heavy release of CO_2 into the atmosphere.

3.2.2 Effects of climate change on the human environment

Having critically examined the potential factors that can trigger climate change, it is important to examine the effect of it on the human environment, to create a direction for the utilization of RES as a medium for combating or coping with global climate change. The effect of climate change on the human environment is far-reaching and has encroached on all areas of human endeavor, including increased energy consumption for domestic homes and agriculture (livestock farming and irrigation). The architectural design of building in recent times is now being tailored toward coping with climate change, which is a key driver of epidemics and diseases across the globe [44, 45]. In the last few years, the spread of global infectious diseases has largely been connected to climate change [46]. It is also known that climate change acts as an agent for the widespread transmission of pathogenic diseases such as vector-borne, water-related, air-borne, and food-borne diseases [47]. Global infectious diseases triggered by the evolution of new virus types in recent years include COVID-19, Ebola, and monkeypox.

It is important to know that food insecurity is one of the direct effects of change in weather patterns. When the global world is food insecure, malnutrition and high morbidity and mortality rates are inevitable. This is further buttressed by the World Health Organization (WHO) that predicts that between 2030 and 2050 climate change will cause 250,000 additional deaths per year, and that the feasibility of this projection stems from malnutrition, the occurrence of malaria, and the prevalence of diarrhea and heat-born stress [48]. Climate change will bring about fluctuations in one or more climatic elements, such as temperature patterns, precipitation patterns, wind patterns, and global sunshine patterns. Figure 3.3 shows a summary of the process of climate change and its consequences.

Climate change is no longer a piece of news; rather the earlier we believe that it is here to stay, the better we can explore or develop appropriate coping strategies within the limits of our nature-endowed resources. As a matter of concern, the WHO in 2021 [48] opined that suitable forms of energy for industrial utilization, transportation, and domestic utilization will increase the chances of reducing GHGs. With an appropriate choice of energy, a better health-care system can be sustained, an improved environment made against air pollution, and abundant food availability for all will be guaranteed. Climate change is believed to have altered the normal pattern of rainfall, sunshine, and the wind speed regime of several locations across the globe. This alteration could be a key advantage to be harnessed when it

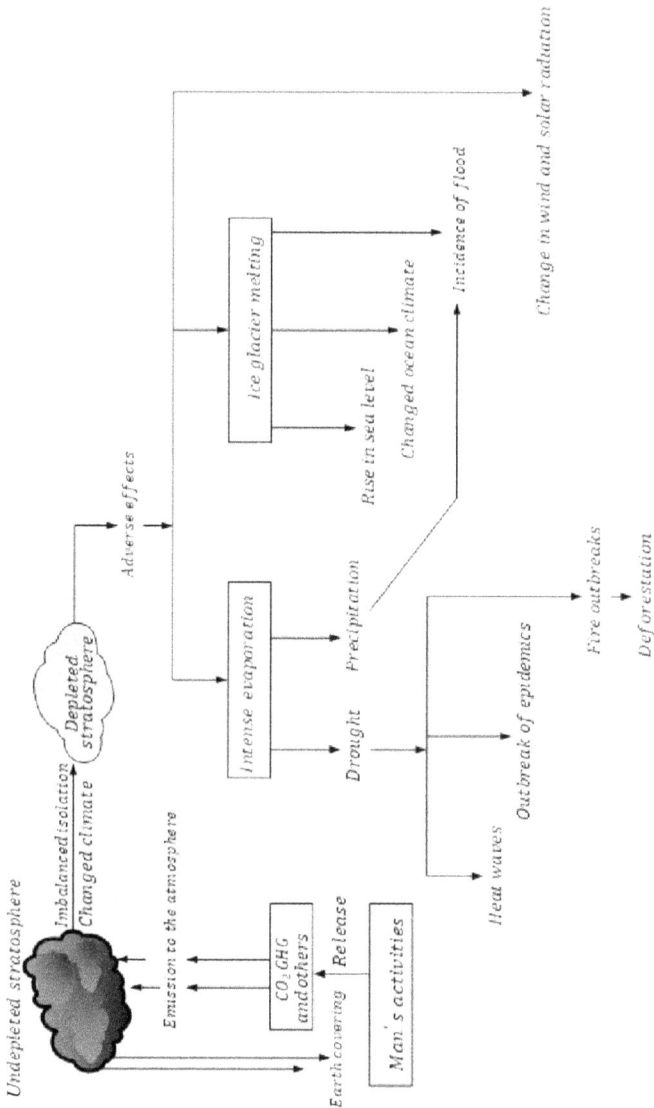

Figure 3.3 Climate change and its environmental consequences.

comes to the adoption of RES. Hence, the main focus of this chapter centers on harnessing the potential of RES in combating climate change.

3.3 POSSIBILITIES WITH RENEWABLE ENERGY RESOURCES

A good number of human activities require electrical energy from the supply grid, and in recent times the majority of these activities have been conveniently carried out with an energy supply from renewable energy. Also, the transition from conventional power systems to the new world of renewable energies is occasioned by many concerns, such as the environmental impact which has brought about climate change, the possibility of the exhaustion of fossil-fuel-based materials, fluctuation in the market price of fossil-fuel materials, and most importantly technological advancements in the area of renewable energy materials. There is a school of thought that strongly believes in the endless existence of fossil fuel reserves (the non-depleted nature of fossil fuels), but the growing concern centers on minimizing the damaging effects arising from the combustion of such fossil materials. Also, another point of concern is the unpredictable hike in the price of fossil fuels in the global market while the cost of obtaining energy from RES is undergoing price cuts daily due to geometrical technological advancement in the exploration and exploitation of these resources. Suffice it to know that if the worst impacts arising from climatic alterations are to be circumvented, the world needs to transit to RES. This would guarantee zero carbon footprints; and these sources are continuously replenished via nature-endowed mechanisms. RES are readily available either from natural processes or waste from municipal landfills. The remaining part of this section focuses on a review of the available RES with an emphasis on zero carbon footprints and the potential of biomass energy resources as well as the integration of small hydros for sustainable household utilization.

3.3.1 A conceptual review of the available RES with an emphasis on zero carbon footprints

The world at large is endowed with sufficient RES that can minimize, if not eradicate, the emissions of dangerous ozone-layer-depleting substances. These renewable resources have the potential to cause a paradigm shift in the power-generating sector. Furthermore, going by the executive summary on the global energy perspective authored by McKinsey & Company in April 2022, it was projected that penetration of renewable energy is expected to top the power generation mix by up to 80–90% come 2050 [49]. Also, a critical assessment of the research report of the authors in [49] revealed a growing pattern in the integration of RES for the global electrical energy generation mix. In 2010, about 20% penetration of renewable in the global energy mix was recorded which was found to have increased to 29% as of 2020. Also, renewable energy mix penetration was projected to have increased to 60% by 2035 and by 2050 to 86%.

It is imperative to note that the reduced cost of procurement of renewable energy materials, in addition to carbon-capturing in fossil-based materials, becomes a veritable platform to genuinely support the growing trend regarding the possibility of transiting the power sector toward a zero carbon footprint. Globally, a significant cut-down to the tune of 37 gigatons in the amount of CO_2 emissions is expected to be achieved by 2050 with the intensive direct consumption and generation of energy from RES [50]. In 2019 renewable energy accounted for 26% of world energy generation, and by 2050 it is expected to have risen to 90% [50]. To this end, efforts are being made to explain in detail the possibilities of achieving a zero-carbon footprint with the use of solar photovoltaic (PV) conversion, wind-based energy resources, biomass energy resources, natural gases, waste-to-energy from landfills, geothermal, and carbon-capturing techniques in fossil-based materials.

3.3.2 Solar photovoltaic systems

One of the most abundant forms of energy in nature is that harnessed from the insolation of the sun with the aid of solar cells [51]. This is readily available in any location around the globe [52]. Electricity generated from the sun is the cleanest form of renewable energy with a zero carbon footprint. The process of generating electrical energy from the sun is founded on the principle of the PV effect. The effect is analogous to the photoelectric effect, the common point of convergence of these two concepts centers on the generation of electrons either by electromagnetic radiation or the absorption of light, which is an electromagnetic wave. Generally, direct beam energy and diffuse energy are the two forms of energy radiated. Although at different degrees and intensities, these two forms of energy reach the earth's surface from the sun.

Cloudiness and the sun's position largely determine the magnitude of solar energy in the direct beam. The solar radiation that falls on a cloud suffers scattering thus reducing the amount of energy reaching the surface of the earth [53, 54]. Like conventional power systems, which apply to all areas of human endeavor, evolving technologies in solar energy have expanded its areas of utilization to a large number of applications such as heating applications (water heating and space heating), on-grid or off-grid electrical energy generation, and dedicated industrial electricity generation via high-temperature solar energy [53, 55].

On the trail to eradicating or cutting down net carbon emissions, research findings have shown that PV cells have the capacity to wipe out 4.25 billion tons of carbon emissions on a yearly basis [56]. Also, according to the submission of climate change 2022 mitigation, PVs can reduce emissions of carbon by 4.25 gigatons yearly [56]. In a similar vein, the International

Energy Agency has stipulated that PVs alone can generate 22% of the globe's electricity by 2050 [57], and with this speculation, a significant percentage of the growing emissions of CO_2 from fossil-based materials will be minimized [56]. Also, due to the ever decreasing cost of procurement, solar PVs are expected to account for 50% of the world's electricity generation since their capacity can be scaled up. To buttress this further, as of 2020, PVs added 126 GW per year and by 2050 it is expected to have risen to 444 GW per year [50]. The evolving technology in the development of PV solar cells have recently made it possible to improve the performance of PVs in locations with low solar irradiance [58, 59].

It is important to understand that the increased presence of moisture, aerosols, and particulates in the atmosphere has a greater tendency to decrease solar radiation due to the possibility of the occurrence of cloudy days. However, advancement in energy storage technologies and improvement in PV cell materials have made it possible to sustain the energy requirement from PVs due to improved performance in areas with low solar irradiance. It is equally possible that solar cells could receive a temporary boost in performance even in cloudy environments with the occurrence of the silver-lining phenomenon in heavy clouds [60]. Equally, during cloudy days, solar panels have the capability to produce 10–25% of their rated wattage [60]. In a bid to increase the energy yield, modification of monoracial PV modules has led to the development of bifacial solar modules which increase the energy yield even on cloudy days [61, 62].

Having critically examined the possibilities of PVs in providing net zero carbon emissions, it is pertinent to design a road map regarding the adoption and utilization of this renewable resource. Such road maps entail an awareness campaign starting from the grassroots to urban centers and then enforcing smart-energy homes backed up with government incentives or subsidies. A good number of human activities, such as irrigation systems, town water supplies, street lighting, space heating, and water heating, can be successfully supplied from standalone rooftop solar PVs and solar farms [63]. The massive deployment of these rooftop solar PVs and solar farms will be a giant attempt at decarbonizing the world.

Similarly, in the area of material development, a good number of modern-day roofing sheets can be coated or imprinted with solar cells which will encourage the conservation and preservation of the available space. This could be a possible future direction in material development as there has recently been constant advocacy for the adoption of green buildings. Also, this has a greater chance of reducing pressure on the grid supply. The whole essence of this is to promote the adoption of decentralized or distributed generation (DG) systems. Figure 3.4 presents a solar home with rooftop solar panels and a solar farm built in the car park. Figure 3.4a is a typical three-bedroom flat in a residential building in Surulere, Lagos, Nigeria with its entire roof covered with solar panels for off-grid power generation. This can be scaled up to a point where an exchange of power could occur from a

(a)   (b)

Figure 3.4 Deployment of solar PV: (a) smart energy home in Surulere, Lagos, Nigeria; (b) solar farm in car park in Delta State, Nigeria.

smart-energy home to the grid and the excess energy generated can be metered to the grid or surrounding neighbors who are less privileged regarding the purchase of requisite infrastructures required for self-generation. Figure 3.4b showcases a hostel in Delta State, Nigeria powered by converting the available car-park space into a mini-solar farm. The same approach can be extended to commercial buildings such as shopping malls, restaurants, filling stations, churches, and public buildings so that they can be energy buoyant without necessarily depending on the grid supply.

Recent developments around the globe have made it possible that a municipality could be powered by a dedicated solar farm. A critical look at Figure 3.5a shows the possibility of powering a 115/34.5 kV substation

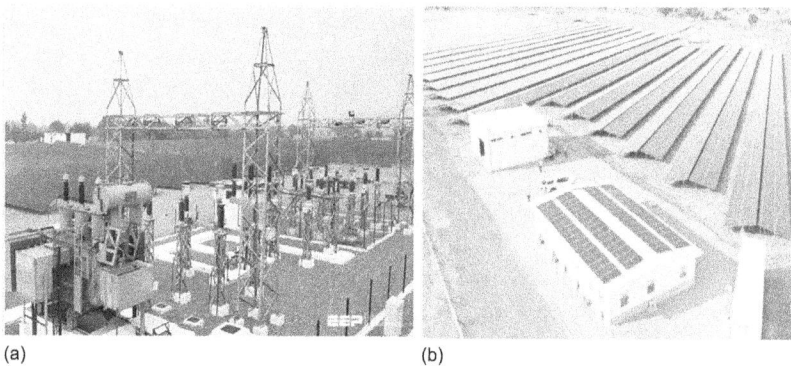

(a)   (b)

Figure 3.5 Grid-connected solar plants: (a) 60 MW grid-tied solar farm; (b) 13 MW grid-tied solar farm [64, 65].

from a 60 MW grid-tied solar farm [64], while Figure 3.5b shows a 13 MW solar farm tied to the grid in Ghana [65]. The configurations described in these figures show that facilities for conventional power systems can still be actively utilized in the transition to net zero carbon emissions. This will only require slight modification by way of incorporating three-phase inverter systems into the configuration to convert the direct current output from the solar farm to alternating current before being connected to the transformer for onward handling over to the transmission system. Similar laudable projects are found around the globe including a 60 MW Malawi solar farm [66], 60 MW of a solar plant was connected to the grid in northern Battambang [67], a 60 MW solar plant in Ivory Coast was connected to the grid [68], and a 180 MW solar PV plant was connected to the grid in Bhutan.

3.3.3 Wind energy

In combating climate change caused by emissions of carbon-related gases, wind energy is a viable option as it is produced from wind which is devoid of emissions of CO_2 and GHGs. The technology that surrounds wind power is fast-expanding globally [69] while the procurement cost of the associated accessories is on a daily basis coming down. Wind power is obtained by converting air movement in the form of kinetic energy, via a wind turbine attached to a generator, into electrical energy [70]. The fundamental principles that explain the concepts of wind and wind power generation are anchored in a thermodynamic principle, the Coriolis effect caused by the rotation of the earth in a counter-clockwise manner, and a principle of electromagnetism [71].

The thermodynamic principle describes the energy output from the wind turbine (with a horizontal or vertical axis), while the electromagnetism principle is concerned with electricity generation by electromagnetic induction in the permanent magnet of the wind turbine. The Coriolis effect describes wind direction in the northern and southern hemispheres of the earth. The viability of wind power generation is largely determined by the prevailing wind regime of such locations. The wind capacity of each location across the globe varies and this variability is one of the key factors in determining the best-suited locations for wind turbine plants. For instance, the wind speed in the northern part of Nigeria is stronger than in any other part of the country [72, 73]. In a similar perspective, the north and west are the windiest parts of the United Kingdom [74], while quite a number of places have the strongest wind speed in the USA, including Boston and Chicago [75]. The height hub, the wind turbine size, and the length of the blade are essentials that largely dictate the energy output from wind farms.

In the last few years, the adoption and utilization of wind energy has expanded faster than expected [76]. A research report presented by the authors in [75, 77] showed a progressive trend in wind energy capacities as illustrated with a bar chart. It was observed that there was an appreciable growth in the global installed capacity from 2002 to 2021, and this is not unconnected to the decline in the cost of procurement of wind turbines and their associated accessories. The last two years from 2020 to 2022 witnessed a massive deployment of wind energy in power generation [78]. The report in [79] shows record-breaking news of the installation of wind power of capacity 93 GW, which raises the total capacity of wind power, both from onshore and offshore, to 743 GW. On the global rankings, China and the USA are leading the rest of the world in the integration of wind energy into national grids [79], while the countries of Denmark, Uruguay, and Ireland respectively have 58.0, 40.4, and 38.0% shares of the electrical energy generation mix from wind farms [80]. The regional growth in wind power capacity over the last decade shows that Asia grew its wind capacity by 300,000 to 350,000 MW; Europe by 200,000 to 250,000 MW; North America by 150,000 to 200,000 MW; the Latin Caribbean by 50,000 to 100,000 MW; and the Pacific, Africa, and Middle East by 0 to 50,000 MW. The trend reported showed that the Pacific, Africa, and Middle East have had the least growth in wind energy technology integration into national grids.

According to the Global Wind Energy Council (GWEC), the year 2020 witnessed unprecedented growth in the wind industry, and as of 2020 wind power capacity worldwide stands at 743 GW. This avoids the production of 1.1 billion tonnes of CO_2 globally [80]. A research report by GWEC stipulated that if net zero carbon emission is to be achieved an average of 280 GW of wind energy should be installed annually from 2030 to 2050 [81]. It is no gainsaying that the installation of wind energy can achieve net zero carbon emissions if its worldwide deployment is enforced.

Wind power technology can be viewed as a mature technology that has undergone a series of metamorphoses in blade structure from horizontal to vertical and recently to vortex bladeless technology [81]. The whole essence of this metamorphosis is to achieve increased energy output from the wind plant [82], which is a promising alternative that can be adopted as a mainstream energy source that is cost-effective and cost-competitive relative to coal-fired and gas-fired power plants [81]. To fully harness the potentiality of wind energy in combating climate change, there is a need to develop a pathway for the utilization and adoption of wind energy. One such road map is shown in Figure 3.6, which entails harvesting wind energy for direct utilization in smart-energy homes, while the excess is imported to the national grid.

Another viable road map is the development of wind farms that can be used to feed regional transmission substations that supply a municipality. One such case is in Kenya where 365 turbines were built as a wind farm on the shores of Lake Turkana in the north. This wind farm was reported to have boosted Kenya's electricity supply by 13% and is the largest wind farm

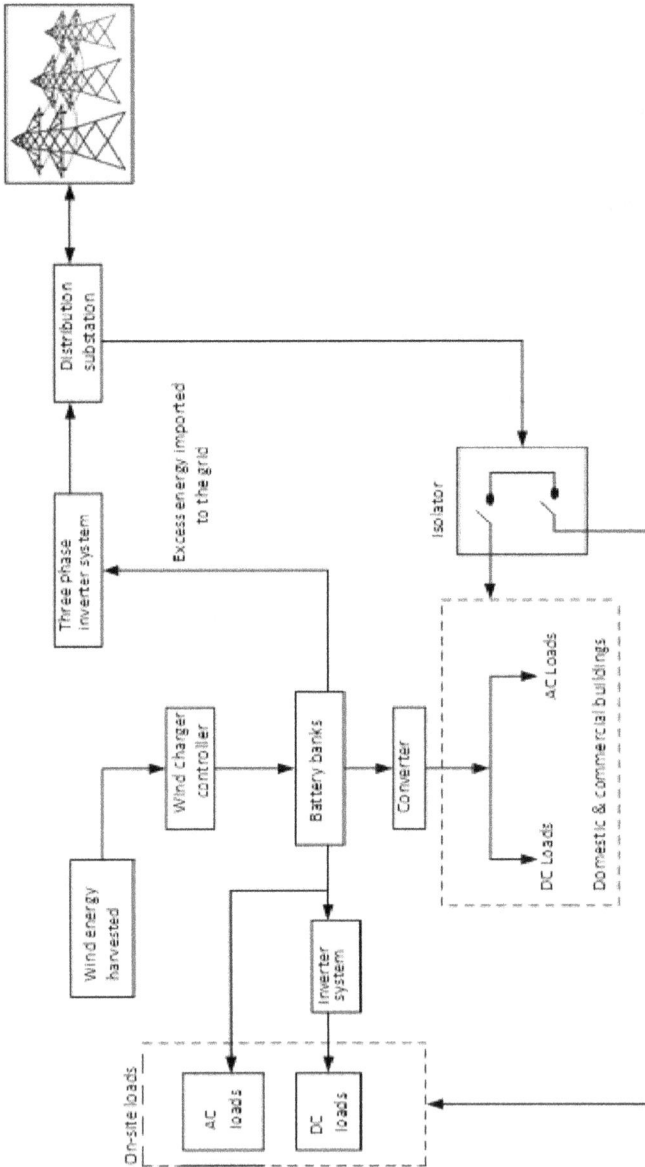

Figure 3.6 Harvesting of wind energy for immediate use and importation to the grid.

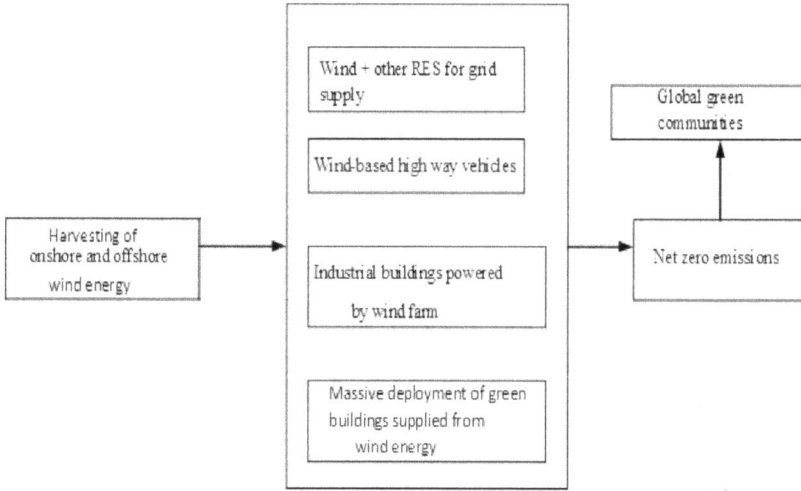

Figure 3.7 Pathways to green global communities via wind energy deployment.

in Africa [83, 84]. Also, community-based electrification from wind farms is equally possible, as announced by the Ethiopian government that is at present on the verge of delivering 100 MW from wind farms to power about 400 homes [85]. Similar projects exist in Jhimpir, Pakistan (500–700 MWh) [86], the Gansu wind farm in China (20 GW) [87], and the USA (122.32 GW), among others. The whole potential of the technology is to drastically reduce the amount of CO_2 emissions in order to mitigate the climate change scenario. The general path to follow begins with individual households, municipalities, and the industrial-based adoption of wind energy as illustrated in Figure 3.7. This pathway, as mentioned above, can potentially cut-down emissions of carbon-related and GHGs that are known to be key causes of climate change.

3.3.4 Natural gas approach

Natural gas is one of the cleanest forms of fossil fuels as it emits less CO_2 during combustion processes. It is a naturally occurring gas formed from the decomposition of microorganisms and the remains of plants and animals beneath the earth's crust. It is the end product of plants and animals that once lived on earth several million years ago. The existence of natural gas is due to intense temperature rises, compression, and great pressure occurring at some depth (swallow or deep) within the earth's crust. Natural gas can take the form of biogenic methane and thermogenic methane [88]. It is biogenic methane if formed from the decomposition of plants and animals deposited on landfills (earth's surface is low in oxygen content) while it is thermogenic methane if produced from the breaking of carbon bonds present in organic matter under intense compression and temperature rise [88].

Natural gas finds application in many human activities which include residential, commercial, industrial, and transportation systems. The industrial application takes the form of separately or centrally owned gas-fired power plants, residential applications for space heating, the powering of air conditioners, lighting outdoors, the powering of washing machines and dryers, while vehicles such as trucks, buses, and cars can equally be fueled using natural gas. From the perspective of global climate change mitigation, coal-fired power plants are gradually being replaced with gas-fired power plants to curtail the emissions of CO_2 and GHG emissions as seen in the US energy sector [89]. As of 2021, gas-fired power stations in the USA supplied 38.3% of the energy generation mix, which is the largest share compared to other energy generation sources (coal, nuclear, and hydro) [89]. It has been postulated that a switch from coal to a natural gas-fired power plant has the propensity to cause an approximate 30% reduction in emissions of CO_2 and GHGs [89]. Natural gas can be used as either a short-term bridge or a mid-term bridge to fuel the low-carbon future that is globally anticipated [90].

Having critically analyzed the possibilities of net zero emissions using natural gas for fueling in the energy sector, it is pertinent to develop pathways for its utilization in curtailing the emission of dangerous climate change gases. A conceptual pathway is shown in Figure 3.8, which tries to mitigate the leakage of methane and carbon capturing during the combustion of natural gas by appropriate treatment before utilization. The conceptual framework proposed in the figure is sustainable, scalable, and encompasses all human activities, which are potential threats to climate change scenarios. Another prominent consideration regarding this pathway revolves around the development, implementation, and enforcement of suitable policies regarding the utilization of wind energy to purposely phase out over-dependency on fossil-based oils in driving the global economy.

3.3.5 Waste-to-energy approach

Waste dumping in unregulated dump sites has negative impacts ranging from impaired aesthetic value and environmental pollution. The dumping of waste in municipal landfills causes the emission of methane gas during natural anaerobic decay processes. Also, when the waste is subjected to burning in incinerators, large tonnes of CO_2 are emitted into the atmosphere [91]. The emission of methane gas and CO_2 has a greater chance of triggering climate change. The authors in [92] assert that methane possesses 21 times the global warming potential of CO_2. In a similar vein, the oxides of nitrogen have 310 times the warming potential of CO_2. The growing global interest in recycling human waste aimed at ensuring a safer society could be one among other factors necessitating the conversion of municipal solid wastes to energy. There is debate as to how to classify the energy generated from

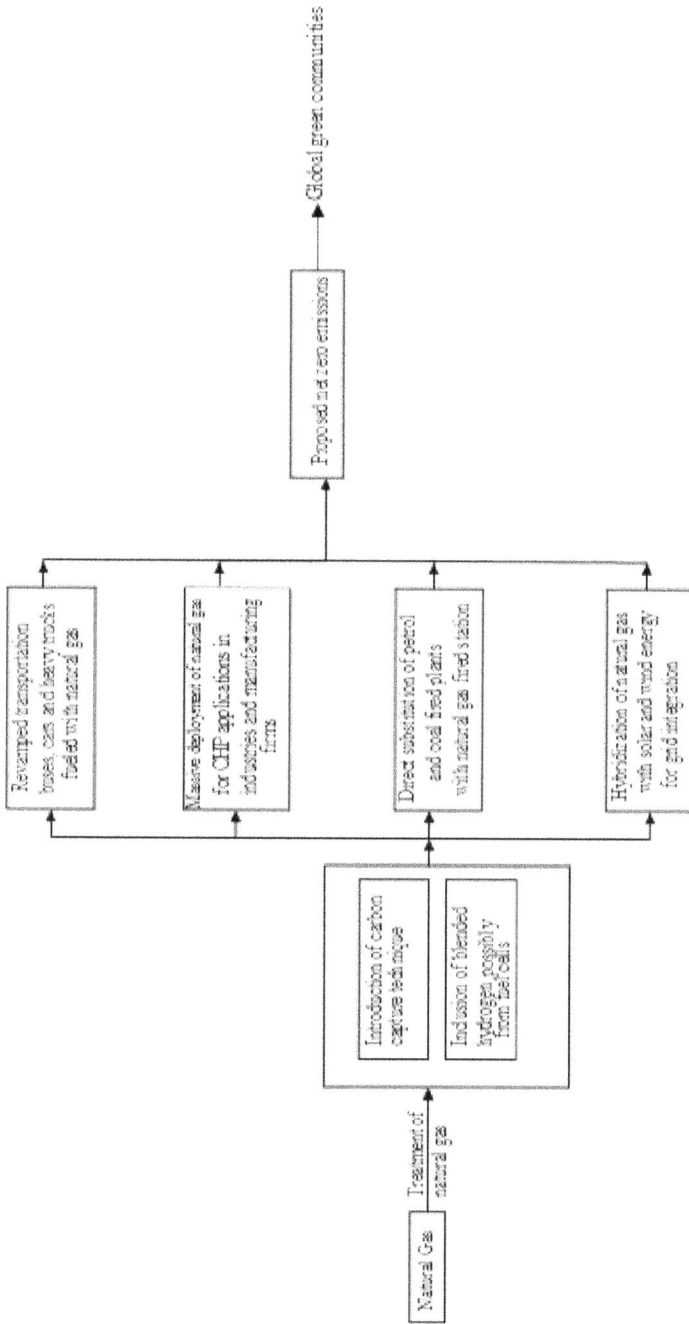

Figure 3.8 Conceptual pathway to green global communities via natural gas.

waste: as either a renewable or non-renewable energy resource. However, its classification as a renewable energy is widely supported considering the many circumstances such as population growth and urbanization that bring about a rise in the consumption level and consequently the availability of sustainable waste. The waste is sustainable since it is constantly being generated by many human activities.

A research report by the author in [93] showed that annual waste generation globally is predicted to upsurge by 73% from what it was in 2020 and rise to approximately 3.88 billion tonnes come 2050. To further buttress this fact of continuous growth in solid waste, 292.4 million tonnes of solid waste was generated in the USA in 2018 [94]. Nigeria alone accounts for 32 million tonnes per year of waste generation [95]. In a similar vein, Italy, particularly Lombardy, generated over 4.8 million tonnes of municipal waste as far back as 2019 [96]. Canada had a sharp increase in municipal solid waste, between 2002 and 2018, when the waste generation rose from 4.8 to 35.6 million tonnes [97]. In Ethiopia, the waste generation per day was estimated to be 9,700 tonnes in 2015, which has significantly risen to 12,200 tonnes per day in 2020 [98]. It is pertinent to review progress in energy generation from municipal waste before designing a pathway for its wide deployment in pursuit of a net zero carbon footprint. For instance, in February 2022, a waste-to-energy (WTE) plant was commissioned in Meadow Lake, Saskatchewan, Canada purposely to power about 5,000 homes in the province [99]; in Washington state, Spokane's WTE plant has successfully generated 22 MW of electrical energy from the combustion of about 800 tonnes of municipal solid waste which can conveniently power 13,000 homes [100]. Also in the United Kingdom, the capacity of the WTE plant increased significantly from 381 MW in 2009 to 1.45 GW in 2021 [101] while that of India stood at 5,690 MW [102].

A critical assessment of the short review in the preceding paragraph shows the sustainable nature of solid waste generation in global communities and the possibility of meeting municipal electrical energy demand from such sources. Also, the rapidly evolving technology in the area of WTE shows the prospect of sustainable electrical energy generation from municipal waste, and a future direction in global energy generation and management. The general approach for extraction of methane gas proceeds from the building of dump sites, the creation of landfill gas wells, gas capture and flaring, passage through the gas pipelines, and the utilization level. Utilization could be for the grid supply via gas-fired power stations, direct consumption for domestic applications, fueling of gas-powered vehicles, space heating, and industrial applications. Over time, energy generated from solid waste is a proven technology that is capable of meeting human desire for electrical energy in the areas of space heating, cooling systems, and industrial processes. It is a proven technology that is capable of displacing fossil-based fuels due to its significantly lower carbon when compared to dumping such wastes in landfills.

3.3.6 Promotion of the potential of biomass energy

It is not gainsaying that humanity is endowed with every necessary resource to combat climate change. Even the natural processes occurring in nature, if fully optimized, are sufficient to bring about a paradigm shift in the recent global warming and climate change scenarios. The global cause of climate change has been attributed to dangerous emissions of oxides of carbon, nitrogen, and GHG. The natural process through which plants process their food (photosynthesis) is capable of absorbing CO_2, which is the chief cause of climate change. The process of obtaining energy from feedstock such as plant wastes and wood is known as the biomass process [103]. The energy obtained from biomass is clean and has the potential to mitigate the threat of climate change.

One unique thing about biomass resources is the fact that they cannot be depleted, just like fossil fuel material, as long as the world continues to exist [104]. To further buttress this argument of the sustainability of biomass feedstock, a good percentage of earth's landmasses are endowed with plants such as trees and perennial grasses which are the key materials for biomass energy. In sum, biomass is relatively abundant in nature, reliable, and a means of reducing waste in landfills, thereby sustaining a cleaner environment. Suffice to know that if energy from biomass is largely promoted, the environmental payback will be great. Biomass energy has been proven to be carbon neutral and hence well thought of as a green energy. In addition, since the anaerobic decay of plants is jettisoned in all their forms, methane emissions arising from this decaying are thus reduced. Also, the utilization of biomass energy cuts across a good number of human energy requirements such as the production of heat and the generation of electricity via the direct burning of biomass feedstock, and indirectly too. Biofuel can be obtained from processed biomass feedstock [104, 105].

Combating climate change with biomass energy has to do with the level of deployment and the extent of its involvement in a country's energy mix. At the level of integration with other resources such as fossil-fuel plants, it is easier to combine coal and biomass resources to primarily fuel coal-fired power plants via the technique of co-firing [104]. With the co-firing technique, the amount of emissions of CO_2 as well as GHGs from coal-fired stations is significantly reduced in addition to the elimination of the cost of building a new biomass plant. Steam-fired plants can maximally benefit from biomass resources as the biomass feedstock, when turned into briquettes, contains a high energy density sufficient to drive a steam turbine to generate the required electricity for the grid supply.

If the global community means to transit to a net zero carbon footprint environment, then transportation systems have to be revolutionized. For instance, in 2020, it has been reported in the USA that 27% of GHG emissions come from transportation systems powered by the burning of gasoline and diesel [106]. Also, across the globe, more than 99% of vehicles – cars, ships, trucks, trains, and buses – work on the principle of the internal combustion of fossil fuel, hence the contribution of transportation systems to the emission of gases causing climate change is colossal.

The findings of the World Economic Forum on climate change as of June 2022 showed that transportation systems alone are responsible for 20% and 27% of global CO_2 and GHG emissions respectively; yet it is a sector that often is widely overlooked when tackling climate change [107, 108].

Having carefully followed the above, there is a need to replace fossil fuels with other alternatives that can sustain the dream of net zero emissions. One such alternative is the utilization of biofuels processed from biomass feedstock [109]. The major forms of biofuels are ethanol and biodiesel. These biofuels can be suitably and efficiently used to power a good number of internal combustion engines used for vehicular movement. If higher efficiency is required, biofuel can be easily fortified to produce increased efficiency. Biomass energy like other RES is the only one that can be processed into liquid fuel, thereby expanding its areas of application. The research findings of [110] show that if biofuel replaces petrol and petroleum products in locomotive engines, the probability of cutting down total GHG emissions stands at 45_80%. Also, progress so far in the deployment of biomass energy in global electricity generation across countries has to be tracked.

In Canada, biomass energy accounted for 1.4% of the total energy generation share [111] in 2021, approximately five quadrillion British thermal units; nearly 5% of aggregate primary energy came from biomass energy in the USA in 2021 [112]. The penetration level of bioenergy in the total energy generation mix is about 15% in Denmark, Finland, and Estonia [113]. As of 2016, South Africa has been able to generate 100 GWh of energy from biomass, though at the moment no record of progress has been found [114, 115]. As far back as 2019, 10% of Germany's share of total energy consumption came from biomass resources and which was particularly applicable in the areas of heating, electricity, and transportation [116]. Biomass energy involvement in California's total energy generation share in 2021 was 2.3% [117], while solid biomass accounted for 61 TWh of energy back in 2019 [118]. The survey on the deployment of biomass energy as illustrated above shows that its adoption rate is growing in leaps and bounds. Biomass could lead the trend of continuous growth currently being experienced in the world of renewable energy penetration in the global energy supply.

In achieving a green global community, the deployment of biomass in the locomotive industries, heat production for industrial and domestic applications, and replacement of coal-fired power plants should be given mass attention. Above all, a green global community is possible, if there were to be stringent environmental policies that promote the adoption and utilization of biomass energy in the place of fossil fuel and coal energy. With this, the global community would be devoid of unnecessary waste accumulation that litters communities, causing dangerous emissions of methane gases during natural decay processes and the release of dangerous CO_2 when such waste undergoes indiscriminate combustion in landfills and unauthorized dump sites.

3.3.7 Small hydropower plants for sustainable household and community-based energy generation

The electrical energy generated from the potential energy of falling water is one of the cleanest forms of energy known with a zero carbon footprint. Large-scale hydropower formed the largest mix of the energy share of many countries across the world, and in the last 20 years its capacity rose significantly to about 70% of global electricity generation [119]. Based on the central aim of this chapter which is a home-based or communities-based approach to combating climate change, it is necessary to consider the sustainability of small hydropower plants in this regard. Such plants have the capacity to generate approximately 30 MW, which is suitable for the off-grid electrification of rural households, a small resort, or irrigation purposes in farm settlements – and with the recent global drive to decarbonize communities, it is imperative to intensify efforts in utilizing untapped water bodies for generating sustainable electrical energy for several hundreds of homes or small communities where grid extension appears difficult.

Small hydro plants can be feasibly sited where there is a possibility of water storage via newly constructed dams, higher chances of returning water to an upper reservoir with the aid of a pump (pumped storage techniques), and the presence of river run-off [120]. One major merit of small hydro plants is that water used in the process is usually returned to its natural course without altering the river flow. Small hydro plants run on zero CO_2 and GHG emissions, their impacts on the environment is limited, and their engineering requirements in term of investment cost and construction duration is comparatively lower than large-scale hydropower plants [121]. To fully optimize the inherent potential and value of small hydropower plant development, there is a need for state governments to craft policies that will not only support but also provide financial incentives for private investors to be alert to off-grid electrification via small hydro plants. The processes involved in granting the proposed incentives in support of the development of small hydropower plants should be extensively reviewed to prevent bottlenecking and funds being diverted or sabotaged.

3.3.8 Geothermal energy

Geothermal energy develops from the heat generated beneath the earth's crust via the decaying of radioactive particles and also from the residual thermal energy resulting from the gravitational effect when the earth was being formed. Geothermal energy is a renewable resource with zero emission of gases triggering climate change. It is constantly being replenished via natural processes within the earth's crust and carries sufficient heat energy that can be used for heating applications and the generation of electrical energy via steam-turbine power plants [122], though it is constantly being neglected. The physical expression of geothermal energy on the earth's surface is the volcanic eruption of molten magma built under intense heat and pressure at some depth in the earth's crust. To support this claim, geothermal energy is often generated in areas with young volcanoes as seen in California and many other countries around the world [123]. Volcanoes are very hot, indicating that geothermal is a huge thermal energy resource that may be employed for electricity generation and other applications that require a heat supply. To harness geothermal energy, holes have to be bored into the geothermal reservoir beneath the earth's crust which consequently permits the massive rapid transfer of either steam or hot water to the surface for onward utilization; it also provides a passage for the return of used geothermal fluids to the sub-surface reservoir.

The energy obtained from geothermal has been successfully deployed for several applications such as heating individual and commercial buildings, as well as for industrial heating and cooling systems. One unique thing with geothermal energy is the possibility of being co-produced with oil and gas; with this unique attribute, GHGs can be heavily minimized; likewise neglected oil and gas field facilities can be revitalized thereby extending the economic life of such fields [124]. Unlike most RES – wind and solar – that suffer from intermittency, geothermal energy resources are known to be nonintermittent in supply [125], making them more reliable and sustainable than other RES. This justifies their utilization for energizing base-load power and heating applications, as well as a sustainable supply to the national grid. In addition to the above-mentioned features, resources for geothermal energy have a high availability factor to the tune of 90% [126, 127].

The technology of geothermal electricity generation is widely employed in 26 countries, while more than 70 countries are currently using geothermal energy for meeting the energy demand of their heating loads [128]. There is no gainsaying as regards the maturity of this technology for deployment in combating climate change. For instance, in 2021 0.4% of US power generation came from geothermal energy [129]. As of 2007, geothermal plants accounted for 1.9% of Indonesia's power generation, which is expected to expand to 5% come 2025 [130]. In the Philippines the geothermal national energy share is 12% [131]. The geothermal energy share in Turkey's power generation is 2.4% as of 2020 [132]. Also in 2020, New Zealand reported that 13% of its power generation comes from geothermal energy [133], while in Mexico, 9% of their total renewable penetration comes from geothermal energy [134]. Also, based on the installed capacity of geothermal power plants, the major leading countries as reported by [135] are the USA (3,714 MW), Indonesia (2,133 MW), the Philippines (1,918 MW), Turkey (1,526 MW), New Zealand (1,005 MW), Mexico (962.7 MW), Italy

(944 MW), Kenya (861 MW), Iceland (755 MW), and Japan (603 MW). The installed capacity of geothermal plants is heavily and rapidly multiplying in Indonesia, Chile, and Turkey, making them leading countries in the utilization and development of geothermal energy as of 2021 [135].

Massive deployment of geothermal energy can be used to combat the crisis of climate change. For instance, it has been projected that by 2050, substantial deployment of geothermal energy should be able to offset more than 500 million metric tons of GHG gases in the electric power sector and more than 1,250 million metric tons in the heating and cooling sector [127]. Also, in line with the submission of experts, heat pumps in geothermal plants can sufficiently reduce possible emissions both from furnaces used in natural gases and also heating systems in fossil fuel-based materials [136]. Unlike other RES, geothermal energy is constantly available and is neither season-dependent nor subject to the control of weather variability. A good number of fossil-fueled vehicles could have their batteries fully charged and so now deployed for transportation [126]. The possibility of replacing a good number of fossil-fueled cars is a giant attempt to cut down the emissions of dangerous gases promoting climate change. Another prominent potential of geothermal energy in combating climate change is the possibility of using geothermal channels to dispose of CO_2 by pumping it down into the deepest part of the earth during carbon capture and storage [137].

The sooner that the massive deployment of geothermal plants across the globe is encouraged and implemented, the earlier will the world reach the goal of net zero carbon footprints, since more CO_2 can be easily disposed of in the deep of the earth's crust via geothermal plant channels. Similarly, geothermal plants are sufficiently adequate in supporting carbon capture and storage technology which is believed to be one of the best solutions to the continual and safe utilization of fossil-fuel-based materials. The pathway to fully realizing the potential of geothermal energy in the transition to a green global community will be heavy investment in the deployment of geothermal technology, since geothermal energy is suitable as the base load, sustainable, and is not weather or season dependent. The building of the most reliable grid can be sustained by geothermal energy since it is more consistent in supply than other RES. Conclusively, energy from geothermal plants occupies a critical gap in achieving the goal of green energy transition and green global communities.

3.4 THE INDISPENSABILITY OF FOSSIL FUELS

The discovery and utilization of fossil fuels such as coal, gas, and oil have indeed revolutionized the world in all facets of life, be it the power sector, transportation systems, or economic systems. It is one of the key drivers that catalyzed the first Industrial Revolution and hence a factor that also contributed immensely to global technological advancement. It is imperative to know that about 80% of the world's energy is supplied by the combustion of fossil fuels [138]. Despite the claim that the continued utilization of fossil fuel-based resources is contributing to global climate change, it is increasingly difficult to abruptly jettison it when looking at the constraints surrounding the development and utilization of RES. These constraints include intermittency for solar, location dependency in terms of a viable wind speed regime for wind energy, and technological know-how for the integration of geothermal energy and biogas. In addition to the above limitations regarding the massive deployment of RES, fossil fuels are extremely excellent fuels with greater efficiency, as a little quantity of them carries sufficient energy density. Also, unlike most RES that are weather dependent, fossil fuel can be conveniently used anywhere, anytime, in the world where the requisite infrastructure exists. To further substantiate the claim of the difficulty in the abandonment of fossil fuel is the level of its penetration into the global economy – it is the main source of energy and a commodity for export exchange – over the last two centuries.

3.4.1 Zeroing the carbon footprint

An excerpt from a piece of research puts it that for any alternative energy resource to be a viable option that can conveniently replace fossil fuels, it must be able to equal its efficiency when used as fuel, it must be as accessible as sustainable energy, and it must enjoy wide societal integration [139]. In the last two centuries, fossil fuels and their utilization have left a pretty mammoth vacuum to be filled, and as it is presently, no alternative source can be adjured as being up to the task despite the negative effects in terms of its contribution to global climate change scenarios. Then the need to adopt a means of curbing the negative effect of fossil fuels on climate change becomes imperative. Fossil fuels are known for their dangerous emissions of large amounts of CO_2 and GHGs which have the potential to trap the heat present in the atmosphere thereby resulting in global warming and climate change events. Over time, technology has made it easy to develop a means of capturing CO_2 before being released into the atmosphere.

One such technology is carbon capture and storage techniques. Carbon capture is the cheapest means of dealing with emissions from heavy industries, where the combustion of fossil fuel materials is inevitable. Efforts are made to explain in detail the concept of carbon capture for effective and continuous utilization of fossil fuel materials. The discussion here includes carbon capture and storage, carbon capture storage and utilization, and bioenergy with carbon capture and storage.

3.4.2 Carbon capture, utilization, and storage

The concept of carbon capture started in the USA back in 1972 when it was deployed for the improved recovery of oil in west Texas [140]. The essential stages involved in carbon capture and storage entails capturing CO_2, compression of the captured CO_2 in liquefied gas, and injection of liquefied CO_2 down into the deepest part of the earth. The CO_2 emitted from the combustion of fossil fuels in power plants and heavy industries are captured at source. It is then compressed into liquefied gas which is thereafter transported via pipelines and discharged into the general mass of the earth. The approach is the safest method of making the combustion of fossil fuels free of the emission of GHGs and CO_2 into the atmosphere. CO_2 can be removed from fossil fuel materials in three ways: it can take place immediately after combustion which is referred to as post-combustion carbon capture, it can take place before the combustion process ends which is known as pre-combustion, and lastly by burning the fuel in a plentiful supply of oxygen which is called oxy-fuel combustion carbon capture. Whichever approach is deployed, the whole essence is to ensure that the combustion of fossil fuels comes with a zero carbon footprint which is suitable for achieving the anticipated global green communities.

The efforts in ensuring safer global communities and sustaining energy transition to green energy gave birth to the process of not only capturing CO_2 but also to its reusability by making new products for meeting the demand of humanity. CO_2 has industrial applications, most especially in the manufacturing, pharmaceutical, polymer, cement, and fertilizer-making industries. With carbon capture utilization and storage, the principal technique is still applicable, except that the captured CO_2 is now recycled for manufacturing products, such as urea fertilizer, plastics, pharmaceutical drugs, accelerating the growth of algae in agriculture, cement, and feedstock for biofuel. From the foregoing massive deployment of carbon capture, utilization and storage are a sure pathway to a net zero carbon footprint.

3.5 CONCLUSION

We have presented in this chapter the possibilities of combating climate change with the massive deployment of RES. The examined RES include solar PV, wind energy, natural gas, waste-to-energy landfill, biomass energy, small hydro, geothermal energy, and carbon capture approaches with the utilization of fossil fuel materials. The pathways towards achieving net-zero carbon footprints with these RES have been fully covered in detail. Also, suitable areas of application of each of the RES have been equally discussed. The progress and integration of each of the RES across major countries around the globe have been equally discussed as was the extent of their contribution towards combating dangerous emissions of CO_2, GHGs, and other gases.

It is imperative to point out that if the paradigm shift towards zero carbon emissions is to be sustained, there is a need for robust policies that will enforce and encourage the utilization of RES for many human daily applications. Similarly, there should be a massive deployment of pilot projects across third-world countries on the forgotten but viable alternative replacement for fossil fuel materials. It is not an overstatement that the government should be ready to invest heavily in providing the requisites awareness of the need to transit to low-carbon energy resources for transportation systems, industrial utilization, and domestic applications. Also, from the foregoing it is important to realize that if the transition to low-carbon energy is a gradual approach, then utilization of fossil fuel materials should be accomplished with the carbon capture technique, which has a way of boosting the growth of other economic sectors, such as agriculture, manufacturing industries, and pharmaceutical companies, to keep our environment safe from dangerous emissions of climate-change triggering gases. In conclusion, RES offer endless possibilities in combating global climate change.

REFERENCES

[1] Lu Y, Khan ZA, Alvarez-Alvarado MS, Zhang Y, Huang Z, Imran M. A critical review of sustainable energy policies for the promotion of renewable energy sources. *Sustainability* 2020; 12:1–30. https://doi.org/10.3390/su12125078

[2] Olabode EO, Ajewole TO, Okawku IK, Ade-Ikuesan OO. Optimal sitting and sizing of shunt capacitor for real power loss reduction on radial distribution system using firefly algorithm: A case study of Nigerian system. *Energy Sources, Part A: Recovery, Utilization, and Environmental Effects* 2019; 1–13. https://doi.org/10.1080/15567036.2019.1673507

[3] Lozano L, Taboada EB. The power of electricity: how effective is it in promoting sustainable development in rural off-grid islands in the Philippines. *Energies* 2021; 14:1–17. https://doi.org/10.3390/en14092705

[4] Okakwu IK, Alayande AS, Olabode OE. Performance and economic analysis of Kainji hydropower plant in Nigeria. *Arid Zone Journal of Engineering, Technology and Environment* 2019; 15(2):461–469.

[5] Mladenov V, Chobanov V, Georgiev A. Impact of renewable energy sources on power system flexibility requirements. *Energies* 2021; 14:1–20. https://doi.org/10.3390/en14102813

[6] IEA 2017 Perspectives for the energy transition: Investment needs for a low-carbon energy system Abu Dhabi: International Renewable Energy Agency Available at: https://www.irena.org/publications/2017/Mar/Perspectivesfor-the-energy-transition-investment-needs-for-a-lowcarbon-energy-system (Assessed on 31 August, 2022).

[7] Olabode OE, Ajewole TO, Okakwu IK, Alayande AS, Akinyele DO. Hybrid power systems for off-grid locations: A comprehensive review of design technologies, applications and future trends. *Scientific African* 2021, 13(2021): e00884. https://doi.org/10.1016/j.sciaf.2021.e00884

[8] Ogundele FO, Ayo O, Taiwo IS. The dilemma of ozone layer depletion, global warming and climate change in tropical countries: a review. *Academic Research International* 2011; 1(2):474–483.

[9] Coffel ED, Mankin JS. Thermal power generation is disadvantaged in a warming world. *Environmental Research Letter* 2020; 16:1–11

[10] Odewale SA, Sonibare JA, Jimoda LA. Electricity sector's contribution to green-house gas concentration in Nigeria. *Management of Environmental Quality: An International Journal* 2017; 28(6):917–929. https://doi.org/10.1108/MEQ-07-2016-0048

[11] Mittal ML. Estimates of emissions from coal fired thermal power plants in India. Available at: https://www3.epa.gov/ttnchie1/conference/ei20/session5/mmittal.pdf. Accessed on 31 August, 2022.

[12] Shi W, Tang W, Qiao F, Sha Z, Wang C, Zhao S. How to reduce carbon dioxide emissions from power systems in Gansu province—analyze from the life cycle perspective. *Energies* 2022; 15:1–15. https://doi.org/10.3390/ en15103560

[13] Crippa M, Guizzardi D, Muntean M, Schaaf E, Lo-Vullo E, Solazzo E, Monforti-Ferrario F, Olivier J, Vignati E. *EDGARv5.0 Greenhouse Gas Emissions; European Commission, Joint Research Center (JRC)*: Brussels, Belgium, 2019; Available online: http://data.europa.eu/89h/488dc3de-f072-4810-ab83-47185158ce2a (Accessed on 31 August, 2022).

[14] Rockström J, Gaffney O, Rogelj J, Meinshausen M, Nakicenovic N, Schellnhuber HJ. A roadmap for rapid decarbonization. *Science* 2017; 355:1269. https://doi.org/10.1126/science.aah3443.

[15] Crippa M, Oreggioni G, Guizzardi D, Muntean M, Schaaf E, Lo-Vullo E, Solazzo E, Monforti-Ferrario F, Olivier JGJ, Vignati E. Fossil CO_2 and GHG emissions of all world countries 2019 report; Publications Office of the European Union: Luxembourg, 2019. https://doi.org/10.2760/687800

[16] Diesel Generators in Developing Countries. Available at: https://www.generatorsource.com/Generators_in_Developing_Countries.aspx. (Accessed on 1 September, 2022).

[17] Oladokun VO, Asemota OC. Unit cost of electricity in Nigeria: A cost model for captive diesel powered generating system. *Renewable and Sustainable Energy Reviews* 2015; 52: 35–40. https://doi.org/10.1016/j.rser.2015.07.028

[18] Khattak MA, Syahir SM, Aiza AA, Mansor SA, Nurhaiza SOS. Feasibility assessment of a diesel power plant: a review. *Journal of Advanced Review on Scientific Research* 2016; 24(1):1–12.

[19] Diesel power generation inventories and black carbon emissions in Nigeria. Available at: https://documents1.worldbank.org/curated/en/853381501178909924/pdf/117772-WP-PUBLIC-52p-Report-DG-Set-Study-Nigeria.pdf (Accessed on 1 September, 2022).

[20] Global energy review: CO_2 emissions in 2021. Available at: https://www.iea.org/reports/global-energy-review-co2-emissions-in-2021-2 (Accessed on 1 September, 2022).

[21] Herbertson AJ. *Outlines of Physiography, an Introduction to the Study of the Earth*; Arnold: London, UK, 1907.

[22] The Climate Is What You Expect; The Weather Is What You Get. Available online: https://quoteinvestigator.com/2012/06/24/climate-vs-weather/(Accessed on 1 September, 2022).

[23] Köppen W. Klassifikation der Klima nach Temperatur, Niederschlag und Jahreslauf. 1918, 64:193–203.

[24] Lamb HH. Climate: Past, Present, and Future, Vol. 1: *Fundamentals and Climate Now*; Methuen: London, UK, 1972. https://doi.org/10.1002/qj.49709941926

[25] National Weather Service Glossary. Available at; https://w1.weather.gov/glossary/index.php?letter=c (Accessed 1 September, 2022).

[26] Climate Glossary by the Climate Prediction Center of the National Weather Service. Available at: https://www.cpc.ncep.noaa.gov/products/outreach/glossary.shtml#C. (Accessed on 1 September, 2022).

[27] Glossary of Meteorology by the American Meteorological Society. Available at: https://glossary.ametsoc.org/wiki/Climate (Accessed 1 September, 2022).

[28] Koutsoyiannis D. Hurst-Kolmogorov dynamics and uncertainty. *JAWRA Journal of the American Water Resources Association* 2011; 47(3):481–495. https://doi.org/10.1111/j.1752-1688.2011.00543.x

[29] Weiler F. Adaptation and health: are countries with more climate-sensitive health sectors more likely to receive adaptation aid? *International Journal of Environmental Research and Public Health* 2019; 16(1353):1–16. https://doi.org/10.3390/ijerph16081353

[30] Nunes LJR, Dias MF. Perception of climate change effects over time and the contribution of different areas of knowledge to its understanding and mitigation. *Climate* 2022; 10(7):1–19. https://doi.org/10.3390/cli10010007

[31] Van-Aalst MK, Cannon T, Burton I. Community level adaptation to climate change: The potential role of participatory community risk assessment. *Global Environmental Change* 2008; 8:165–179. https://doi.org/10.1016/j.gloenvcha.2007.06.002

[32] Valipour M, Bateni SM, Jun C. Global surface temperature: a new insight. *Climate* 2021; 9(5):81–100. https://doi.org/10.3390/cli9050081

[33] Li Y, Xie Z, Qin Y, Xia H, Zheng Z, Zhang L, Pan Z, Liu Z. Drought under global warming and climate change: an empirical study of the Loess Plateau. *Sustainability* 2019; 11(1281):1–16. https://doi.org/10.3390/su11051281

[34] Leitold R, Garschagen M, Tran V, Revilla DJ. Flood risk reduction and climate change adaptation of manufacturing firms: Global knowledge gaps and lessons from Ho Chi Minh City. *International Journal of Disaster Risk Reduction* 2021; 61:102351. https://doi.org/10.1016/j.ijdrr.2021.102351

[35] Chisale HLW, Chirwa PW, Babalola FD, Manda SOM. Perceived effects of climate change and extreme weather events on forests and forest-based livelihoods in Malawi. *Sustainability* 2021; 13(11748):1–15. https://doi.org/10.3390/su132111748

[36] Wortmann M, Duethmann D, Menz C, Bolch T, Huang S, Tong J, Kundzewicz ZW, Krysanova V. Projected climate change and its impacts on glaciers and water resources in the headwaters of the Tarim River, NW China/Kyrgyzstan. *Climate Change* 2022; 171:257–260. https://doi.org/10.1007/s10584-022-03343-w

[37] Kraaijenbrink PDA, Bierkens MFP, Lutz AF, Immerzeel WW. Impact of a global temperature rises of 1.5 degrees Celsius on Asia's glaciers. *Nature* 2017; 549:257–260. https://doi.org/10.1038/nature23878

[38] Siddique R, Mejia A, Mizukami N, Palmer RN. Impacts of Global Warming of 1.5, 2.0 and 3.0 °C on Hydrologic Regimes in the Northeastern U.S. *Climate* 2021; 9(1):1–15. https://doi.org/ 10.3390/cli9010009

[39] Teegavarapu RSV. Changes and trends in precipitation extremes and characteristics, trends and changes in hydroclimatic variables. *Trends and Changes in Hydroclimatic Variables* 2019; 91–148. https://doi.org/10.1016/B978-0-12-810985-4.00002-5

[40] Stefanidis K, Varlas G, Papadopoulos A, Dimitriou E. Four decades of surface temperature, precipitation, and wind speed trends over lakes of Greece. 2021, 13(17): 9908 https://doi.org/10.3390/su13179908

[41] Fourth National climate assessment; chapter two; our changing climate. Available at: https://nca2018.globalchange.gov/chapter/2/ (Accessed on 6 September, 2022).

[42] USGS science for a changing world; Volcanoes can affect climate. Available at: https://www.usgs.gov/programs/VHP/volcanoes-can-affect-climate (Accessed on 6 September, 2022).

[43] Union of Concerned Scientists. Tropical deforestation and global warming. Published Jul 27, 2008, Updated Nov 10, 2021. Available at: https://www.ucsusa.org/resources/tropical-deforestation-and-globalwarming#:~:text=When%20forests%20are%20cut%20down,percent%20of%20global%20warming%20pollution (Accessed on 4 September, 2022).

[44] Earth Eclipse: How does deforestation affect climate change? Available at: https://eartheclipse.com/environment/climate-change/how-does-deforestation-affect-climate-change.html (Accessed on 6 September, 2022).

[45] Mazzocchi F. The covid-19 pandemic and the climate crisis: a call to question the mindset of modernity. *Challenges* 2022; 13(33):1–11. https://doi.org/10.3390/challe13020033

[46] Mora C, McKenzie T, Gaw IM, Dean JM, Hammerstein HV, Knudson TA, Setter RO, Smith CZ, Webster KM, Patz JA, Franklin, EC. Over half of known human pathogenic diseases can be aggravated by climate change. *Nature Climate Change* 2022; 1–9. https://doi.org/10.1038/s41558-022-01426-1

[47] Wu X, Lu Y, Zhou S, Chen L, Xu B. Impact of climate change on human infectious diseases: Empirical evidence and human adaptation. *Environment International* 2016; 14–23. https://doi.org/10.1016/j.envint.2015.09.007

[48] World Health Organization (2021). Climate change and health. Available online: https://www.who.int/news-room/fact-sheets/detail/climate-change-and-health (Accessed on 8 September, 2022).

[49] McKinsey & Company: Global Energy Perspective 2022 Executive Summary. Pages 1–28. https://www.mckinsey.com/~/media/McKinsey/Industries/Oil%20and%20Gas/Our%20Insights/Global%20Energy%20Perspective%202022/Global-Energy-Perspective-2022-ExecutiveSummary.pdf. (Accessed on 11 September, 2022).

[50] International renewable Energy Agency (2022). World Energy Transitions Outlook 2022. Pages 1–20. Available at: https://www.irena.org//media/files/irena/agency/publication/2022/mar/irena_weto_summary_202.2.pdf?la=en&hash=1da99d3c3334c84668f5caae029bd9a076c10079 (Accessed on 11 September, 2022).

[51] Adegoke CW, Oluwafisoye PA, Ajiboye OK, Ajewole TO. Operational design parameters in multi-lamp module solar street light system. *Nigerian Journal of Solar Energy* 2010; 21:1–8.

[52] Oluwafisoye PA, Adegoke CW, Ajiboye OK, Ajewole TO. Development of a 3.5 kW solar power-pack dedicated to laboratory (ultra violet) equipment at UNIOSUN. *Nigerian Journal of Solar Energy* 2010; 21:9–14.

[53] Goswami DY (2015). *Principle of solar engineering.* 3rd Edition, CRC Press, Taylor & Francis, 1–790.

[54] Omoigui MO, Ajewole TO, Ariyo FK. (2011). Investigation of the transient behaviour of a grid-connected photovoltaic power generator. *Nigerian Society of Engineers Technical Transactions* 2011; 46(1):97–107.

[55] Ajewole TO, Babalola OS, Omoigui MO. Stability impact of grid-tied photovoltaic plant on the distribution network. *International Journal of Scientific and*

Engineering Research 2015; 6(12):787–790. https://doi.org/10.1429 9/ijser. 2015.12.011

[56] Solar the energy workhorse in latest gloomy IPCC verdict. PV magazine. April, 5th 2022. Available at: https://www.pv-magazine.com/2022/04/05/solar-the-energy-workhorse-in-latest-gloomy-ipcc-verdict/ (Accessed on 10 September, 2022).

[57] Nelson J, Gambhir A, Ekins-Daukes, N. Solar power for co_2 mitigation. *Imperial College London Grantham Institute for Climate Change* 2014; 1–16.

[58] International Energy Agency (2021). Net Zero by 2050 A Roadmap for the Global Energy Sector. Available online: https://iea.blob.core.windows. net/assets/deebef5d-0c34-4539-9d0c-10b13d840027/NetZeroby2050-ARoadmapfortheGlobalEnergySector_CORR.pdf (Accessed on 13 September, 2022).

[59] Lawal MO, Bada OM, Ajewole TO. An experimental approach towards pv-based solar system sizing for an engineering laboratory. *Tanzania Journal of Engineering and Technology* 2022; 41(2):98–108. https/doi.org/10.52339/tjet. v40i2.736

[60] Solar panels performance: clouds, rain and snow. Available at: https://www. solaris-shop.com/blog/solar-panel-performance-clouds-rain-and-snow/ (Accessed on 13 September, 2022).

[61] Ecogeneration: Bifacial shines as PV power booster in cloudy conditions. Available at: https://www.ecogeneration.com.au/bifacial-shines-as-pv-power-booster-in-cloudy-conditions/ (Accessed on 13 September, 2022).

[62] Shoukry I, Libal J, Kopecek R, Wefringhaus E, Werner J. Modelling of bifacial gain for stand-alone and in-field installed bifacial PV modules. *Energy Procedia* 2016; 92:600–608. https://doi.org/10.1016/j.egypro.2016.07.025

[63] Ajewole TO, Olabode OE, Alawode KO, Lawal MO. Small-Scale Electricity Generation through Thermal Harvesting in Rooftop Photovoltaic Picogrid using Passively Cooled Heat Conversion Devices. *Environmental Quality Management* 2020; 29(4):95–102, http://doi.org/10.1002/tqem.21696

[64] Electrical Engineering Portal: Design of 60MW grid tied solar power plant with 115kV/34.5kV substation. Available at: https://electrical-engineering-portal. com/download-center/books-and-guides/alternative-energy/solar-power-plant (Accessed on 14 September, 2022).

[65] ESI Africa newsletter (2022). Grid-tied solar plant successfully commissioned in Ghana. Available at: https://www.esi-africa.com/solar/grid-tied-solar-plant-successfully-commission-in-ghana/ (Accessed on 14 September, 2022).

[66] InfraCo AFRICA: Malawi's President inaugurates the 60MW Salima Solar plant 17th Nov 2021. Online: https://infracoafrica.com/malawis-president-inaugurates-the-60mw-Salima-solar-plant/ (Accessed on 14 September, 2022).

[67] Available at: https://www.phnompenhpost.com/business/60mw-solar-station-battambang-hooked-grid. (Accessed on 14 September, 2022).

[68] Ten IPPs in run for developing 60 MW solar PV plants in Ivory Coast. Available at: https://solarquarter.com/2021/12/01/ten-ipps-in-run-for-developing-60-mw-solar-pv-plant-in-ivory-coast/ (Accessed on 14 September, 2022).

[69] Ajewole TO, Lawal MO, Omoigui MO. Development of a lab-demo facility for wind energy conversion systems. *International Journal on Energy Conversion* 2016; 4(1):1–6, https://doi.org/10.15866/irecon.v4i1.8572

[70] Ajewole TO, Kayode O, Omoigui MO, Oyekanmi WA. (2016). Demonstration of power factor improvement in wind energy conversion systems using wind turbine emulator. *Nigerian Society of Engineers Technical Transactions* 2016; 50(1):108–123.

[71] The geography of transportation systems: Global wind pattern. Available at: https://transportgeography.org/contents/chapter1/transportation-and-space/ global-wind-patterns/ (Accessed on 14 September, 2022).

[72] Idris WO, Ibrahim MZ, Albani A. The status of the development of wind energy in Nigeria. *Energies* 2020; 13(6219):1–16

[73] Ajewole TO, Craven PMR, Kayode O, Babalola OS. (2018). Simulation of load-sharing in standalone distributed generation system. *Proceedings of the 7th International Conference on Clean and Green Energy*, February 7–9, 2018 Paris, France, pp. 108–115, https://doi.org/10.1088/1755-1315/154/1/012014

[74] Where are the windiest parts of the UK? Available at: https://www.metoffice. gov.uk/weather/learn-about/weather/types-of-weather/wind/windiest-place-in-uk#:~:text=The%20windiest%20places%20in%20the,Summit%20on%20 20%20March% (Accessed on 14 September, 2022).

[75] Current results in weather and science facts. Available at: https://www. currentresults.com/Weather-Extremes/US/windiest-cities.php (Accessed on 14 September, 2022).

[76] Ajewole TO, Alawode KO, Omoigui MO, Oyekanmi WA. Design validation of a laboratory-scale wind turbine emulator. *Cogent Engineering* 2017; 4:1–13, https://dx.doi.org/10.1080/23311916.2017.1280888

[77] Global Wind Energy Council. Global Wind Report. 2021. Available online: https://gwec.net/global-wind-report-2021/ (Accessed on 14 September, 2022).

[78] Ajewole TO, Oyekanmi WA, Babalola AA, Omoigui MO. RTDS modeling of a hybrid-source autonomous electric microgrid. *International Journal of Emerging Electric Power Systems* 2017; 18(2):1–11. https://doi.org/10.15 15/ ijeeps-2016-0157

[79] Renewables 2021 global status report. Available at: https://www.ren21.net/wp-content/uploads/2019/05/GSR2021_Full_Report.pdf (Accessed on 14 September, 2022).

[80] Global wind report 2021. Available at: https://gwec.net/global-wind-report-2021/ (Accessed on 14 September, 2022).

[81] Changing winds: emerging wind turbine technologies. Available at: https:// www.powermag.com/changing-winds-emerging-wind-turbine-technologies/ (Accessed on 14 September, 2022).

[82] Next-generation wind technology. Available at: https://www.energy.gov/eere/ wind/next-generation-wind-technology (Accessed on 14 September, 2022).

[83] https://nzebnew.pivotaldesign.biz/knowledge-centre/renewable-energy/wind/ (Accessed on 15 September, 2022).

[84] Africa's largest wind power project opens in northern Kenya. Available at: https:// www.reuters.com/article/us-kenya-energy-environment-idUSKCN1UE2GM (Accessed on 15 September, 2022).

[85] New wind farm for Ethiopia. Available at: https://iclg.com/alb/15359-new-wind-farm-for-ethiopia (Accessed on 15 September, 2022).

[86] Operation of Jhimpir wind turbines offers a ray of hope in power gloom. Available at: https://www.dawn.com/news/1032576(Accessed on 15 September, 2022).

[87] The world's biggest wind farms. Available at: https://www.nesfircroft.com/blog/2021/11/the-worlds-biggest-wind-farms?source=google.com (Accessed on 15 September, 2022).

[88] Natural gas. Available at: https://education.nationalgeographic.org/resource/natural-gas (Accessed on 16 September, 2022).

[89] U.S energy information administration. Available at: https://www.eia.gov/outlooks/aeo/retrospective/ (Accessed on 16 September, 2022).

[90] Just 5 percent of power plants release 73 percent of global electricity production emissions. Available at: https://www.smithsonianmag.com/smart-news/five-percentpower-plants-release-73-percent-global-electricity-production-emissions-180978355/ (Accessed on 16 September, 2022).

[91] Elehinafe FB, Okedere OB., Ayeni AO, Ajewole TO. (2022). Hazardous organic pollutant from open burning of municipal wastes in southwest Nigeria. *Journal of Ecological Engineering* 2022; 23(9):288–296. https://doi.org/10.12911/22998993/15064

[92] Department of Energy and Environmental Protection. Available at: https://portal.ct.gov/DEEP/Reduce-Reuse-Recycle/Climate-Change/Climate-Change-and-Waste (Accessed on 19 September, 2022).

[93] The world bank on solid waste management. Available at: https://www.worldbank.org/en/topic/urbandevelopment/brief/solid-waste-management (Accessed on 19 September, 2022).

[94] United States Environmental Protection Agency. Available at: https://www.epa.gov/facts-and-figures-about-materials-waste-and-recycling/national-overview-facts-and-figures-materials. (Accessed on 19 September, 2022).

[95] Solid waste management in Nigeria. Available at: https://www.bioenergyconsult.com/solid-waste-nigeria/ (Accessed on 19 September, 2022).

[96] Municipal waste generated in Italy 2019, by region. Available online: https://www.statista.com/statistics/683093/municipal-waste-generated-in-italy-by-region/ (Accessed on 21 September, 2022).

[97] National solid waste diversion and disposal. Available at: https://www.canada.ca/en/environment-climate-change/services/environmental-indicators/solid-waste-diversion-disposal.html (Accessed on 21 September, 2022).

[98] Solid waste management and recycling in Ethiopia. Available at: https://aaeafrica.org/home/solid-waste-management-and-recycling-in-ethiopia/ (Accessed on 21 September, 2022).

[99] Canada waste-to-energy market–growth,trends,covid-19 impact,and forecasts (2022–2027). Available at: https://www.mordorintelligence.com/industry-reports/canada-waste-to-energy-arketindustry#:~:text=Canada's%20waste%2Dto%2Denergy%20marketdue%20to%20supply%20chain%20issue (Accessed on 21 September, 2022).

[100] Solid waste service spokenecity. Available at: https://my.spokanecity.org/solidwaste/waste-to-energy/ (Accessed on 21 September, 2022).

[101] Energy from waste cumulative installed capacity in the United Kingdom (UK) 2009–2021. Available at: https://www.statista.com/statistics/1097210/cumulative-waste-to-energy-capacity-united-kingdom/ (Accessed on 21 September, 2022).

[102] Government of India, Ministry of new and renewable energy. Available at: https://mnre.gov.in/waste-to-energy/current-status (Accessed on 21 September, 2022).

[103] Ajewole TO, Alawode KO, Abidoye LK, Oluwasanmi BO. Determination of thermo-electric potentials of some local sawdust as energy source feedstock for electric power generation. *UNIOSUN Journal of Science* 2017; 2(2):112–119.

[104] Biomass energy. Available at: https://education.nationalgeographic.org/resource/biomass-energy (Accessed on 22 September, 2022).

[105] Ajewole TO, Elehinafe FB., Okedere OB, Somefun TE. Agro-Residues for Clean Electricity: A Thermo-Property Characterization of Cocoa and Kolanut Wastes Blends. *Heliyon* 2021; 7(9):1–9. https://doi.org/10.1016/j.heliyon. 20 21.e08055

[106] United States Environmental Protection Agency: "Carbon Pollution from Transportation" Available at: https://www.epa.gov/transportation-air-pollution-and-climate-change/carbon-pollution-transportation (Accessed on 22 September, 2022).

[107] World economic forum on climate change. Available at: https://www.weforum. org/agenda/2022/06/green-transport-and-cleaner-mobility-are-key-to-meeting-climate-goals/ (Accessed on 23 September, 2022).

[108] United States Environmental Protection Agency- Fast facts on transportation greenhouse gas emission. Available at: https://www.epa.gov/greenvehicles/fast-factstransportation-greenhouse-gas-emissions (Accessed on 23 September, 2022).

[109] Ajewole TO, Aworinde AK, Okedere OB, Somefun TE. Agro-Residues for Clean Electricity: In-Lab Trial of Power Generation from Blended Cocoa/Kolanut Wastes. *Heliyon* 2022, 8(3):1–6. https://doi.org/10.1016/j.heliyon.2022. e09091

[110] Environmental benefits of biomass. Available at: https://www.eubia.org/cms/wiki-biomass/employment-potential-in-figures/environmental-benefits/ (Accessed on 23 September, 2022).

[111] Available at: https://www.nrcan.gc.ca/our-natural-resources/energy-sources-distribution/renewable-energy/about-renewable-energy/7295 (Accessed on 23 September, 2022).

[112] Biomass-renewable energy from plants and animals. Available at: https://www.eia.gov/energyexplained/biomass/ (Accessed on 23 September, 2022).

[113] IEA Bioenergy Countries' Report-Update 2021. Available at: https://www.ieabioenergy.com/blog/publications/iea-bioenergy-countries-report-update-2021/ (Accessed on 23 September, 2022).

[114] Akinbami OM, Oke SR, Bodunrin MO. The state of renewable energy development in South Africa: an overview. *Alexandria Engineering Journal* 2021; 60:5077–5093. https://doi.org/10.1016/j.aej.2021.03.065.

[115] Production and Operating Capacity – The Renewable Energy Data and Information Service. http://redis.energy.gov.za/power-production/ (accessed on 23 September, 2022).

[116] IEA Bioenergy – Implementation of bioenergy in Germany – 2021 update. Available at: https://www.ieabioenergy.com/wp-content/uploads/2021/11/Country Report2021 (Accessed on 23 September, 2022).

[117] California Energy Commission-2021 Total System Electric Generation. Available at: https://www.energy.ca.gov/data-reports/energy-almanac/california-electricity-data/2021-total-system-electric-generation (Accessed on 23 September, 2022).

[118] IEA Bioenergy-Implementation of bioenergy in China. Available at: https://www.ieabioenergy.com/wp-content/uploads/2021/11/CountryReport2021_China_final.pdf (Accessed on 23 September, 2022).

[119] Hydropower special market report analysis and forecast to 2030. Available at: https://iea.blob.core.windows.net/assets/83ff8935-62dd-4150-80a8c5001b740e21/ (Accessed on 26 September, 2022).

[120] Ajewole TO, Ajewole OT, Fagbamiye MO, Omoigui MO. A study on the hydro-electric potential of olumirin waterfall at erin-ijesa in Osun State of Nigeria. *UNIOSUN Journal of Science* 2018; 3(1):64–72.

[121] Konak N, Sungu-Eryilmaz Y. Does small run-of-river hydro power development in turkey deliver on its sustainability premise. *Society and Natural Resources* 2015; 29(7):807–821. https://doi.org/10.1080/08941920.2015.1086459

[122] Environmental and Energy Study Institute-Geothermal. Available at: https://www.eesi.org/topics/geothermal/description#:~:text=Heat%20from%20below%20the%20earth's,gases%20that%20cause%20climate%20change (Accessed on 27 September, 2022).

[123] Volcanoes and geothermal energy. Available at: https://www.openaccessgovernment.org/volcanoes-geothermal-energy/14541/ (Accessed on 27 September, 2022).

[124] U.S Department of Energy – Geothermal energy production with co-produced and geopressuredresource9os. Available at: https://www1.eere.energy.gov/geothermal/pdfs/low_temp_copro_fs.pdf/ (Accessed on 27 September, 2022).

[125] Longa FD, Nogueira LP, Limberger J, Wees JV, Zwaan BV. Scenarios for geothermal energy deployment in Europe. *Energy* 2020; 206:1–10. https://doi.org/10.1016/j.energy.2020.118060

[126] Bromley CJ, Mongillo M, Hiriart G, Goldstein B, Bertani R, Huenges E, Tester AJ, Muraoka H, Zui V. Contribution of geothermal energy to climate change mitigation: the IPCC renewable energy report. *Proceedings World Geothermal Congress* 2010 Bali, Indonesia, 25–29 April 2010.

[127] Office of Energy Efficiency & Renewable Energy-Geothermal energy's role in addressing the climate crisis. Available at: https://www.energy.gov/eere/geothermal/geothermal-energys-role-addressing-climate-crisis#: (accessed on 27 September, 2022).

[128] Geothermal power. Available at: https://en.wikipedia.org/wiki/Geothermal_poweracc Assessed on 27 September, 2022).

[129] U.S. Energy Information Administration: Electricity in the United States. Available at: https://www.eia.gov/energyexplained/electricity/electricity-in-the-us.php (accessed on 27 September, 2022).

[130] Geothermal power in Indonesia. Available at: https://en.wikipedia.org/wiki/Geothermal_power_in_Indonesia (Accessed on 27 September, 2022).

[131] The Manila Times: PH seen as major geothermal market. Available at: https://www.manilatimes.net/2022/07/16/business/green-industries/ph-seen-as-major-geothermalmarkatt/1851052#:~:text=The%20Philippines%20is%20one%20of,nearly%20double (Accessed on 27 September, 2022).

[132] Turkish Association of Geothermal Investors. Available at: https://www.thinkgeoenergy.com (Accessed on 27 September, 2022).

[133] Electricity sector in New Zealand. Available at: https://en.wikipedia.org/wiki/Electricity_sector_in_New_Zealand (Accessed on 27 September, 2022).

[134] Source of renewable energy in Mexico. Available at: https://en.wikipedia. org/wiki/Renewable_energy_in_Mexico#:~:text=Geothermal%20is%20the %20second%20most,the%20total%20global%20renewable%20energy. Accessed on 27 September, 2022).

[135] ThinkGeoEnergy's Top 10 Geothermal Countries 2020 – installed power generation capacity. Available at: https://www.thinkgeoenergy.com/thinkgeo energys-top-10-geothermal-countries-2020-installed-power-generation-capacity-mwe/ (Accessed on 28 September, 2022).

[136] Geothermal Bubbles Up as Another Way to Fight Climate Change. Available at: https://www.pewtrusts.org/en/research-and-analysis/blogs/stateline/2022/ 09/09/geothermal-bubbles-up-as-another-way-to-fight-climate-change (Accessed on 28 September, 2022).

[137] Renewable Energy Magazine: Iceland Using Geothermal to Remove CO_2. Available at: https://www.renewableenergymagazine.com/emily-folk/iceland-using-geothermal-to-remove-co2-20200910 (Accessed on 28 September, 2022).

[138] Client Earth: Fossil fuels and climate change: the facts. Available at: https:// www.clientearth.org/latest/latest-updates/stories/fossil-fuels-and-climate-change-the-facts/ (Accessed on 29 September, 2022).

[139] University of Arkansas Sustainability Blog: Will we ever be able to stop using fossil fuels? Available at: https://wordpressua.uark.edu/sustain/3-reasons-we-are-still-using-fossil-fuels/ (Accessed on 29 September, 2022).

[140] The basics of carbon capture and storage. Available at: https://www.farmweek now.com/general/the-basics-of-carbon-capture-and-storage/article_77977b6e-8926-11ec-85f4-d3153749d493.html (Accessed on 29 September, 2022).

Chapter 4

Quantum neural networks for machine learning applied to the tracking and control of the dynamics of stochastic transmission lines

Harish Parthasarathy
Netaji Subhash University of Technology, New Delhi, India

Arti Vaish
Department of computer science, School of Engineering,
O.P. Jindal University, Raigarh, India

CONTENTS

4.1 INTRODUCTION

A quantum neural network is based on the Schrödinger equation evolving in accordance with a self-adjoint Hamiltonian operator and hence the state of such a system is governed by unitary dynamics which implies that the wave function modulus square will at all times be a probability density if it is so at the initial time. Moreover, by controlling a set of classical parameters in the Hamiltonian operator, we can control the wave function dynamics in such a way that the corresponding probability density tracks a given family of probability density functions evolving in time [1, 2]. In this way, an entire family of evolving PDFs can be compressed into simple finite parameter trajectories which can later on be used for synthesis. Thus, data compression is the first achievement of a quantum neural network (QNN). The second advantage is learning. We take up the example of a transmission line driven by a source process and connected to a load and that is subject to random line loading. The stochastic differential equations for the line voltage and current lead to Fokker–Planck or Chapman–Kolmogorov equations for the joint line voltage and current PDF with time [3]. These partial difference equations only contain information about the line distributed parameters but not about the source process or the load. The boundary conditions satisfied by the line PDF in order to match the source and load constraints give us extra terms in the cost function for tracking a given joint PDF [4]. Alternatively, we can train the QNN to track the line voltage and current joint PDF without taking into account the source and load constraints, that is, by just matching, using the gradient algorithm, the QNN PDF with the true line PDF by adapting the parameters in the Schrödinger Hamiltonian. This forms the training stage. The trained parameters of the QNN reflect only the statistics of the line loading noise. Now in the testing stage, given a new line with the same loading noise statistics but different source and load characteristics, we start with the trained parameters and then adjust them slightly by matching the source and load constraints. Alternatively, we can take into account the source and load information by encoding this information into the adaptation constant in the gradient algorithm by computing the error energy between the source process and the average of the voltage process at the source end, obtained using the QNN wave function and also the error energy corresponding to the load constraints. If these error energies are large, we accordingly increase the adaptation constant and, if they are small, we decrease it at the next time instant. This chapter presents a methology for carrying out this programme. Further, since the number of sample spatial points along the transmission line is very large, the corresponding joint PDF is of a very large dimension and hence, in order to use a QNN to track this, we must use a Schrödinger equation with a very large number of position variables. This is not feasible in practice. However, quantum field theoretic models naturally incorporate a very large number of independent position variables; indeed the canonical position field can be developed into a Fourier series and each Fourier coefficient can be regarded as an independent position variable, and the Schrödinger equation for the wave functional of the field can be viewed as a Schrödinger equation for the wave function of the field's Fourier coefficients. The Hamiltonian for such a Schrödinger equation is derived from the Lagrangian density of the field and hence an appropriate large dimensional QNN can be designed for tracking the large dimensional line PDF. In view of the increasing recent interest in string theory, we also include in this chapter a section on how to use the Feynman path integral for the string field interacting with control classical fields as well as with a quantum gauge field, so that the quantum gauge field has a given transition probability from one state to the other. This section is based on computing probabilities in quantum mechanics using Feynman's path integral approach rather than Schrödinger's equation, as the former is better suited to quantum field theoretic problems [5, 6]. This chapter also includes some aspects of classical and quantum filtering theory especially in situations when we do not have access to the line probability density but only to the line conditional probability density given noisy measurements, or when we wish to control the quantum filtered state for tracking based on the Hudson–Parthasarathy quantum stochastic calculus and the Belavkin filter.

4.2 TRACKING THE TRANSMISSION LINE PDF USING THE MANY PARTICLE SCHRÖDINGER EQUATION

Statement of the problem: Consider a transmission line of length d carrying a voltage $v(t, z)$ and a current $i(t, z)$ with $t \geq 0, 0 \leq z \leq d$. Taking into account random line loading, the dynamics of such a line are given by the stochastic partial differential equations

$$-\partial_z v(t, z) = L(z) \partial_t i(t, z) + R(z) i(i, z) + \sigma_v W_v(t, z),$$
$$-\partial_z i(t, z) = C(z) \partial_t v(t, z) + G(z) v(t, z) + \sigma_i W_i(t, z)$$

where W_v, W_i represent respectively random voltage and current loading along the line. These equations can in principle be solved given the terminal conditions, namely the source and load conditions

$$v(t, 0) + i(t, 0) R_s = v_s(t), v(t, d) = i(t, d) R_L$$

The source voltage $v_s(t)$ is assumed to be a non-random process. This follows by solving these equations of which we can in principle calculate the joint probability density of the line voltage and current at time-space points $(t_1, z_1, ..., t_k, z_k)$. We denote this joint PDF by

$$p(v_1, ..., v_k, i_1, ..., i_k \mid t_1, z_1, ..., t_k, z_k)$$

Our aim is to estimate this family of joint PDFs by using the fact that it is highly dependent upon the source process $v_s(t)$. To simplify matters, we shall be primarily interested in the situation where the joint PDF of the current and voltage at different spatial points along the line are at a fixed time which varies, that is, we are interested in the joint PDF $p(v_1, ..., v_k, i_1, ..., i_k | t, z_1, ..., z_k)$ of $v(t, z_j), i(i, z_j), j = 1, 2, ..., k$. In a QNN designed to solve this problem, we use the Schrödinger equation in $2k$ variables:

$$i \partial_t \psi(t, v_1, ..., v_k, i_1, ..., i_k) = (-1/2) \sum_{m=1}^{k} \left(\partial_{v_m}^2 + \partial_{i_m}^2 \right) \psi + V(t, \theta(t)) \psi$$

where $\theta(t)$ is a set of time varying parameters in the Schrödinger potential that is adapted so as to ensure that the Schrödinger PDF $|\psi(t, .)|^2$ tracks the given PDF $p_t = p(. |t, z_m, m = 1, 2, ..., k)$. The fact that p_t depends upon the source voltage $v_s(t)$ and its past values means that the mean values of the voltage and current obtained using the Schrödinger PDF must be highly correlated with the source voltage $v_s(.)$. In order to make use of this information, we construct these means at each time instant, predict these means using a nonlinear predictor based on the present and past values of the source voltage, and make the adaptation constant increase with the increase in the magnitude of the corresponding prediction error energy. The training of this QNN can be done in two ways, one requiring information about the Tx line PDF and the other not requiring this information and requiring only information about some statistical moments of the line variables. In the former, we aim at minimizing $|p_t - |\psi_t|^2|^2$ by adapting $\theta(t)$ for a given source voltage process $v_s(t)$. After this training is over, we will not be given the line PDF for the new line. We will be given only the source voltage and the adapated parameter history. Then, we will synthesize the line PDF by starting with the adapted line parameters and adjusting these parameters further so that the mean of the Schrödinger PDF has a minimum prediction energy when prediction is carried out based on the given source voltage. In the second method, we directly adapt the Schrödinger potential parameters so as to minimize the error between the statistical moments of the line voltage and current and those computed using the Schrödinger PDF. This scheme does not involve any training. It is based on the premise that the Schrödinger equation represents the natural way in which the random line loading generates the line variable PDF given the source process history.

In order to carry out this programme, we first calculate the joint PDF of the voltage and current from the transmission line model as follows, Fourier transforming in time the Tx line equations gives us

$$-\partial_z V(\omega, z) = \left(R(z) + j\omega L(z)\right) I(\omega, z) + W_v(\omega, z),$$

$$-\partial_z I(\omega, z) = \left(G(z) + j\omega C(z)\right) V(\omega, z) + \sigma_i W_i(\omega, z)$$

which can be expressed in vector notation as

$$-\partial_z \xi(\omega, z) = A(\omega, z) \xi(\omega, z) + W(\omega, z)$$

where

$$\xi(\omega, z) = \left[V(\omega, z), I(\omega, z)\right]^T, \quad A(\omega, z) = \begin{pmatrix} 0 & Z(\omega, z) \\ Y(\omega, z) & 0 \end{pmatrix},$$

$$W(\omega, z) = \left[W_v(\omega, z), W_i(\omega, z)\right]^T$$

Let $\Phi(\omega, z, z')$ denote the state transition matrix associated with the above system of vector differential equations, that is,

$$\partial\Phi(\omega, z, z') = -A(\omega, z)\Phi(\omega, z), z \geq z', \Phi(\omega, z', z') = I_2$$

Then the solution to the above equation is

$$\xi(\omega, z) = -\int_0^z \Phi(\omega, z, z')W(\omega, z')dz' + \Phi(\omega, z, 0)\xi(\omega, 0)$$

Taking the inverse Fourier transform w.r.t. ω gives us

$$\xi(t,z) = \left[v(t,z), i(t,z)\right]^T = -\int \phi(t - s, z, z')w(s, z')dsdz'$$
$$+ \int \phi(t - s, z, 0)\xi(s, 0)ds$$

where

$$\phi(t, z, z') = (2\pi)^{-1}\int \Phi(\omega, z, z')\exp(j\omega t)d\omega$$

In order to obtain $\xi(t, 0)$ or $\xi(\omega, 0)$, we use the source and load boundary conditions, which assuming the load to be resistive, gives

$$[1, R_s]\xi(\omega, 0) = v_s(\omega), [1, -R_L]\xi(\omega, d) = 0$$

These conditions give us two equations for the two variables in $\xi(\omega, 0)$ which are

$$[1, R_s]\xi(\omega, 0) = v_s(\omega), \int_0^z [1, R_L]\Phi(\omega, z, z')W(\omega, z')dz'$$

$$+ [1, -R_L]\Phi(\omega, z, 0)\xi(\omega, 0) = 0$$

which are easily solved for $\xi(\omega, 0)$ in terms of $v_s(\omega)$, $W(\omega, .)$.

There is a special case in which the Tx line stochastic differential equations directly yield a PDF for the joint PDF of $v(t, .)$, $i(t, .)$. This occurs when W_v, W_i are expressible as linear combinations of differentials of the Brownian motion process, that is, we write

$$W_v(t, z) = v(t, z) = \sum_n f_n(z)dB_n(t)/dt, W_i(t, z) = \sum_n g_n(z)dB_n(t)/dt$$

where the B_n's are independent Brownian motion processes. In this case, we choose a set of orthonormal basis functions $\phi_n(z)$, $n = 1, 2, \ldots$ along the line and expand

$$v(t,z) = \sum_n v_n(t)\phi_n(z), i(t,z) = \sum_n i_n(t)\phi_n(z)$$

Substituting these expressions into the line equations gives us

$$\sum_n v_n(t)\phi_{n'}(z) + \sum_n L(z)\phi_n(z)i_{n'}(t) + \sum_n R(z)\phi_n(z)i_n(t)$$
$$= \sum_n f_n(z)dB_n(t)/dt$$

$$\sum_n i_n(t)\phi_{n'}(z) + \sum_n C(z)\phi_n(z)i_{n'}(t) + \sum_n G(z)\phi_n(z)i_n(t)$$
$$= \sum_n g_n(z)dB_n(t)/dt$$

From these, we derive

$$i_{n'}(t) + \sum_m v_m(t)\int \phi_{m'}(z)\phi_n(z)dz/L(z) + \sum_m i_m(t)\int R(z)\phi_m(z)\phi_n(z)dz/L(z)$$
$$= \sum_m \left(\int f_m(z)\phi_n(z)dz/L(z) \right) dB_m(t)/dt$$

$$v_{n'}(t) + \sum_m i_m(t)\int \phi_{m'}(z)\phi_n(z)dz/C(z) + \sum_m v_m(t)\int G(z)\phi_m(z)\phi_n(z)dz/C(z)$$
$$= \sum_m \left(\int g_m(z)\phi_n(z)dz/C(z) \right) dB_m(t)/dt$$

These equations should be interpreted as either Ito or Stratonovich stochastic differential equations. Since the transition from one to the other interpretation is well known, we shall use the Ito formalism and express the above as

$$di_n(t) = \left(\sum_m \left(A_1(n,m)v_m(t) + A_2(n,m)i_m(t) \right) \right) dt + \sum_m A_3(n,m)dB_m(t),$$

$$dv_n(t) = \left(\sum_m \left(A_4(n,m) i_m(t) + A_5(n,m) v_m(t) \right) \right) dt + \sum_m A_6(n,m) dB_m(t),$$

where

$$A_1(n,m) = -\int \phi'_m(z) \phi_n(z) dz/L(z), \quad A_2(n,m) = -\int R(z) \phi_m(z) \phi_n(z) dz/L(z),$$

$$A_3(n,m) = \int f_m(z) \phi_n(z) dz/L(z) \quad A_4(n,m) = -\int \phi'_m(z) \phi_n(z) dz/C(z),$$

$$A_5(n,m) = -\int G(z) \phi_m(z) \phi_n(z) dz/C(z), \quad A_6(n,m) = \int g_m(z) \phi_n(z) dz/C(z)$$

Equivalently, in vector notation,

$$di(t) = \left(A_1 v(t) + A_2 i(t) \right) dt + A_3 dB(t),$$
$$dv(t) = \left(A_4 i(t) + A_5 v(t) \right) dt + A_6 dB(t)$$

where the infinite dimensional vectors $i(t)$, $v(t)$, $B(t)$ have the obvious meanings as also do the infinite dimensional matrices A_k, $k = 1, \ldots, 6$. Assuming that we have discrete spatial measurements $v(t, z_k)$, $i(t, z_k)$, $k = 1, 2, \ldots, M$, we can use these to approximately calculate

$$v_n(t) = \int v(t,z) \phi_n(z) dz \approx \sum_k v(t, z_k) \phi_n(z_k)(z_{k+1} - z_k)$$

$$i_n(t) = \int i(t,z) \phi_n(z) dz \approx \sum_k i(t, z_k) \phi_n(z_k)(z_{k+1} - z_k)$$

and then obtain, using these, the empirical distribution of the truncated versions of $v(t)$, $i(t)$. It should be noted that $i(t)$, $v(t)$ has a joint PDF $p(t, i, v)$ satisfying the Fokker–Planck equation

$$\partial_t p(t,i,v) = -\nabla_v . \left(\left(A_1 v + A_2 i \right) p(t,i,v) \right) - \nabla_i . \left(\left(A_4 i + A_5 v \right) p(t,i,v) \right)$$

$$+ (1/2) Tr \left[\begin{pmatrix} \nabla_i \\ \nabla_v \end{pmatrix} . \left(\nabla_i^T, \nabla_v^T \right) p(t,i,v) \right]$$

By solving this Fokker–Planck equation, we obtain $p(t, i, v)$ and use this to train the Schrödinger equation by adapting the potential parameters so that the modulus square of the wave function tracks this density. Specifically, discretizing the above Fokker–Planck equation in time gives us

$$p(t+h,i,v) = p(t,i,v) + h.Lp(t,i,v)$$

where L is the above Fokker–Planck operator:

$$Lf(i,v) = -\nabla_v.\left((A_1v + A_2i)f(i,v)\right) - \nabla_i.\left((A_4i + A_5v)f(i,v)\right)$$
$$+(1/2)Tr\left[\begin{pmatrix}\nabla_i\\\nabla_v\end{pmatrix}.(\nabla_i^T,\nabla_v^T)f(i,v)\right]$$

Writing the discretized Schrödinger equation as

$$\psi(t+h,i,v) = \psi(t,i,v) - ih.H(\theta(t))\psi(t,i,v)$$

where

$$H(\theta) = (-1/2)\nabla_{(i,v)}^2 + V(\theta)$$

is the Hamiltonian operator, we train the parameters θ so that $|\psi(t, i, v)|^2$ tracks $p(t, i, v)$ as follows:

$$\theta(t+h) = \theta(t) - \mu(t).\frac{\partial}{\partial\theta(t)}\left(\int\left(p(t+h,i,v) - |\psi(t+h,i,v)|^2\right)^2 didv\right.$$
$$= \theta(t) - \mu(t).\frac{\partial}{\partial\theta(t)}\left(\int\left(p(t+h,i,v) - |\psi(t+h,i,v) - ih.H(\theta(t))\psi(t,i,v)|^2\right)^2 didv\right.$$

This iteration is to be combined with the iterations

$$p(t+h,i,v) = p(t,i,v) + h.Lp(t,i,v), \psi(t+h,i,v)$$
$$= \psi(t,i,v) - ih.H(\theta(t))\psi(t,i,v)$$

to obtain the parameter trajectory $\theta(t)$ that represents the family of PDFs $p(t, i, v), t = 0, h, 2h, \ldots$ of the line current and voltage. The aim of the QNN is however to estimate $p(t, i, v)$ for a given source voltage process $v_s(t)$. So far, we have not made use of the information in the PDF about the source voltage and the load. To see how this enters into the picture, we use the boundary conditions

$$v(t,0) + R_s i(t,0) = v_s(t), v(t,d) = i(i,d)R_L$$

where the load R_L is assumed to be resistive for simplicity. We note that

$$v(t,0) = \sum_b v_n(t)\phi_n(0), i(t,0) = \sum_n i_n(t)\phi_n(0), v(t,d)$$
$$= \sum_n v_n(t)\phi_n(d), i(t,d) = \sum_n i_n(t)\phi_n(d)$$

So incorporating these constraints into the Fokker–Planck equation, we obtain the PDF $p(t, i, v)$. Actually, the line stochastic differential equations are not valid near the source and near the load. So, in a spatially discretized scenario, we assume that the line differential equations are valid for $z \in [\delta, d - \delta]$ and for $0 < z < \delta$ and $d - \delta < z < d$; the line voltage and currents are determined from the source and load conditions.

We now explore applications of the extended Kalman filter in this QNN. Suppose we take discrete noisy measurements of the line voltage and currents. These measurements are represented by an equation of the form

$$dZ(t) = h(i(t), v(t))dt + dV(t)$$

This is because $v(t, z)$, $i(t, z)$ at any point z along the line can be represented as a superposition of $i(t) = (i_n(t))$ and $v(t) = (v_n(t))$. Using the extended Kalman filter (EKF), we estimate the conditional density of the state $p(i, v | Z_t)$, $Z_t = \{z(s) : s \le t\}$ from the stochastic dynamics of $i(t)$, $v(t)$. The EKF gives the dynamics of $p(i, v | Z_t)$ and the Schrödinger equation based QNN can be used to calculate the potential parameter adaptation so that $|\psi(t, i, v)|^2$ tracks $p(t, i, v | Z_t)$. We can go a step further with this. We take noisy measurements of an observable $X = h(i, v)$ on the Tx line at different times. These noisy measurements have the form

$$dZ(t) = h(i(t), v(t))dt + dV(t)$$

and we assume that these noisy measurements actually represent the quantum average $\int h(i, v) |\psi(t, i, v)|^2 didv$. We then incorporate noise in Schrödinger's equation and treating the wave function at each time as the state vector, we apply the EKF to the Schrödinger equation with the measurement model

$$dZ(t) = \int h(i,v) |\psi(t,i,v)|^2 didv + dV(t)$$

to estimate the wave function and hence the probability density of (i, v) at each time. Ideally speaking, using the Fokker–Planck equation for the line PDF, we should represent our measurements $h(i(t), v(t))$ as

$$dZ(t) = \left(\int p(t,i,v)h(i,v)didv \right)dt + dV(t)$$

and apply the noisy Schrödinger equation to these measurements to calculate the wave function estimate on a real time basis. In fact, we can, even
without using the Schrödinger equation, directly tackle the problem of estimating the line PDF using the Fokker–Planck equation for it based on the
noisy measurements of a set of moments of the line voltage and current at
different times. Indeed, suppose we write down the Fokker–Planck equation
for the line as

$$\partial p(t,i,v)/\partial t = \mathcal{L}p(t,i,v)$$

where $\mathcal{L}$ is the Fokker–Planck operator introduced above. We add noise
to this equation and regard $P_t = \{p(t, i, v) : (i, v) \in \mathbb{R}^N\}$ as our current state
vector with $N/2$ denoting the number of terms in the generalized Fourier
expansion to which the series has been truncated. Thus, our state model is

$$dP_t = \mathcal{L}P_t dt + \sigma.dB(t)$$

Let $h(i,v) = \left(h_k(i,v)\right)_{k=1}^{M}$ denote the observables on the line whose statistical moments have been measured. Thus, our measurement model is

$$dZ(t) = dt \int h(i,v)p(t,i,v)didv + dV(t)$$

which can be expressed as

$$dZ(t) = \langle h, P_t \rangle + dt + dV(t)$$

We can then use the results of standard infinite dimensional Kalman filter
theory to estimate the state P_t on a real time basis.

Now we are also in a position to incorporate quantum noise into the
QNN. Consider the following Heisenberg quantum stochastic differential
equation (QSDE) in the sense of Hudson and Parthasarathy for the evolving
observable $j_t(X)$:

$$dj_t(X) = j_t\left(L(\theta(t))X\right)dt + j_t\left(\theta_1(X)\right)dA(t) + j_t\left(\theta_2(X)\right)dA(t)^*$$

where $L(\theta)$ is the Lindblad operator:

$$L(\theta)(X) = i\left[H(\theta), X\right] - (1/2)\left(MM^*X + XMM^* - 2MXM^*\right)$$

In the presence of non-demolition measurements

$$dY(t) = j_t\left(M + M^*\right)dt + dA(t) + dA(t)^*$$

we can use the Belavkin filter to estimate the quantum system state and also the parameter θ as a function of time on a real time basis. Having obtained an estimate of the evolving system state, we can calculate the PDF of any observable $X = h(i, v)$ in that state. More generally, once we know the line PDF family $p(t, i, v)$, we choose the Lindblad operator M so that the quantum system observable $M + M^*$ represents $h(i, v)$ and then we apply the Belavkin filter to estimate the quantum state and hence the PDF of any observable $g(i, v)$ as a function of time. Note that the quantum formalism also gives us in this case an estimate of the joint density $p(t, i, v)$ as $|\psi(t, i, v)|^2$ since the i, v forms a set of commuting observables which form the position representation in the Schrödinger equation.

Now we can talk of synthesis, that is, testing of the trained QNN. The training is carried out using known values of the PDF family of the Tx line. Training results in a parameter trajectory $\theta(t)$, $t \geq 0$. The corresponding PDF family also depends on the source voltage process $v_s(t)$ and the load R_L. Now $v_s(t)$ equals $v(t, 0)$ while $R_L = v(t, d)/i(t, d)$. Given we now have a new line with the same distributed parameters and line loading noise statistics but a different source voltage process $v_s(t)$ and different load R_L, we wish to obtain an estimate of the line PDF family using the QNN. Since the trained QNN parameter trajectory $\theta(t)$ does not depend upon the source process and the load, we retain this set of parameters as a first guess. Then, we calculate, using the Schrödinger equation PDF, the mean values of $v(t, 0) = \Sigma_n v_n(t)\phi_n(0)$, $i(t, 0) = \Sigma_n i_n(t)\phi_n(0)$, $v(t, d) = \Sigma_n v_n(t)\phi_n(d)$ and $i(t, d) = \Sigma_n i_n(t)\phi_n(d)$. Denote these quantum averages by $< . >$. We then readjust the trained parameter $\theta(t)$ so that these mean values provide a better match for $v_s(t)$ as $<v(t, 0) > - R_s < i(t, 0)>$ and R_L as $<v(t, d) > / < i(t, d)>$ at the next time iteration. Specifically, if $\theta(t)$ is the trained parameter based on the line stochastic differential equation (SDE) PDF, that is, the parameter process $\theta(.)$ depends only on the statistics of the line loading, and if we wish to update this parameter further (as we do in transfer learning) to $\phi(t)$, then we would use

$$\phi(t) = \theta(t) + \delta\theta(t)$$

where

$$\phi(t+h) = \theta(t) - \mu_1 \cdot \frac{\partial}{\partial\theta(t)} \times [w_1(v_s(t+h)$$
$$- \int \sum_n (v_n - R_s i_n) \times \phi_n(0) \left|\psi(t+h, i, v)\right|^2 didv)^2$$
$$+ w_2 \left(\int \sum_n (R_L i_n - v_n)\phi_n(d) \times \left|\psi(t+h, i, v)\right|^2\right)^2\right]$$

and where, of course, as earlier, we substitute

$$\psi\left(t+h,i,v\right)=\psi\left(t,i,v\right)-iH\left(\theta\left(t\right)\right)\psi\left(t,i,v\right)$$

and then carry out the differentiation w.r.t. $\theta(t)$.

Suppose now that the line loading noise is non-Gaussian white, that is, the time derivative of a Levy process. In that case the state vector $\xi(t) = (i(t), v(t))$ will be Markovian but not necessarily Gaussian. Let K denote the infinitesimal generator of this Markov process with kernel $K(\xi, \eta)$. In other words,

$$E(\phi\left(\xi\left(t+h\right)\right)|\xi\left(t\right))=\phi\left(\xi\left(t\right)\right)+K\phi\left(\xi\left(t\right)\right)h+o\left(h\right)$$

where

$$K\phi\left(\xi\right)=\int K\left(\xi,\eta\right)\phi\left(\eta\right)d\eta$$

with the integration being carried out in infinite dimensional space. The probability density $p(t, \xi)$ now satisfies the Chapman–Kolmogorov equation, or the forward equation

$$\partial p\left(t,\xi\right)/\partial t = K^{*}p\left(t,\xi\right)$$

This follows from the composition rule

$$p(t_3,\xi_3 \mid t_1,\xi_1)=\int p(t_3,\xi_3 \mid t_2,\xi_2)p(t_2,\xi_2 \mid t_1,\xi_1)d\xi_2$$

Multiplying both sides of this equation by $p(t_1, \xi_1)$ and integrating w.r.t. ξ_1 results in

$$p\left(t_3,\xi_3\right)=\int p(t_3,\xi_3 \mid t_2,\xi_2)p\left(t_2,\xi_2\right)d\xi_2$$

Differentiating both sides w.r.t. t_3 and setting $t_3 = t_2$ with the observation that

$$\partial p(t,\xi \mid s,\eta)/\partial t \mid_{t=s} = K\left(\eta,\xi\right)$$

results in the forward equation stated above:

$$\partial p\left(t,\xi\right)=\int K\left(\eta,\xi\right)p\left(t,\eta\right)d\eta = K^{*}p\left(t,\xi\right)$$

After discretizing this equation in time, we get

$$p\left(t+h,\xi\right)=p\left(t,\xi\right)+h.Kp\left(t,\xi\right)$$

This equation describes the stochastic dynamics of the line due to loading effects only and is independent of the source and load. It carries information only about the line distributed parameters and the statistics of the line loading noise. When we train our QNN to follow these dynamics, we also desire the PDF to carry information about the source and load. This is accomplished by assuming that the averages of $v(t, 0) = \Sigma_n v_n(t)\phi_n(0)$, $v(t, d) = \Sigma_n v_n$ $(t)\phi_n(d)$ and $i(t, d) = \Sigma_n i_n(t)\phi_n(d)$ are required to approximately satisfy the boundary conditions

$$\langle v(t,0) \rangle - R_s \langle i(t,0) \rangle = v_s(t), < v(t,d) - R_L \langle i(t,d) \rangle = 0$$

Keeping this in mind, our adaptive update equation for the QNN parameters would be

$$\theta(t+h) = \theta(t) - \mu . \frac{\partial}{\partial \theta(t)} E(t, \theta(t))$$

where

$$E(t,\theta) = \int \left(p(t+h,i,v) - \left| \psi(t+h,i,v) \right|^2 \right)^2 didv$$

$$+ w(1) \left(v_s(t+h) - \sum_n (v_n - R_s i_n)\phi_n(0) \left| \psi(t+h,i,v) \right|^2 didv \right)^2$$

$$+ w(2) \left(\int \sum_n (R_L i_n - v_n)\phi_n(d) \left| \psi(t+h,i,v) \right|^2 didv \right)^2$$

with

$$\psi(t+h,i,v) = \psi(t,i,v) - iH(\theta)\psi(t,i,v)$$

substituted.

4.3 NONLINEAR SIGMA MODEL APPLIED TO THE DESIGN OF A QNN FOR TRACKING THE TX LINE PDF

In most situations, it is very hard to design a quantum system with a large number of independent position variables to track the line PDF. The nonlinear sigma model is a quantum field theoretic model that naturally provides a large number of degrees of freedom as required by the nature of the Tx line problem. The idea is that physical systems on the quantum scale that naturally have the nonlinear sigma dynamics occur and these systems have control parameters appearing in the nonlinear metric that can be adapted at will. For example, if string theory is correct, then the dynamics of the string is decided by the metric of the D-dimensional space–time string or equivalently by the background gravitational field; it can also be controlled by a gauge field. The idea is that the field can be expanded in Fourier modes within a box and each Fourier component of the field behaves as an independent position field which appears in the Schrödinger equation for the truncated wave functional of the field. A quantum field naturally carries a large number of position variables in contrast to a system of several particles. That is why a field theoretic approach to the QNN is much better and more natural than a particle theoretic approach.

The Lagrangian of a nonlinear sigma model is given by

$$L(\phi) = g^{\mu v}(\phi)\phi_{,\mu}\phi_{,\mu} - V(\phi)$$

A specialization of this model is provided in string theory:

$$L = g_{\mu v}(X(\tau,\sigma))h^{ab}(\tau,\sigma)\sqrt{h(\tau,\sigma)}X^{\mu}_{,a}X^{v}_{,b}$$

where X^{μ} are functions of τ, σ with a, b running over the indices 0, 1 where $\sigma^0 = \tau$, $\sigma^1 = \sigma$. After applying a reparametrization and conformal transformation of the world sheet metric, this Lagrangian reduces to

$$g_{\mu v}(X)\eta_{ab}X^{\mu}_{,a}X^{v,b}, \eta_{00} = -\eta_{11} = 1, \eta_{01} = \eta_{10} = 0$$

We consider the field in the nonlinear sigma model to be enclosed within a box and expand the field in spatial Fourier modes:

$$\phi(t,r) = \sum_k \phi_k(t).\exp(ik.r)$$

where

$$k = (2\pi/L)(n_1, n_2, n_3), n_1, n_2, n_3 \in Z$$

We assume the following form of the metric:

$$g^{00} = 1, g^{0r} = 0, r = 1, 2, 3$$

Substituting this into the Lagrangian gives, after assuming a metric expansion,

$$g^{\mu v}(\phi) = g^{\mu v}(0) + \sum_{r \geq 1} g^{\mu v}(r)\phi^r$$

$$= g^{\mu v}(0) + \sum_r g^{\mu v}(r)\phi_{k_1}(t)...\phi_{k_r}(t)\exp(i(k_1 + ... + k_r).r)$$

Thus, the Lagrangian density can be expressed as

$$L = g^{\mu\nu}(\phi)\phi_{,\mu}\phi_{,\nu} = \phi_{,0}^2 + g^{rs}\phi_{,r}\phi_{,s}$$
$$= \phi'_{k_1}(t)\phi'_{k_2}(t)\exp\left(i(k_1 + k_2).r\right) - g^{rs}(m)\phi_{k_1}(t)\phi_{k_2}\phi_{k_3}(t)...\phi_{k_{m+2}}(t)k_{1r}k_{2s}$$

so that the Lagrangian given by the spatial integral of the Lagrangian density becomes

$$2M(\phi) = 2\int L(\phi)d^3r =$$
$$\delta[k_1 + k_2]\phi_{k_1}'(t)\phi_{k_2}'(t) + g^{rs}(0)k_{1r}k_{2s}\delta[k_1 + k_2]\phi_{k_1}(t)\phi_{k_2}(t)$$
$$-\sum_{m\geq 1}g^{rs}(m)\delta[k_1 +...+ k_{m+2}]k_{1r}k_{2s}\phi_{k_1}(t)\phi_{k_2}\phi_{k_3}(t)...\phi_{k_{m+2}}(t)$$

Noting that the reality of the field implies $\phi_{-k}(t) = \phi_k(t)*$, this expression can be put into the form

$$M(\phi) = \sum_k \left(\left|\phi_{k'}(t)\right|^2 - A_2[k]\left|\phi_k(t)\right|^2\right) - \sum_{k_1,...,k_m,m\geq 3} A_m[k_1,...,k_m]\phi_{k_1}(t)...\phi_{k_m}(t)$$

This Lagrangian is a potential perturbation of a harmonic oscillator Lagrangian, one complex oscillator being associated with each mode. Writing

$$X_k(t) = Re\left(\phi_k(t)\right), Y_k(t) = Im\left(\phi_k(t)\right)$$

we can express this Lagrangian as

$$M(\phi) = M\left(X_k, Y_k, X_{k'}, Y_{k'}, k = 1, 2,...\right)$$
$$= \sum_k \left(X_{k'}(t)^2 + Y_{k'}(t)^2 - A_2(k)\left(X_k(t)^2 + Y_k(t)^2\right)\right) - V\left(X_k(t), Y_k(t), k \geq 1\right)$$

where V is a polynomial in X_k's and Y_k's which contain only cubic and higher terms. Now we can set up the Schrödinger equation for this system, or more precisely calculate the perturbations in the time-evolving wave function caused by the cubic and higher perturbation terms and hope to control the corresponding PDF using the parameters in this perturbation. The parameters of the metric $g^{\mu\nu}(\phi|\theta)$ naturally appear in the frequencies $A_2(k)$ and in the higher order potential V; and these can be used to control the wave function. A simple Legendre transform theoretic analysis of this Lagrangian yields the Schrödinger Hamiltonian

$$H(\theta) = (-1/2)\sum_k \left(\partial^2/\partial X_k^2 + \partial^2/\partial Y_k^2\right)$$
$$+ (1/2)\sum_k A_2(k \mid \theta)\left(X_k^2 + Y_k^2\right) + V(X_k, Y_k, k = 1, 2,... \mid \theta)$$

which can be used to update the wave function so that its modulus square tracks the evolving PDF of the line.

Another approach to fine tuning of the line is based on encoding the information contained in the source and load into the adaptation constant: this means that the source pdf and load are also to be tuned so that the line voltage and current pdf are close to the QNN pdf. From the QNN pdf, we reconstruct the line voltage and load and if these are large in distance from the true line voltage and load, then we increase the size of the adaptation constant, and vice versa.

Statistical performance analysis of the algorithm based on large deviation theory is as follows. The inputs to the QNN are the measured values of the line PDF as well as the source voltage process and the load. While measuring these quantities, errors will be involved. We wish to develop a first-order perturbation theoretic analysis of the statistics of the parameter fluctuation errors in terms of those measurement errors. We first observe that the parameter update equations can be expressed as

$$\theta(t+h) = \theta(t) - \mu \frac{\partial}{\partial \theta(t)} \left(\left\| P_{t+h} - \left| \psi(t+h) \right|^2 \right\|^2 \right)$$

with

$$\psi(t+h) = \psi(t) - ih.H(\theta(t))\psi(t)$$

Since all partial differential operators are discretized during the process of implementation, it follows that $H(\theta(t))$ is to be regarded as a matrix valued function of the parameter $\theta(t)$.

4.4 DESIGN OF QNNs BASED ON SPONTANEOUS SYMMETRY BREAKING

Let $V(\phi)$ be the quantum effective action. In the case where the action and path measure are both invariant under the gauge group, it can be shown that the quantum effective action is also invariant under the gauge group, that is, if T is any generator of the gauge group, that is, T is a gauge Lie algebra element, then

$$\sum_{n,m} \partial(V(\phi)/\partial\phi_n) T_{nm}\phi_m = 0$$

for all ϕ. It follows then that if ϕ_0 is a ground state, that is,

$$\partial V(\phi_0)/\partial\phi_n = 0$$

then another differentiation of the above equation gives us

$$\sum_{n,m}\left(\partial^2 V(\phi_0)/\partial\phi_k\partial\phi_n\right)T_{nm}\phi_{0m} = 0$$

Now if the ground state ϕ_0 is not invariant under a given element T of the gauge Lie algebra, or equivalently under the elements $exp(tT)$ of the gauge group, then it follows that $T_{nm}\phi_{0m} \neq 0$, that is, $T\phi_0 \neq 0$, and we get $T\phi_0$, which is a non-zero eigenvector of the mass matrix $((\partial V(\phi_0)/\partial\phi_n\partial\phi_m))$ with a zero eigenvalue, that is, the state $T\phi_0$ has zero mass. Thus, denoting by

$$W = \{T\phi_0 : T \in \mathfrak{g}\}$$

the subspace of the ground states, it follows that every non-zero element of W is a zero mass eigenvector of the mass matrix. If $T\phi_0 \neq 0$, then it follows that T is the broken symmetry of the ground state ϕ_0 and hence yields a zero mass eigenvector. Thus, $dim\,W$ can be interpreted as the number of zero mass particles produced by the broken symmetry of the ground state. This means that there is one massless particle, called a Goldstone boson, corresponding to each broken degree of gauge symmetry. Now let H be the stability group of the ground state ϕ_0, that is,

$$H = \{g \in G : g\phi_0 = \phi_0\}$$

Equivalently, the Lie algebra of H denoted by $\mathfrak{h}$ is given by

$$\mathfrak{h} = \{T \in \mathfrak{g} : T\phi_0 = 0\}$$

Then, we write

$$L(\phi) = (1/2)\partial_\mu\phi.\partial^\mu\phi - V(\phi)$$

with V being G-invariant. Writing

$$V_0(\delta\phi) = V(\phi_0 + \delta\phi)$$

we see that the Lagrangian in terms of $\delta\phi$ becomes, since ϕ_0 is independent of x,

$$L_0(\delta\phi) = (1/2)\partial_\mu\delta\phi.\partial^\mu\delta\phi - V_0(\delta\phi)$$

and it is immediately seen that L_0 is only H-invariant and not G-invariant. By controlling the ground state ϕ_0, we can control, that is, vary, the unbroken subgroup H. So the problem is the following. Let H be a given subgroup of G. We wish to control the ground state ϕ_0 with time so that H always remains unbroken and such that a given time varying probability distribution is well approximated by the Schrödinger equation for the wave functional ψ of the broken field $\delta\phi$. To do so, we choose a basis β of generators of the broken Lie algebra $\mathfrak{h} = Lie(H)$. Let $T_1, \ldots, T_r$ be the basis elements of $\mathfrak{h}$ and define the matrix

$$A_H = \begin{pmatrix} T_1 \\ T_2 \\ \ldots \\ T_r \end{pmatrix}$$

Let $N = N(A_H)$, the null-space of A_H. Choose a basis for $N = R\left(A_H^*\right)^\perp$ and denote this by $\{\phi_1, \ldots, \phi_m\}$. Then any ϕ_0 that is $\mathfrak{H}$-invariant can be expressed as

$$\phi_0 = \sum_{k=1}^m c(k)\phi_k$$

The idea is then to control the $c(k)'s$ with time in the potential

$$V_0\left(t, \delta\phi(x)\right) = V\left(\sum_{k=1}^m c_k(t)\phi_k + \delta\phi(x) \right)$$

of Schrödinger's wave equation for the wave functional $\psi(\delta\phi)$ so as to cause the PDF to track a given one.

4.5 APPLICATION OF STRING THEORY TO CONTROL THE TRANSITION PROBABILITY DISTRIBUTION OF A QNN

We consider a quantum string gauge field along with a classical gauge field and a classical gravitational field that interacts with a quantum string. We control the classical gravitational field as well as the classical gauge field so that the string propagator gets controlled in such a way that the quantum gauge field which interacts with the string has an appropriate transition probability between two states. Once again, since the quantum gauge field is typically an infinite dimensional field, its probability distribution will be of a very large dimension and hence can be used in tracking the transmission line voltage and current joint probability distribution at different spatial points.

The Lagrangian density of the string is

$$L(X) = (1/2) g_{\mu\nu}(X) \eta_{ab} X^{\mu}_{,a} X^{\nu}_{,b} + (1/2) B_{\mu\nu}(X) \varepsilon(ab) X^{\mu}_{,a} X^{\nu}_{,b}$$

The equations of motion

$$\delta_X \int L(X) d^2\sigma = 0$$

result in

$$-\left(g_{\mu\nu}(X) \eta_{ab} X^{\nu}_{,b} \right)_{,a} + (1/2) g_{\rho\nu,\mu}(X) \eta_{ab} X^{\rho}_{,a} X^{\nu}_{,b}$$

$$-\left(B_{\mu\nu}(X) \varepsilon(ab) X^{\nu}_{,b} \right)_{,a} + (1/2) B_{\rho\nu,\mu}(X) X^{\rho}_{,a} X^{\nu}_{,b} = 0$$

Note that

$$\varepsilon(ab) X^{\nu}_{,ab} = 0$$

and so the above simplifies to

$$-g_{\mu\nu}(X) \ddot{X}^{\nu} - \left(g_{\mu\nu,\rho}(X) - (1/2) g_{\rho\nu,\mu}(X) \right) \eta_{ab} X^{\rho}_{,a} X^{\nu}_{,b}$$

$$-\left(B_{\mu\nu,\rho}(X) - (1/2) B_{\rho\nu,\mu}(X) \right) \varepsilon(ab) X^{\rho}_{,a} X^{\nu}_{,b} = 0$$

or equivalently

$$-g_{\mu\nu}(X) \ddot{X}^{\nu} - \left(\Gamma_{\mu\rho\nu}(X) \eta_{ab} - H_{\mu\rho\nu}(X) \varepsilon(ab) \right) X^{\rho}_{,a} X^{\nu}_{,b} = 0$$

where

$$\Gamma_{\mu\nu\rho} = (1/2) \left(g_{\mu\nu,\rho} + g_{\mu\rho,\nu} = g_{\rho\nu,\mu} \right),$$

$$H_{\mu\rho\nu} = (1/2) \left(g_{\mu\rho,\nu} + g_{\rho\nu,\mu} = B_{\nu\mu,\rho} \right)$$

The string propagator equations are:

$$\Delta^{\mu\nu}(\tau,\sigma \mid \tau',\sigma') = \left\langle T\left(X^{\mu}(\tau,\sigma) X^{\nu}(\tau',\sigma') \right) \right\rangle$$
$$= \theta(\tau - \tau') \left\langle X^{\mu}(\tau,\sigma) . X^{\nu}(\tau',\sigma') \right\rangle$$
$$+ \theta(\tau' - \tau) \left\langle X^{\nu}(\tau',\sigma') . X^{\mu}(\tau,\sigma) \right\rangle$$

Thus

$$\Delta^{\mu\nu}(\tau,\sigma \mid \tau',\sigma') = \delta(\tau-\tau')\langle[\partial_0 X^\mu(\tau,\sigma), X^\nu(\tau,\sigma')]\rangle$$
$$+\langle T(F^\mu(\tau,\sigma).X^\nu(\tau',\sigma'))\rangle = -i\delta(\tau-\tau')\delta(\sigma-\sigma')\eta^{\mu\nu}$$
$$-\langle T(F^\mu(\tau,\sigma).X^\nu(\tau',\sigma'))\rangle$$

where

$$F^\mu(\tau,\sigma) = (\Gamma^\mu_{\rho\nu}(X)\eta_{ab} - H^\mu_{\rho\nu}(X)\varepsilon(ab))X^\rho_{,a}X^\nu_{,b} = 0$$
$$= X^\rho_{,a}X^\nu_{,b}\left[\sum_{\rho,\nu,a,b,\mu_1,\dots,\mu_r} C(\mu,\rho,\nu,a,b,\mu_1,\dots,\mu_r)X^{\mu_1}\dots X^{\mu_r}\right]$$

Suppose we now have an external control field $K_{\mu\nu ab}(\theta(t))$ dependent only on a set of parameters $\theta(t)$ that can vary with time and that interacts with the string field in accordance with the Lagrangian density

$$\Delta L(\tau,\sigma,X) = K_{\mu\nu ab}(\theta(\tau))X^\mu_{,a}(\tau,\sigma)X^\nu_{,b}(\tau,\sigma)$$

The Schrödinger transition kernel from time zero to time t can be expressed as a Feynman path integral

$$U(t,X(t),X(0)) = \int \exp(iS_0(X)+i\Delta S(X))\Pi_{\tau\le t,\sigma\in[0,d]}dX(\tau,\sigma)$$

which under the assumption that $\Delta S(X)$ is small, can be approximated as

$$\int \exp(iS_0(X))(1+i\Delta S(X))\Pi_{\tau\le t,\sigma}dX(\tau,\sigma)$$

This path integral is computed by keeping the initial and final string fields as fixed $X(0,.) = X(0)$, $X(t,.) = X(t)$; the wave functional at time t given by

$$\psi(t,X) = \int U(t,X,X(0))\psi(0,X(0))dX(0)$$

can be controlled by adjusting the parameters in $K_{\mu\nu ab}(\theta(t))$. It should be noted that since we are constraining the terminal points in the path integral, we cannot use the string propagator to evaluate the term corresponding to the quadratic form $\Delta S(X) = \int \Delta L d\tau d\sigma$. However, there is one special case in which this can be used, namely the case in which the string interacts with the classical gravitational field, a classical string gauge field as above, and, in addition, with a quantum string gauge field $\Delta B_{\mu\nu}(X)$ having a Lagrangian $L_0(\Delta B)$ with the interaction Lagrangian density being $\Delta B_{\mu\nu}(X)\varepsilon(ab)X^\mu_{,a}X^\nu_{,b}d^2\sigma$. Assuming that this quantum gauge field is weak, it suffices to evaluate it at the coordinates of the centre of the string which are ordinary real valued functions of τ, σ in contrast to the string field perturbations of this centre which is an operator valued function of τ, σ. We assume that the quantum gauge field $\Delta B_{\mu\nu}(X)$ has a Lagrangian density

$$L_0(\Delta B) = \Delta H_{\mu\nu\rho}(X)\Delta H^{\mu\nu\rho}(X),$$

where

$$\Delta H_{\mu\nu\rho}(X) = \sum_{(\mu\nu\rho)}\Delta B_{\mu\nu,\rho}(X)$$

and where the sum is a cyclic sum. The action functional of this field in the absence of interactions with the string is given by

$$\int L_0(\Delta B(X))d^dX = S_B(\Delta B)$$

In the presence of interactions with an isolated string, the total action of the string becomes

$$S_0(X,g,B) + S_B(\Delta B) + S_I(X,\Delta B)$$

where $S_0(X, g, B)$ is the action of the string interacting with the classical gravitational and gauge field as described above and

$$S_I(X,\Delta B) = \int \Delta B_{\mu\nu}(X)\varepsilon(ab)X^\mu_{,a}X^\nu_{,b}d^2\sigma$$
$$\approx \int \Delta B_{\mu\nu}(\tau,\sigma)\varepsilon(ab)X^\mu_{,a}(\tau,\sigma)X^\nu_{,b}(\tau,\sigma)d\tau d\sigma$$

where, in the last step, we have made the above mentioned approximation of evaluating ΔB at the string centre. The transition amplitude for the quantum gauge field ΔB making a transition from an initial state $|phi_0\rangle$ to a final state $|\phi_1\rangle$ under this interaction with the isolated string with the string remaining in the vacuum state is given by

$$A(\phi_1,\phi_0,g,B) = \int_{\Delta B=\phi_0,\phi_1} \exp\left(iS_B\times(\Delta B)+iS_I(X,\Delta B)+iS_0(X,g,B)\right)$$
$$\times\left(\Pi_Y d\Delta B(Y)\right).DX$$

with the path integration over X being unrestricted and that over ΔB being over the region in which initially, that is, at $Y^0 = 0$, the state of the field is ΔB is ϕ_0 and finally, that is, at $Y^0 = \infty$, its state is ϕ_1. Using second-order perturbation theory, this path integral approximates to

$$(-1/2) \int_{\Delta B = \phi_0, \phi_1} \exp\left(iS_B\left(\Delta B\right) + iS_0\left(X, g, B\right)\right) S_I\left(X, \Delta B\right)^2 \left(\Pi_Y d\Delta B(Y)\right) DX$$

$$= (-1/2) \int_{\Delta B = \phi_0, \phi_1} \exp\left(iS_B\left(\Delta B\right)\right) D\Delta B. \left(\int \exp\left(iS_0\left(X, g, B\right)\right) S_I\left(X, \Delta B\right)^2 DX\right)$$

More generally, if we do not make any approximations, then we can also write

$$A\left(\phi_1, \phi_0, g, B\right) = \int_{\Delta B = \phi_0, \phi_1} \exp\left(iS_B\left(\Delta B\right)\right) F\left(\Delta B, g, B\right) D\Delta B$$

where

$$F\left(\Delta B, g, B\right) = \int \exp\left(iS_0\left(X, g, B\right)\right). \exp\left(iS_I(X, \Delta B)\right) DX$$
$$\approx (-1/2) \int \exp\left(iS_0\left(X, g, B\right)\right) S_I\left(X, \Delta B\right)^2 DX$$

Note that g, B are classical fields. The path integral over X can be expressed in terms of the propagator of the string X in the presence of the control classical fields g, B. In fact, for fixed ΔB, $S_I(X, \Delta B)$ with ΔB evaluated at the centre of the string, as explained above, is a quadratic functional of X and its square is therefore a homogeneous fourth-degree functional of X. The resulting path integral over X is therefore expressible as a product of two terms, each of which is linear in the string propagator and in ΔB. Specifically,

$$S_I\left(X, \Delta B\right)^2 = \int \varepsilon\left(ab\right)\varepsilon\left(cd\right) \Delta B_{\mu\nu}\left(\tau, \sigma\right) \Delta B_{\alpha\beta}\left(\tau', \sigma'\right)\left(X^{\mu}_{,a}(\tau, \sigma) X^{\nu}_{,b}\left(\tau, \sigma\right)\right)$$

$$X^{\alpha}_{,c}\left(\tau', \sigma'\right) X^{\beta}_{,d}\left(\tau', \sigma'\right) d^2\sigma. d^2\sigma'$$

where

$$d^2\sigma = d\tau. d\sigma, d^2\sigma' = d\tau' d\sigma'$$

This gives us

$$F\left(\Delta B, g, B\right) \approx \int \varepsilon\left(ab\right)\varepsilon\left(cd\right) \Delta B_{\mu\nu}\left(\tau, \sigma\right) \Delta B_{\alpha\beta}\left(\tau', \sigma'\right)$$

$$\times \left\langle T\left(X^{\mu}_{,a}\left(\tau, \sigma\right) X^{\nu}_{,b}\left(\tau, \sigma\right) X^{\alpha}_{,c}\left(\tau', \sigma'\right) X^{\beta}_{,d}\left(\tau', \sigma'\right)\right)\right\rangle d^2\sigma. d^2\sigma'$$

with

$$\left\langle T\left(X^{\mu}_{,a}\left(\tau,\sigma\right)X^{\nu}_{,b}\left(\tau,\sigma\right)X^{\alpha}_{,c}\left(\tau',\sigma'\right)X^{\beta}_{,d}\left(\tau',\sigma'\right)\right)\right\rangle$$

$$= \int \exp(iS_0\left(X,g,B\right)X^{\mu}_{,a}\left(\tau,\sigma\right)X^{\nu}_{,b}\left(\tau,\sigma\right)X^{\alpha}_{,c}\left(\tau',\sigma'\right)X^{\beta}_{,d}\left(\tau',\sigma'\right)d^2\sigma d^2\sigma'$$

$$= \Delta\left(\mu\nu\alpha\beta,abcd,g,B\right)$$

say. So we can also write

$$F(\Delta B, g, B) \approx \varepsilon(ab)\varepsilon(cd)\int \Delta B_{\mu\nu}(\tau,\sigma)\Delta B_{\alpha\beta}(\tau',\sigma')\Delta(\mu\nu\alpha\beta,abcd,g,B)d^2\sigma.\,d^2\sigma'$$

It is now clear that in the expression for the transition probability of the quantum gauge field ΔB from the initial state ϕ_0 to the final state ϕ_1 given by the expression

$$P(\phi_1;\phi_0 \mid g,B) = A\left(\phi_1,\phi_0,g,B\right)\big|^2 = \Bigg| \int_{\Delta B=\phi_0,\phi_1} \exp\left(iS_B\left(\Delta B\right)\right)F\left(\Delta B,g,B\right)D\Delta B\Bigg|^2$$

we can control g, B so as to achieve a specified transition probability. This means that by allowing the quantum gauge field to interact with a quantum string whose dynamics are controlled by classical gravitational and gauge fields g, B, we can achieve almost any transition probability functional in gauge field space. This idea suggests the application of string theory to probability density tracking. It is the string theoretic analogue of the quantum field theoretic problem of how to control Fermion transition probabilities between Fermion states by allowing the Fermionic field to interact with the sum of a classical and a quantum electromagnetic field $A_{o\mu}$, ΔA_{μ}, where the quantum electromagnetic field ΔA_{μ} remains in the vacuum state. In this latter case, however, if we approximate the exponential of the interaction action

$$\int \bar{\psi}\left(x\right)\gamma^{\mu}\psi\left(x\right)\Delta A_{\mu}\left(x\right)d^4x$$

by a second-order term, we end up with only quadratic terms in the quantum electromagnetic (EM) field and fourth-degree terms in the Fermionic field. By controlling the classical EM field so as to yield a desired form of the propagator of the quantum EM field, we can influence the transition probabilities for the Fermions.

4.6 CONCLUSION

In this chapter, we have explained how to control the parameters in the Hamiltonian or in the Lagrangian of a quantum dynamical system for particles or fields, so that the resulting PDF or functional will be able to track the joint probability density of the line voltage along a transmission line at different spatial locations. In the Hamiltonian method, we control the parameters of the potential or the Lindblad operators if the quantum system is an open system, that is, connected to the bath while in the Lagrangian method, and we control classical fields so that the transition probability of the main field between two times calculated using the Feynman path integral has a specific pattern. By "bath", we mean the quantum noisy environment within which the quantum system is kept. Coupling of the system to the noisy bath causes the system dynamics to become modified in the master equation. This equation is also referred to as the noisy Schrödinger equation. In the process of doing this, we discussed many stochastic aspects of transmission line theory especially how to derive Fokker–Planck and Chapman–Kolmogorov-like equations for the PDF of the line voltage in the presence of random line loading.

REFERENCES

1. Andrew Jazwinsky, *Stochastic processes and filtering theory*, Academic Press.
2. Leonard Schiff, *Quantum mechanics*.
3. Steven Weinberg, *The quantum theory of fields*, Vols. I and II, Cambridge University Press.
4. P.A.M. Dirac, *The principles of quantum mechanics*, Oxford University Press.
5. M. Green, J. Schwarz and E. Witten, *Superstring Theory*, Cambridge University Press.
6. K.R. Parthasarathy, *An introduction to quantum stochastic calculus*, Birkhauser.

Chapter 5

Power grid adaptive and block processing control based on the extended Kalman filter and large deviation theory

Harish Parthasarathy

Netaji Subhash University of Technology, New Delhi, India

Arti Vaish

Department of computer science, School of Engineering,
O.P. Jindal University, Raigarh, India

CONTENTS

5.1 INTRODUCTION

In the case when noise is non-Gaussian and white, that is, contains both Gaussian and Poisson white noise components but the noise amplitude is weak, we explain how, using the theory of large deviations, the approximate log likelihood function can be constructed and hence the unknown parameters of the model can be estimated using a block processing approach. We then discretize the spatial components of the dynamics and end up with a system of nonlinear state variable stochastic differential equations (SDEs) where the state vector comprises the voltage field at each discretized spatial pixel as well as its time derivative. We formulate the extended Kalman filter (EKF) for estimating the state from noisy measurements of some functions of the state process, such as the power at each spatial pixel, on a real time basis [1, 2]. This is important when we cannot obtain accurate measurements of the state. Using this estimated state, we address the trajectory tracking control problem, namely the problem of controlling the voltage and current field to follow a desired trajectory, by means of an error feedback controller based on the instantaneous error between the desired state trajectory and the EKF estimated trajectory. We then linearize the system of three SDEs,

DOI: 10.1201/9781003436461-5

one for the state dynamics, another for the desired state dynamics, and the third for the EKF state estimator/observer dynamics, thereby resulting in a set of coupled linear SDEs for the state estimation error and the trajectory tracking error. We then use the large deviation method to estimate the controller coefficients so as to minimize the probability of deviation of some appropriate combination of the two error energies by an amount more than a threshold over the given time duration [3].

5.2 STATEMENT OF THE PROBLEM AND DISCUSSION

Let $X_1(t), ..., X_N(t)$ be the voltages delivered at centres $r_1, ..., r_N$ in a city. The state vector $\left(X_i(t)\right)_{i=1}^{N}$ satisfies a nonlinear SDE with inputs $u_1(t), ..., u_M(t)$. The state vector $\left(X_i(t)\right)_{i=1}^{N}$ satisfies usually a non-linear SDE with inputs $u_i(t), i = 1, 2, ..., M$. These inputs are the voltages produced by generators located at $s_1, ..., s_M$. The reason why X_i's should satisfy SDEs can be attributed to transmission line theory in which the cables that carry the voltage and current signals from the generators to the respective regions are distributed parameter networks characterized by resistance, inductance, capacitance, and conductance per unit line length and hence a combination of Kirchoff's current Law (KCL), Kirchoff's voltage law (KVL) and characteristic elemental relations between the voltage and current for distributed elements leads to an SDE [4]. Noise in such SDEs is produced by line loading effects. More generally, we can consider a stochastic field $X(t, r)$ in space–time corresponding to the voltage at r at time t. From basic transmission line theory, we know that this voltage field and the corresponding current field $Y(t, r)$ will in view of the KCL and KVL jointly satisfy first-order space–time partial differential equations (PDEs) and hence, eliminating the current, the voltage field will satisfy second-order space–time PDEs [5, 6]. For example, if there is a line connecting r and $r + dr$, $dr = \delta.\xi$ along the direction defined by the unit vector ξ, and if this line carries a current $Y(t, r, \xi)$ at r and has distributed parameters $R(\xi), L(\xi), C(\xi), G(\xi)$, then we have the KVL

$$X(t,r) - X(t,r+dr) = X(t,r) - X(t,r+\delta.\xi)$$
$$= L(\xi)\partial_t Y(t,r,\xi) + R(\xi).Y(t,r,\xi)$$

Likewise, if there is a parallel line near r separated by the vector $\delta.\eta$, where η is a unit vector and $Y(t, r, \eta)$ is the current from the line along ξ at r along the direction η, then the KCL gives

$$Y(t,r,\xi) - Y(t,r+\delta.\xi,xi) =$$
$$\sum_{\eta}\left[G(\eta)\left(X(t,r) - X(t,r+\delta.\eta)\right) + C(\eta)\partial_t\left(X(t,r) - X(t,r+\delta.\eta)\right)\right]$$

The entire network of transmission lines is thus characterized by this system of equations for the voltage field $X(t, r)$ and the current field $Y(t, r, \eta)$. Taking the limit $\delta \to 0$, PDEs for X, Y can be obtained from these. If in addition there is line loading, the right sides of the above equation will contain noise source terms $W_X(t, r)\delta$ and $W_Y(t, r, \xi).\delta$; this system can be analysed using the theory of infinite dimensional stochastic PDEs. Eliminating the current field after making appropriate approximations, we can generally derive second-order PDEs satisfied by the voltage field $X(t, r)$. This has the form

$$\partial_t^2 X(t,r) = F\left(X(t,r), \partial_t X(t,r), \partial_{x_i} X(t,r), \partial_{x_i x_j} X(t,r)\right), u(t,r)\right) +$$
$$\sum_{k \geq 1} F_k\left(X(t,r), \partial_t X(t,r), \partial_{x_i} X(t,r), \partial_{x_i x_j} X(t,r), u(t,r)\right) W_k(t,r) \quad (5.1)$$

where $u(t, r)$ is the source field. The boundary conditions in conventional transmission line theory are that, at the source end, we specify the relationship between the source voltage and the line voltage and current at that end while, at the load end, we specify the relationship between the line voltage and current in terms of the load. Likewise in our generalized continuous space distributed transmission line theory, at certain stipulated points $s_1, \ldots, s_M$ in space, we specify a relationship between the voltage field and its first-order partial derivatives w.r.t. space–time arguments, corresponding to the presence of either a source or a load at those points. Thus, our boundary conditions are of the form

$$G_k\left(X(t, s_k), \partial_t X(t, s_k), \nabla_r X(t, s_k), u_k(t)\right) = 0, k = 1, 2, \ldots, M$$

We usually assume in such stochastic models that the line loading noise fields $W_k(t, r)$ are mutually uncorrelated in space but not necessarily in time. Thus, the noise correlation structure is

$$\mathbb{E}\left(W_k(t,r) W_m(t',r')\right) = \rho_{km}(t,t') \delta^3(r - r')$$

We "diagonalize" the kernel $\rho_{km}(t, t)$ using standard Karhunen–Loeve/spectral theory:

$$\sum_m \int \rho_{km}(t,t') \chi_{sm}(t') dt' = \lambda_s \chi_{sk}(t)$$

so that $\chi_{sk}(t)$ are orthonormal in the sense that

$$\sum_k \int \chi_{sk}(t) \chi_{lk}(t) dt = \delta_{sl}$$

We choose an orthonormal basis $\phi_n(r)$, $n = 1, 2, \ldots$ for $L^2(D)$ where D is the spatial region of the distributed network. Thus, we have

$$\delta\left(r - r'\right) = \sum_n \phi_n\left(r\right)\phi_n\left(r'\right)$$

We can now expand the noise field as

$$W_k\left(t, r\right) = \sum_{sn} w\left(sn\right)\chi_{sk}\left(t\right)\phi_n\left(r\right)$$

or equivalently

$$w\left(sn\right) = \sum_k \int W_k\left(t, r\right)\chi_{sk}\left(t\right)\phi_n\left(r\right)dtd^3r$$

We get that $\{w(sn)\}$ is an orthogonal family of random variables:

$$\mathbb{E}\left(w\left(sn\right)w\left(s'n'\right)\right) = \sum_{kl} \int \mathbb{E}(W_k\left(t, r\right)W_l\left(t', r'\right)\chi_{sk}\left(t\right)\chi_{s'l}\left(t'\right)\phi_n\left(r\right)\phi_{n'}\left(r'\right)dtdt'd^3rd^3r'$$

$$= \sum_{kl} \int \rho_{kl}\left(t, t'\right)\delta^3\left(r - r'\right)\chi_{sk}\left(t\right)\chi_{s'l}\left(t'\right)\phi_n\left(r\right)\phi_{n'}\left(r'\right)dtdt'd^3rd^3r'$$

$$= \left(\sum_{k,l} \int \chi_{sk}\left(t\right)\rho_{kl}\left(t, t'\right)\chi_{s'l}\left(t'\right)dtdt'\right)\delta_{nn'}$$

$$= \lambda_s \delta_{ss'}\delta_{nn'}$$

and we have the Karhunen–Loeve (KL) expansion for the noise process:

$$W_k\left(t, r\right) = \sum_{sn} w\left(sn\right)\chi_{sk}\left(t\right)\phi_n\left(r\right)$$

We also expand the voltage field $X(t, r)$ using a complete set of orthonormal basis functions $\eta_s(t)$, $s = 1, 2$, in the time domain:

$$X\left(t, r\right) = \sum_{sn} x\left(sn\right)\eta_s\left(t\right)\phi_n\left(r\right), x\left(sn\right) = \int X\left(t, r\right)\eta_s\left(t\right)\phi_n\left(r\right)dtd^3r$$

Note that

$$\sum_s \chi_{sk}\left(t\right)\chi_{sl}\left(t'\right) = \delta_{kl}\delta\left(t - t'\right), \sum_s \eta_s\left(t\right)\eta_s\left(t'\right) = \delta\left(t - t'\right)$$

and also the input fields are

$$u(t,r) = \sum_{sn} u(sn)\eta_s(t)\phi_n(r)$$

The dynamics (5.1) of the network field can now be expressed as

$$\sum_{sn} x(sn)\eta''_s(t)\phi_n(r)F\left(\sum_{sn} x(sn)\eta_s(t)\phi_n(r), \sum_s x(sn)\right.$$
$$\chi'_s(t)\phi_n(r), \sum_{sn} x(sn)\eta_s(t)\partial_i\phi_n(r), \sum_{sn} x(sn)$$
$$\eta_s(t)\partial_{ij}\phi_n(r), \sum_{sn} u(sn)\eta_s(t)\phi_n(r)) + \sum_k F_k\left(\sum_{sn} x(sn)\eta_s(t)\right.$$
$$\phi_n(r), \sum_{sn} x(sn)\eta'_s(t)\phi_n(r), \sum_{sn} x(sn)\eta_s(t)\partial_i\phi_n(r), \sum_{sn} x(sn)$$
$$\left.\sum_{sn} x(sn)\eta_s(t)\partial_{ij}\phi_n(r), \sum_{sn} u(sn)\eta_s(t)\phi_n(r)\right) \cdot \sum_{sn} w(sn)\chi_{sk}(t)\phi_n(r)$$

$$(5.2)$$

From (5.2), we deduce, on multiplying both sides by $\eta_{s'}(t)\phi_{n'}(r)$ and then integrating over time and space, a non-linear non-causal difference equation of the form

$$x(sn) = H_0(s,n,\{x(pq)\},\{u(pq)\} \mid \theta)$$
$$+ \sum_{s'n'} H(s,n,s',n',\{x(pq)\},\{u(pq)\} \mid \theta)w(s'n') \qquad (5.3)$$

When we assume that $W_k(t, r)$ are jointly zero-mean Gaussian fields, it follows that the random vectors (RVs) $w(sn)$ are jointly zero-mean orthogonal Gaussian RVs with variance λ_s. These equations can be expressed in vector notation:

$$\mathbf{x} = H_0(\mathbf{x}),\mathbf{u} \mid \theta) + H_1(\mathbf{x},\mathbf{u} \mid \theta)\mathbf{w} \qquad (5.4)$$

where $\mathbf{w}$ is a white Gaussian random vector. The same model will also work in the case when the processes $W_k(t, r)$ have an arbitrary space–time correlation structure. In this case, we must construct, using the KL method, an orthonormal basis in space–time for the correlation kernel, that is,

$$\mathbb{E}\left(W_k(t,r)W_m(t',r')\right) = \rho_{km}(t,r,t',r')$$

with the orthonormal basis eigenfunctions $\phi_{sk}(t, r)$ satisfying

$$\sum_m \int \rho_{km}(t,r,t',r')\phi_{sm}(t',r')dt'd^3r' = \lambda_s\phi_{sk}(t,r)$$

We then have the orthonormality relations

$$\sum_k \int \phi_{sk}(t,r)\phi_{s',k}(t,r)dtd^3r = \delta_{ss'}$$

and the noise process now has the spectral representation

$$W_k(t,r) = \sum_s w(s)\phi_{sk}(t,r), w(s) = \sum_k \int W_k(t,r)\psi_{sk}(t,r)dtd^3r$$

so that

$$\mathbb{E}\big(w(s)w(s')\big) = \delta_{ss'}$$

Then we aim to construct the approximate likelihood function for the parameters θ which are assumed to be present in the functions F, F_k and hence also in H, H_k from (5.4). Equation (5.4), after introducing a perturbation parameter δ into the noise term, is

$$\mathbf{x} = H_0(\mathbf{x},\mathbf{u}\,|\,\theta) + \delta.H_1(\mathbf{x},\mathbf{u}\,|\,\theta)\mathbf{w}$$

We solve this using perturbation theory:

$$\mathbf{x} = \sum_{n\geq 0}\delta^n\mathbf{x}(n)$$

Equating the coefficients of δ^0, δ^1 gives us respectively

$$\mathbf{x}(0) = H_0(\mathbf{x}(0),\mathbf{u}\,|\,\theta), \mathbf{x}(1) = H_{0'}(\mathbf{x}(0),\mathbf{u}\,|\,\theta)\mathbf{x}_1 + H_1\big(\mathbf{x}(0),\mathbf{u}\big)\mathbf{w}$$

Equating the coefficients of δ^2 gives us

$$\mathbf{x}(2) = H_0'(\mathbf{x}(0),\mathbf{u}\,|\,\theta),\mathbf{x}(2) + (1/2)H_0''(\mathbf{x}(0),\mathbf{u}\,|\,\theta)(\mathbf{x}_1 \otimes \mathbf{x}_1)$$
$$+ H_{1'}(\mathbf{x}(0),\mathbf{u}\,|\,\theta)\big(\mathbf{x}(1)\otimes\mathbf{u}\big)$$

From these expressions, we see that, up to $O(\delta^2)$, $\mathbf{x}$, given the inputs $\mathbf{u}$ and the parameter vector θ, can be expressed as a quadratic polynomial in the

Gaussian vector $\mathbf{w}$, that is, in the following form. We have, under the appropriate conditions of the invertibility of matrices,

$$\mathbf{x}_1 = (I - \mathbf{H}_0'(\mathbf{x}(0), \mathbf{u} \mid \theta))^{-1} G_1(\mathbf{x}_0, \mathbf{u}) \mathbf{w},$$

$$\begin{aligned} \mathbf{x}_2 &= (I - \mathbf{H}_0'(\mathbf{x}(0), \mathbf{u} \mid \theta))^{-1} \Big((1/2) \mathbf{H}_0''(\mathbf{x}(0), \mathbf{u} \mid \theta)(\mathbf{x}_1 \otimes \mathbf{x}_1) \\ &\quad + \mathbf{H}_1'(\mathbf{x}(0), \mathbf{u} \mid \theta)(\mathbf{x}(1) \otimes \mathbf{u}) \Big) = (1/2)(I - \mathbf{H}_{0'}(\mathbf{x}(0), \mathbf{u} \mid \theta))^{-1} \mathbf{H}_0''(\mathbf{x}(0), \\ &\quad \mathbf{u} \mid \theta)[(I - \mathbf{H}_0'(\mathbf{x}(0), \mathbf{u} \mid \theta))^{-1} \otimes (I - \mathbf{H}_0'(\mathbf{x}(0), \mathbf{u} \mid \theta))^{-1}) \\ &\quad (G_1(\mathbf{x}_0, \mathbf{u}) \otimes G_1(\mathbf{x}_0, \mathbf{u}))(\mathbf{w} \otimes \mathbf{w}) + \mathbf{H}_1'(\mathbf{x}(0), \mathbf{u} \mid \theta) \\ &\quad \Big[(I - \mathbf{H}_{0'}(\mathbf{x}(0), \mathbf{u} \mid \theta))^{-1} G_1(\mathbf{x}_0, \mathbf{u}) \otimes \mathbf{u} \Big] \mathbf{w} \end{aligned}$$

The probability distribution of a linear quadratic function of a Gaussian random vector is well known in terms of its moment generating function/characteristic function. In fact, if ξ is a zero-mean Gaussian RV with covariance matrix R, then the moment generating function of the RV

$$c_0 + c_1 \xi + c_2(\xi \otimes \xi)$$

can be computed by using the formula for an RV η having $N(m, R)$ distribution:

$$\mathbb{E}(\exp(\eta^T B \eta)) = K_0 \int \exp\Big(\eta^T B \eta - (\eta - m)^T R^{-1}(\eta - m)/2\Big) d\eta$$

which is evaluated by completing the squares and using the formula

$$\int \exp(-\eta^T C \eta / 2) d\eta = (2\pi)^{N/2} \det(C)$$

The moment generating function of $\mathbf{x}$, $\mathbb{E}\Big(\exp(t^T x)\Big)$ can thus be evaluated using this formula and numerically inverted to obtain the approximate probability density function (PDF) of $\mathbf{x}$ from which the parameters θ can be estimated using the maximum likelihood method. However, when noise $\mathbf{w} = w(\varepsilon)$, parametrized by a small parameter $\varepsilon \to 0$, is weak, with a rate function obtained as the Legendre transform of the limiting scaled logarithmic moment generating function

$$I_w(z) = \sup_f \Big(f^T z - \Lambda(f) \Big), \Lambda(f) = \lim_{\varepsilon \to 0} \varepsilon \log \Big(\mathbb{E}\Big(\exp(\varepsilon^{-1} f^T w(\varepsilon)) \Big) \Big)$$

then the rate function of $\mathbf{x} = x(\varepsilon)$ can be calculated using the method of Dawson and Gartner as:

$$I_x(x \mid \theta) = \inf \Big(I_w(z) : H_0(\mathbf{x}, \mathbf{u} \mid \theta) + H_1(\mathbf{x}, \mathbf{u} \mid \theta)z = \mathbf{x} \Big)$$

and then, using large deviation theory, the probability that $\mathbf{x}$ will take the measured value x is given asymptotically as $\varepsilon \to 0$ by

$$P\left(\mathbf{x}(\varepsilon) = x\right) \approx \exp\left(-\varepsilon^{-1} I_x(x \mid \theta)\right)$$

So the maximum likelihood estimate of θ in the asymptotic limit is given by

$$\hat{\theta} = \operatorname{argmin}_\theta I_x(x \mid \theta)$$

This asymptotic formula, obtained using large deviation theory, does not require the noise to have any specific distribution like the Gaussian law. It works whenever the family of noise RVs has a rate function. In the particular case when w has the same size as x and H_1 is invertible, we find that the rate function of $\mathbf{x}$ is given by

$$I_x(x \mid \theta) = I_w\left(H_1(x, u \mid \theta) - 1\left(x - H_0(x, u \mid \theta)\right)\right)$$

and the approximate maximum likelihood estimator (MLE) of θ can be obtained by minimizing this w.r.t. θ. We can also design a block processing controller based on the large deviation principle by incorporating the control forces into the dynamics of $X(t, r)$, so that the resulting relation between x and w becomes

$$x = H_0(x, u \mid \theta) + H_1(x, u \mid \theta)w + K\left(x_d - x\right)$$

where x_d is the desired state (i.e., the generalized Fourier coefficient of the state) and K is a feedback gain matrix. Usually x_d will satisfy the noiseless dynamics

$$x_d = H_0(x_d, u \mid \theta)$$

and hence the controller design will involve calculating K to minimize the effect of noise on the system. Writing

$$\delta x = x - x_d$$

we find on linearization that up to the linear orders in

$$\delta x = H_0'(x_d, u \mid \theta)\delta x + H_1(x_d, u \mid \theta)w - K\delta x$$

where

$$H_0'(x_d, u \mid \theta) = \partial H(x_d, u \mid \theta) / \partial x$$

the Jacobian matrix of H_0 w.r.t. x. We note that δx can be solved for yielding

$$\delta x = \left(I - H_0'(x_d, u \mid \theta) + K\right) - 1H_1(x_d, u \mid \theta)w$$

so that if $f_w(w)$ is the PDF of w, then the PDF of δx under the assumption of invertibility of H_1 is given by

$$f_x(\delta x \mid \theta) = \left| \det\left(\left(I - H_0'(x_d, u \mid \theta) + K\right)^{-1} H_1(x_d, u \mid \theta) \right) \right|^{-1}$$

$$f_w\left(\left(\left(I - H_0'(x_d, u \mid \theta) + K\right)^{-1} H_1(x_d, u \mid \theta) \right) - 1\delta x \right)$$

from which the MLE of θ is calculated first for $K = 0$ and then K is determined so that

$$P\left(\left| \delta x \right| > a \right) = \int_{|\delta x| > a} f_x(\delta x \mid \theta) d\delta x$$

is minimized, that is, the controller is designed so that the probability of the state deviating from the desired state by an amount greater than the threshold a is at a minimum. Alternatively, if $I_w(z)$ is the rate function of the noise w, we get the rate function of δx as

$$I(\delta x \mid \theta, K) = I_w\left(H_1(x, u \mid \theta)^{-1} \left((I + K)\delta x - H_0(x, u \mid \theta) \right) \right)$$

and the asymptotic deviation probability

$$\exp\left(-\varepsilon^{-1} \inf(I(\delta x \mid \theta, K) : |\delta x| > a) \right)$$

is minimized, or equivalently,

$$\inf(I(\delta x \mid \theta, K) : |\delta x| > a)$$

is maximized.

Now we shall look at the dynamics of our power distribution problem from the standpoint of state vector SDEs and describe the controller design, both based on an adaptive real time approach as well as on a block processing based approach using large deviation theory. For this analysis, we shall assume that the noise processes $W_k(t)$ obtained by spatial discretization is a mixture of Gaussian white noise and Poissonian white noise. The Gaussian noise component responds to the central limit law, namely a superposition of a very large number of very small random effects, which results after appropriate scaling into a Gaussian process. This is a consequence of atmospheric effects like temperature, humidity, pressure, and winds in the atmosphere acting along the length of the transmission line. The Poissonian component is the discrete contribution and can be attributed to random tapping of the line voltage at a discrete set of spatial points and at a discrete set of times. The model for noise is therefore

$$W_k(t) = \sum_j c_1(k,j) dB_j(t)/dt + \sum_j c_2(k,j) dN_j(t)/dt$$

where $B_j s$ are independent standard Brownian motion processes and $N_j s$ are independent Poisson processes with rates λ_j. Since the noise processes are assumed to be weak and the rates of the Poissonian components are large when the line tapping rate is large, we use a scaled model of the above:

$$W_k(t,\varepsilon) = \sum_j c_1(k,j) \sqrt{\varepsilon} B'_j(t) + \sum_j c_2(k,j) N'_j(t/\varepsilon)$$

The logarithmic moment generating functional of the process $\sqrt{\varepsilon} B'_j(t)$ is

$$\log \mathbb{E} \exp\left(\int_0^T \sqrt{\varepsilon} f(t) B'_j(t) dt \right) = \exp\left(\varepsilon \int_0^T f^2(t) dt \right) = \Lambda_\varepsilon(f)$$

and hence the limiting scaled logarithmic moment generating functional of this process is

$$\lim_{\varepsilon \to 0} \varepsilon.\Lambda_\varepsilon\left(\varepsilon^{-1} f\right) = (1/2) \int_0^T f^2(t) dt$$

So the rate function of this family of processes, obtained by taking the Legendre transform of the above, is

$$I(w) = (1/2) \int_0^T w(t)^2 dt$$

Again, the logarithmic moment generating functional of the process $N_{j'}(t/\varepsilon)$ is

$$\log\ \mathbb{E}\ \exp\left(\int_0^T f(t)N'_j(t/\varepsilon)dt\right) = \log\ \mathbb{E}\left[\exp\left(\varepsilon\int_0^{T/\varepsilon} f(\varepsilon t)N_{j'}(t)dt\right)\right]$$

$$= \exp\left(\int_0^{T/\varepsilon} \lambda_j\left(\exp(\varepsilon f(\varepsilon t))-1\right)dt\right)$$

$$= \exp\left(\varepsilon^{-1}\int_0^T \lambda_j\left(\exp(\varepsilon f(t))-1\right)dt\right) = \Lambda_\varepsilon(f)$$

Thus, the limiting scaled logarithmic moment generating function of this process is

$$\lim_{\varepsilon\to 0}\varepsilon.\Lambda_s\left(\varepsilon^{-1}f\right) = \int_0^T \lambda_j\left(\exp(f(t))-1\right)dt$$

Combining these two results and using the independence of the processes B_j, n_k, j, $k = 1, 2, \ldots$ results in the following formula for the limiting logarithmic moment generating functional of the processes $W_k(t)$, $k = 1, 2, \ldots, t \in [0, T]$:

$$\lim_{\varepsilon\to 0}\varepsilon.\log\left(\mathbb{E}\left[\exp\left(\varepsilon^{-1}\int_0^T f_k(t)W_k(t,\varepsilon)dt\right)\right]\right)$$

$$= (1/2)\sum_{j,k} c_1(k,j)^2\int_0^T f_k(t)^2\ dt + \sum_{k,j}\lambda_j\int_0^T\left((\exp(c_2(k,j)f_k(t))-1\right)$$

$$dt = \Lambda\left(f_1,f_2,\ldots\right))$$

and the rate functional of this family of processes is given by

$$I_W\left(w_k, k=1,2,\ldots\right) = \sup_{f_k,k=1,2,\ldots}\left[\sum_k\int_0^T f_k(t)w_k(t)dt - \Lambda\left(f_k, k=1,2,\ldots\right)\right]$$

Thus, after spatial discretization, if we express the power distribution dynamics taking into account feedback control as:

$$X'(t) = F_0(X(t),u(t)\,|\,\theta) + \sum_k F_k(X(t),u(t)\,|\,\theta)W_k(t,\varepsilon) + K\left(X_d(t)-X(t)\right)$$

where the state vector $X(t)$ stands for the aggregate of spatially discretized processes $X(t, r_j)$, $\partial_t X(t, r_j)$, $j = 1, 2, \ldots$, then we can evaluate the rate functional of the vector valued process $X(t)$, $t \in [0, T]$ as

$$I_X(x) = I_X(x(t), t \in [0, T] \mid \theta, K) = \inf(I_W(w_k, k = 1, 2, \ldots):$$
$$x'(t) - F_0(x(t), u(t) \mid \theta) - \sum_k F_k(x(t), u(t) \mid \theta) w_k(t) - K(X_d(t) - x(t))$$
$$= 0, t \in [0, T])$$

and we can then apply the approximate maximum likelihood method to estimate the first θ by setting $K = 0$ from measurements of $X(.)$ and then determining K by minimizing $\inf(I_X(x) : |x - X_d|_\infty > a)$. In practice, we must devise a numerical scheme of search optimization for arriving at this block processing based controller design. It should be noted that if the parameters vary slowly with time, and we take noisy measurements on some function of the state process at different times, then we can use the EKF to estimate the parameters on a real time basis from these measurements. The contoller design can also be carried out adaptively in this framework by a least mean square (LMS)-like algorithm:

$$K(t + dt) = K(t) - \mu(t) \cdot \frac{\delta}{\delta K(t)} \mid X_d(t + dt) - X(t)$$
$$-F_0(X(t), u(t) \mid \theta) dt - K(t)(X_d(t) - X(t))\mid^2$$

with $X(t + dt)$ obtained from the noisy plant dynamics using the controller $K(t)$:

$$X(t + dt) = X(t) + F_0(X(t), u(t) \mid \theta) dt + \sum_k F_k(X(t), u(t) \mid \theta) W_k(t, \varepsilon) dt$$
$$+ K(X_d(t) - X(t)) dt$$

This method works well provided that we are able to measure the state $X(t)$ at each time instant accurately. If, however, we are able only to measure accurately some noise corrupted function $h(X(t))$ of the state process at each time, then denoting this measurement by $dY(t) = h(X(t)) dt + V(t) dt$, the adaptive controller design would be based on first calculating using the EKF the state estimate $\widehat{X}(t + dt)$ using the new measurement $Y(t + dt)$ with the state dynamics obtained using the controller coefficient $K(t)$, then calculating $K(t + dt)$ by minimizing

$$\mid X_d(t + 2dt) - \widehat{X}(t + dt) - F_0\left(\widehat{X}(t + dt), u(t + dt) \mid \theta\right) dt$$
$$- K(t + dt)\left(X_d(t + dt) - \widehat{X}(t + dt)\right) \mid^2$$

This procedure enables us to do simultaneous state estimation and control for trajectory tracking. A more rigorous method of controller design is to start by writing down the approximate linearized dynamics for both the state estimation error and the trajectory tracking error using the controlled dynamics of the state, of the EKF state estimator, and then of the desired state. Noting that the state is not directly observable and hence the controller should be based on feedback of the error between the desired state and the EKF state estimate, these are respectively

$$dX(t)/dt = F_0\big(X(t), u(t)\big) + \sum_k F_k\big(X(t), u(t)\big)W_k(t)$$
$$+ K(t)\big(X_d(t) - \widehat{X}(t)\big), dY(t) = h\big(X(t)\big)dt + dV(t)$$
$$d\widehat{X}(t) = F_0\big(\widehat{X}(t), u(t)\big)dt + L(t)\big(dY(t)/dt - h\big(\widehat{X}(t)\big)\big),$$
$$dX_d(t)/dt = F_0\big(X_d(t), u(t)\big)$$

Where $L(t)$, the Kalman gain, is given by

$$L(t) = P(t)h'\big(\widehat{X}(t)\big)^T R_V^{-1}$$

supplemented with the Riccati equation for the state estimation error covariance matrix

$$dP(t)/dt = F'\big(\widehat{X}(t), u(t)\big)P(t) + P(t)F'\big(\widehat{X}(t), u(t)\big)^T +$$
$$\sum_k F_k\big(\widehat{X}(t), u(t)\big)R_{Wk}(t)F'\big(\widehat{X}(t), u(t)\big)^T - P(t)h'\big(\widehat{X}(t)\big)^T R_V^{-1}h'\big(\widehat{X}(t)\big)P(t)$$

and then defining the state estimation error

$$e(t) = X(t) - \widehat{X}(t)$$

and the trajectory tracking error

$$f(t) = \widehat{X}(t) - X_d(t)$$

we get obtain the linearized dynamics as

$$de(t)/dt = F_{0'}\big(\widehat{X}(t), u(t)\big)e(t) + \sum_k F_k\big(\widehat{X}(t), u(t)\big)W_k(t) - K(t)f(t),$$

$$df(t)/dt = F_{0'}\left(\widehat{X}(t),u(t)\right)f(t)+L(t)\left(h'\left(\widehat{X}(t)\right)e(t)+V'(t)\right)$$

Since these are linearized equations, we can replace the random quantity $\hat{X}(t)$ by the non-random quantity $X_d(t)$ and then defining the non-random functions of time

$$F_0(t) = F_{0'}\left(X_d(t),u(t)\right), F_k(t) = F_k\left(X_d(t),u(t)\right), H(t) = h'\left(X_d(t)\right)$$

we get

$$de(t)/dt = F_0(t)e(t)-K(t)f(t)+\sum_k F_k(t)W_k(t), df(t)/dt$$

$$= F_0(t)f(t)+L(t)\left(H(t)e(t)+V'(t)\right)$$

or equivalently, in matrix notation, as

$$\frac{d}{dt}\begin{pmatrix} e(t) \\ f(t) \end{pmatrix} = \begin{pmatrix} F_0(t) & -K(t) \\ L(t)H(t) & F_0(t) \end{pmatrix}\begin{pmatrix} e(t) \\ f(t) \end{pmatrix} + \begin{pmatrix} \sum_k F_k(t)W_k(t) \\ L(t)V'(t) \end{pmatrix}$$

We observe that $W_k(t)$, $V'(t)$ are independent white noise processes with the former being a superposition of Gaussian and Poissonian white noise processes and the latter being a white Gaussian noise process. Denote by

$$I_{WV}\left(w_1,w_2,...,v\right)$$

the joint rate functionals of the processes W_k, $k = 1, 2, ..., V$ over the time interval $[0, T]$. We have seen above that this rate functional is of the form

$$(1/2)\int v(t)^T R_v^{-1}v(t)dt +$$

$$\sup_{f_1,f_2,...}\left(\begin{array}{c} \sum_k \int_0^T f_k(t)w_k(t)dt - (1/2)\sum_{j,k} c_1(k,j)^2\int_0^T f_k(t)^2\,dt \\ -\sum_{k,j}\lambda_j\int_0^T \left(\exp\left(c_2(k,j)f_k(t)\right)-1\right)dt \end{array}\right)$$

Then the rate functional of the error processes $e(t)$, $f(t)$, $t \in [0, T]$ is now easily evaluated using the above linearized equations and hence we can calculate the controller coefficients K so as to minimize the threshold deviation probability or equivalently maximize the infimum of the rate function over the threshold deviation region. This constitutes a block processing approach and works only if we take $K(t) = K$ to be a fixed constant matrix. If, however, we wish to design the controller adaptively, then we would use a stochastic gradient algorithm of the form

$$dK(t)/dt = -\mu(t)\frac{\delta}{\delta K(t)}$$

$$\left[a(1)\left|e'(t) - F_0(t)e(t) - K(t)f(t)\right|^2 + a(2)\right.$$
$$\left. \times \left|f'(t) - F_0(t)f(t) - L(t)H(t)e(t)\right|^2\right]$$

Yet another way of block processing based controller design is to solve the above linearized error dynamics using standard state transition theory and design the controller K based on large deviation theory.

$$\xi(t) = \left[e(t)^T, f(t)^T\right]^T = \int_0^t \Phi(t,\tau \mid K)W(\tau)d\tau$$

where

$$\partial_t\Phi(t,\tau \mid K) = \begin{pmatrix} F_0(t) & -K(t) \\ L(t)H(t) & F_0(t) \end{pmatrix}\Phi(t,\tau \mid K), \Phi(\tau,\tau \mid K) = I$$

and

$$W(t) = \begin{pmatrix} \sum_k F_k(t)W_k(t) \\ L(t)V'(t) \end{pmatrix}$$

$W(t)$ is a white noise process and, hence, we can write

$$\mathbb{E}\left(W(t)W(s)^T\right) = R_W(t)\delta(t-s)$$

This gives

$$\mathbb{E}\left(\xi(t)\xi(t)^T\right) = \int_0^t \Phi(t,\tau \mid K)R_W(\tau).\Phi(t,\tau \mid K)^T d\tau$$

and we can design K so that

$$\mathbb{E}\int_0^T \xi(t)^T Q(t)\xi(t)dt = \int_{0<\tau<t<T} Tr\left(Q(t)\Phi(t,\tau \mid K)R_W(\tau)\Phi(t,\tau \mid K)^T\right)dtd\tau$$

is a minimum. Yet another method of calculating the controller coefficients K is to start with the expression

$$\xi(t) = \int_0^t \Phi(t,s \mid K)W(s)ds, t \geq 0$$

for the joint error process and, using the rate function $I_W(w)$ of the $W(.)$ process expressible in the form

$$I_W(w) = \int_0^T \Lambda^*\left(w(t)\right)dt$$

(because of the whiteness of the $W(.)$ process), calculate the rate function $I_{\xi|K}$ of the $\xi(.)$ process and then choose K so that $inf\{I(\xi|K) : |\xi| > a\}$ is maximized – for that would indeed amount to minimizing the asymptotic probability of deviation of the error process from the stability zone.

5.3 DESCRIPTION OF THE POWER GRID FIELD IN TERMS OF THE STOCHASTIC ELECTROMAGNETIC FIELDS IN CURVED BACKGROUND SPACE–TIME

The voltage field $X(t, r)$ at each space–time point (t, r) and the current field $Y(t, r, \xi)$ at the space–time point (t, r) along the direction ξ can equivalently be described by an electromagnetic field that satisfies Maxwell's equations. In fact, if $H(t, r)$ is the magnetic field and $E(t, r)$ the electric field, then we consider a parallelogram in space having vertices at positions $r, r + \delta_1\xi, r + \delta_2\eta, r + \delta_1\xi + \delta_2\eta$ and express the integral form of the Maxwell equation

$$curlH = J + \varepsilon.\partial_t E,$$

namely,

$$\int_\Gamma H.dr = \int_S J.ndS + \varepsilon.\partial_t \int_S E.ndS$$

over the parallelogram surface after discretization of space as

$$H(t,r).\xi.\delta_1 + H(t,r + \delta_1\xi).\eta.\delta_2 - H(t,r + \delta_2\eta).\xi.\delta_1 - H(t,r).\eta.\delta_2 =$$
$$\varepsilon.\partial_t E(t,r).\xi \times \eta.\delta_1\delta_2 + Y(t,r,\xi \times \eta \,/\, |\xi \times \eta|).|\xi \times \eta|\delta_1\delta_2$$

taken in conjunction with the equation that relates the magnetic vector potential A and the voltage field X (which is the electric potential)

$$E(t,r) = -\nabla X(t,r) - \partial_t A(t,r)$$

which when integrated along the path joining r to $r + \delta$. ξ yields

$$E(t,r).\xi.\delta = -X(t,r + \delta\xi) + X(t,r) - \partial_t A(t,r).\xi.\delta$$

and the equation that relates the magnetic field to the magnetic vector potential

$$H(t,r) = curlA(t,r)/\mu$$

which in integral form for a closed curve Γ enclosing the open surface S reads after applying Stokes' theorem,

$$\mu\int_S H(t,r).ndS = \int_\Gamma A(t,r).dr$$

or equivalently in the discrete spatial domain

$$\mu H(t,r).(\xi \times \eta)\delta_1\delta_2 = A(t,r).\xi.\delta_1 + A(t,r + \delta_1.\xi).\eta.\delta_2$$

$$-A(t,r + \delta_2\eta).\xi.\delta_1 - A(t,r).\eta\delta_2$$

These three spatially discretized equations give us a one-to-one correspondence between the voltage and current fields on the one hand and the electric and magnetic fields on the other, or equivalently, the voltage/scalar potential and the magnetic vector potential. By solving the last, that is, the third, equation to express A in terms of H in the discrete spatial domain and then substituting this into the second gives us a relationship between E and H in terms of X. The first equation gives us another discrete spatial relationship between E and H by now in terms of the current field Y completing thereby the one-to-one correspondence between (E, H) and (X, Y). It should be noted that in this spatially discretized model, the electric and magnetic fields are defined on the lattice generated by the vectors $\delta_1\xi$ and $\delta_2\eta$ with ξ and η varying over a discrete set. We now consider the alternative description of the electromagnetic field in terms of the potential field (A, X) which satisfies the Lorentz gauge condition

$$divA = -\partial_t X$$

with A satisfying the wave equation with source

$$\partial_t^2 A(t,r) - \nabla^2 A(t,r) = \mu J(t,r)$$

Integrating this equation over the surface of the above parallelogram gives us the discretized equation

$$\partial_t^2 \left(A(t,r).\xi \times \eta \right) - \nabla^2 \left(A(t,r).\xi \times \eta. \right) = \mu Y(t,r,\xi \times \eta /| \xi \times \eta)$$

and these equations provide an alternative one-to-one correspondence between Y and (A, X) under the assumption of the Lorentz gauge. Thus, we consider a noisy line loaded version of the differential equation satisfied by A.

Taking into account noisy line loading and also distributed charges, we have the following differential equations for A, X:

$$\partial_t^2 A(t,r) - \nabla^2 A(t,r) = -\mu \sum_\xi Y(t,r,\xi)\xi ---(a)$$

$$\partial_t^2 X(t,r) - \nabla^2 X(t,r) = -\rho(t,r)/\varepsilon ---(b)$$

with the Lorentz gauge condition

$$\partial_t X + divA = 0 ---(u)$$

which is consistent with (a) and (b) provided that charge is conserved:

$$div \sum_\xi Y(t,r,\xi)\xi + \partial_t \rho = 0$$

We also recall the discretized line field equations for the voltage and current taking into account inhomogeneity and anisotropy in the distributed line parameters and also accounting for line loading:

$$Y(t,r,\xi) - Y(t,r+\delta.\xi,\xi) = \delta \sum_\eta \left(C(r,\eta)\partial_t \left(X(t,r) - X(t,r+\delta\eta) \right) \right.$$

$$\left. + G(r,\eta)\left(X(t,r) - X(t,r+\delta\eta) \right) \right) + W_Y(t,r,\xi) ---(c)$$

$$X(t,r) - X(t,r+\delta\xi) = \delta[R(r,\xi)Y(t,r,\xi) + L(r,\xi)\partial_t Y(t,r,\xi)]$$
$$+ W_X(t,r) - - - (d)$$

(c) and (d) are simultaneously solved to give the fields X, Y. Only those solutions are retained which are consistent with the gauge condition (u). Having thus determined X, the charge density ρ in space can be calculated from (b) and A can be determined from Y using (a). Alternatively, given the charge and current fields (ρ, Y), we solve (a) and (b) for A, X using the retarded potential formula and substitute X, Y into (c) and (d) and calculate the model distributed parameters C, G, L, R by a least squares fitting method. Now consider the equations (c) and (d) for X, Y assuming knowledge of the distributed parameters R, L, G, C. We solve these to obtain X, Y as a function of the source and load conditions at a discrete set of spatial points and also of the line loading noise processes W_X, W_Y. The rate functional for A and hence of E, B can thus be evaluated by using the relationships between (X, Y) and (E, B). From this rate functional, we can calculate the asymptotic probability of the electromagnetic fields deviating away from the stability zone thereby causing radiation hazards. To minimize this deviation probability and hence radiation hazards, we can adjust the distributed parameters of the network field appropriately, subject to the constraint that the voltage and current fields X, Y fall within a desired range so that smooth operation of the power system network is not hampered.

Now we take a look at the effects of space–time curvature on the electromagnetic fields and hence on the rate function of these fields. Assume a choice of coordinates so that the metric becomes of synchronous form, that is,

$$d\tau^2 = dt^2 - \gamma_{rs}(t,r)dx^r dx^s$$

(Such a coordinate system always exists since the four functions that specify the change of coordinates can be chosen so as to make the four functions $g_{0\mu}$ vanish.) Let A_μ denote the electromagnetic four potential. Then, we define the electric and magnetic field components by

$$E_r = F_{0r} = A_{r,0} - A_{0,r}, B_r = -(1/2)\varepsilon(rsk)F_{rk} = -(1/2)\varepsilon(rsk)(A_{k,r} - A_{r,k})$$

or equivalently, in component form

$$E_1 = A_{1,0} - A_{0,1}, E_2 = A_{2,0} - A_{0,2}, E_3 = A_{3,0} - A_{0,3},$$

$$B_1 = -F_{23} = -A_{3,2} + A_{3,2}, B_2 = -F_{31} = -A_{1,3} + A_{3,1}, B_3 = -F_{12} = -A_{2,1} + A_{1,2}$$

These definitions agree with the flat space–time definitions of the electromagnetic field

$$E = -\nabla A^0 - A_{,0}, B = curl A$$

in terms of the four vector potentials, with the latter being given by the contravariant vector

$$A^0 = A_0, A^r = -A_r, r = 1, 2, 3$$

It is easy to see that the above definitions of the electromagnetic field satisfy the usual homogeneous Maxwell equations

$$curlE = -\partial_t B, div B = 0, E = (E_r, r = 1, 2, 3), B = (B_r, r = 1, 2, 3)$$

which are consequences of the tensor equation

$$F_{\mu\nu:\sigma} + F_{\nu\sigma:\mu} + F_{\sigma\mu:\nu} = F_{\mu\nu,\sigma} + F_{\nu\sigma,\mu} + F_{\sigma\mu,\nu} = 0$$

The inhomogeneous Maxwell equations in general relativity are the tensor equations

$$F^{\mu\nu}_{:\nu} = \mu.J^{\mu}$$

which are equivalent in the synchronous system to

$$\left(F^{\mu\nu} \sqrt{\gamma} \right)_{,\nu} = \mu_0 J^{\mu} \sqrt{\gamma}$$

Defining the modified electric and magnetic field components as

$$E^r = -F^{0r} = F^{r0}, B^r = (-1/2)\varepsilon(rsk)F^{sk}, r = 1, 2, 3$$

so that

$$F^{rs} = -\varepsilon(rsk)B^k$$

or equivalently, in terms of components,

$$E^r = \gamma^{rs}F_{s0}, B^r = (-1/2)\varepsilon(rsk)\gamma^{sm}\gamma^{kp}F_{mp}$$

which on noting that

$$F_{mp} = -\varepsilon(mpk)B_k$$

can, equivalently, be expressed as

$$E^r = \gamma^{rs}E_s, B^r = (1/2)\varepsilon(rsk)\varepsilon(mpq)\gamma^{sm}\gamma^{kp}B_k$$

the inhomogeneous Maxwell equations can be expressed as

$$\left(E^r\sqrt{\gamma}\right)_{,\gamma} = div\left(\tilde{E}\sqrt{\gamma}\right) = \mu.J^0\sqrt{\gamma}$$

and

$$\left(E^r\sqrt{\gamma}\right)_{,0} - \varepsilon\left(rsk\right)\left(B^k\sqrt{\gamma}\right)_{,s} = -\mu.J^r\sqrt{\gamma}$$

or equivalently,

$$\partial_t\left(\sqrt{\gamma}\tilde{E}\right) + \mu\sqrt{\gamma}J = curl\left(\sqrt{\gamma}\tilde{B}\right)$$

where

$$\tilde{E} = \left(E^r\right)_{r=1}^3, \tilde{B} = \left(B^r\right)_{r=1}^3$$

In summary, we have a set of four inhomogeneous linear second PDEs for the four components of the electromagnetic four potential $Z(t, r) = (A_\mu(t, r))$. These four PDEs can be cast in the general form

$$\partial_t^2 Z\left(t,r\right) + A_0\left(t,r\right)\partial_t Z\left(t,r\right) + A_{1k}\left(t,r\right)\partial_k Z\left(t,r\right) + A_{2k}\left(t,r\right)\partial_t\partial_k Z\left(t,r\right)$$
$$+A_{3km}\left(t,r\right)\partial_k\partial_m Z\left(t,r\right) = J\left(t,r\right) + W\left(t,r\right)$$

where J is a four vector source field and $W(t, r)$ is a four vector noise field. It should be noted that the 4×4 matrix valued functions $A_0, A_{1k}, A_{2k}, A_{3km}$ of space–time are expressible in terms of the 3×3 metric $\gamma_{rs}(t, r)$. It should also be noted that if we change A_μ to $A_\mu + \partial_\mu\phi$ where ϕ is an arbitrary function of space–time, the electromagnetic field does not change. Therefore we can adopt a general relativistic Lorentz gauge condition of the form

$$\partial_t\left(A^0\sqrt{\gamma}\right) + \partial_r\left(A^r\sqrt{\gamma}\right) = \partial_\mu\left(A^\mu\sqrt{\gamma}\right) = 0$$

which is equivalent to the condition

$$A_\mu^\mu = 0$$

and, with this gauge, considerable simplification of the dynamical equations can be achieved. The problem of control of the electromagnetic field can now be formulated in terms of the Kalman filter along the following lines. Introduce state variables

$$X\left(t\right) = \left(Z\left(t,r_m\right), \partial_t Z\left(t,r_m\right)\right)_{m=1}^N$$

where N is the number of discrete spatial pixels. By replacing partial derivatives w.r.t. the spatial variables by finite differences, we can approximate the above field equation for Z in linear state variable form:

$$X'(t) = A_0(t)X(t) + G(t)J(t) + W(t)$$

The spatially discretized electromagnetic field is measured at a finite sparse set of spatial pixels. By noting the linear relationship between the electromagnetic four potential $Z(t, r)$, its time derivative $\partial_t Z(t, r)$, and the electromagnetic field $E(t, r)$, $B(t, r)$ in terms of spatial derivatives of the former, we can express the measurement model in the form

$$dY(t) = HX(t)dt + dV(t)$$

where H is a sparse matrix obtained by spatial discretization of the equations

$$E_r(t,r) = A_{r,0}(t,r) - A_{0,r}(t,r), B_r(t,r) = -(1/2)\varepsilon(rsk)(A_{k,s} - A_{s,k})$$

Here, the vector $Y(t)$ has components given by the electromagnetic field at a finite sparse set of spatial pixels followed by measurement noise corruption. The Kalman filter for estimating $X(t)$ and hence the electromagnetic field at all the pixels on a real time basis can now be formulated from this state and measurement model and hence feedback control forces based on the estimate of $X(t)$ and a desired field trajectory so as to modify the dynamics for trajectory tracking can be designed. Specifically, if $E(t) = CX(t)$ expresses the relationship between the spatially discretized electromagnetic field vector $E(t)$ and the electromagnetic four potential vector, and if its time derivative $X(t)$ and $E_d(t)$ is the desired electromagnetic (EM) field vector, then our feedback control law would be

$$X'(t) = A_0(t)X(t) + G(t)J(t) + K(t)\Big(E_d(t) - C\hat{X}(t)\Big) + W(t)$$

The Kalman filter (KF) is given by

$$d\hat{X}(t)/dt = A_0(t)\hat{X}(t) + G(t)J(t) + L(t)\Big(Y'(t) - H\hat{X}(t)\Big)$$

with

$$L(t) = P(t)H^T R_V^{-1}$$

where $P(t)$ satisfies the Riccati equation

$$P'(t) = A_0(t)P(t) + P(t)A_0(t)^T + R_W - P(t)H^T R_V^{-1}HP(t)$$

When the process noise is non-Gaussian white, this filter is not optimal in the likelihood sense, although it is optimal in the minimum mean square sense. In that case, we must regard the process $X(t)$ as a non-Gaussian Markov process with specified infinitesimal generator K_t and derive the EKF by approximating the Kushner–Kallianpur nonlinear stochastic filter. The Kushner–Kallianpur (KK) filter is

$$d\pi_t(\phi) = \pi_t(K_t\phi)dt + \left(\pi_t(h\phi) - \pi_t(\phi)\pi_t(h)\right)^T R_V^{-1}\left(dY(t) - \pi_t(h)dt\right)$$

where the state process $X(t)$ is Markovian with generator K_t and the measurement model is

$$dY(t) = h\big(X(t)\big)dt + dV(t)$$

where $V'(t)$ is white Gaussian with

$$dV(t)dV(t)^T = R_V dt$$

and

$$\pi_t(\phi) = \mathbb{E}(\phi\big(X(t)\big) \mid Y(s) : s \le t)$$

We now take $\phi(X) = X_a$ and compute

$$K_t\phi(X) = \int K_t(X,y)y_a dy = K_t(X_a)$$

Write

$$X = \hat{X} + e$$

and make the second-order approximation

$$\begin{aligned} K_t(X,y) = K_t\big(\hat{X} + e, y\big) &\approx K_t\big(\hat{X}, y\big) + K_{t'}\big(\hat{X}, y\big)e \\ &+ (1/2)K_{t''}\big(\hat{X}, y\big)(e \otimes e) \end{aligned}$$

to get

$$\pi_t\big(K_t(X_a)\big) \approx F_{ta}\big(\hat{X}\big) + (1/2)F''_{ta}\big(\hat{X}\big)Vec(P)$$

where

$$P = \text{cov}(e \mid Y_t),$$

$$F_{ta}(X) = \int K_t(X, y) y_a dy$$

Note that, in general, we have the following second-order approximation. For any observable ϕ,

$$\pi_t\left(K_t(\phi)\right) = \pi_t\left(\int K_t(\hat{X} + e, y)\phi(y)dy\right)$$

$$\approx \int K_t(\hat{X}, y)\phi(y)dy + (1/2)\left(\int K_{t''}(\hat{X}, y)\phi(y)dy\right)Vec(P)$$

Again,

$$\pi_t(hX_a) = pi_t\left(h(X)X_a\right) \approx \pi_t\left(h(\hat{X} + e)(\hat{X}_a + e_a)\right)$$
$$\approx h(\hat{X})\hat{X}_a + h'(\hat{X})\pi_t(ee_a) = h(\hat{X})\hat{X}_a + h'(\hat{X})P_a$$

where P_a is the ath column of P. Further,

$$\pi_t(h)\pi_t(X_a) = \pi_t\left(h(\hat{X} + e)\right)\pi_t\left(\hat{X}_a + e_a\right) \approx h(\hat{X})\hat{X}_a$$

So

$$\pi_t(hX_a) - \pi_t(h)\pi_t(X_a) \approx h'(\hat{X})P_a$$

Likewise, taking $\phi(X) = X_a X_b$, we get

$$K_t(\phi) = K_t(X_a X_b) = \int K_t(x, y) y_a y_b dy$$

so

$$\pi_t\left(K_t(X_a X_b)\right) = \pi_t\left(\int K_t(\hat{X} + e, y) y_a y_b dy\right) \approx$$
$$\int K_t(\hat{X}, y) y_a y_b dy + (1/2)\left(\int K_{t''}(\hat{X}, y) y_a y_b dy\right)Vec(P)$$
$$\pi_t(hX_a X_b) = \pi_t\left(h(\hat{X} + e)(\hat{X}_a + e_a)(\hat{X}_b + e_b)\right) \approx$$
$$h(\hat{X})\hat{X}_a \hat{X}_b + h'(\hat{X})\left(P_a \hat{X}_b + P_b \hat{X}_a\right) + h(\hat{X})P_{ab}$$

and

$$\pi_t(b)\pi_t(X_a X_b) = \pi_t\left(b\left(\hat{X} + e\right)\right)\left(\hat{X}_a \hat{X}_b + P_{ab}\right)$$
$$\approx b(\hat{X})\left(\hat{X}_a \hat{X}_b + P_{ab}\right)$$

Thus,

$$\pi_t\left(b X_a X_b\right) - \pi_t(b)\pi_t(X_a X_b) \approx b'\left(\hat{X}\right)\left(P_a \hat{X}_b + P_b \hat{X}_a\right)$$

These equations can be substituted into the KK equation, thereby obtaining a set of coupled differential equations for $\hat{X}(t), P(t)$.

5.4 CONCLUSIONS

In this chapter, we have introduced a generalization of one-dimensional transmission line theory to space–time theory in which the transmission line network can extend along all the three spatial dimensions. We formulated the stochastic dynamics of the line in the case when there is random noise in the source as well as random line loading along the length of the line. The noise can be the time derivative of any Levy process, that is, a linear combination of time derivatives of Brownian motion and Poisson processes so that the line dynamics describes a Markov process. When noisy measurements of the line voltage are taken at a discrete set of spatial points, we have explained how to apply the results of standard real time stochastic filtering theory to estimate the line voltage and how to use this real time estimate to design an error feedback force into the transmission line dynamics in order to ensure that the line voltage tracks a desired line voltage. We also proposed a large deviation based method for optimal design of the controller for accurate trajectory tracking.

REFERENCES

1. Andrew Jazwinsky, *Stochastic processes and filtering theory*, Academic Press.
2. Leonard Schiff, *Quantum mechanics*.
3. Steven Weinberg, *The quantum theory of fields*, Vols. I and II, Cambridge University Press.
4. P.A.M. Dirac, *The principles of quantum mechanics*, Oxford University Press.
5. M. Green, J. Schwarz and E. Witten, *Superstring Theory*, Cambridge University Press.
6. K.R. Parthasarathy, *An introduction to quantum stochastic calculus*, Birkhauser.

Chapter 6

Energy-aware power control scheme for IoT applications

Merin Susan Philip and Poonam Singh
National Institute of Technology Rourkela, Rourkela, India

CONTENTS

6.1 INTRODUCTION

The Internet of Things (IoT) has advanced over a decade in various sectors such as industry 4.0, smart agriculture, and smart cities. The IoT is a network where devices or machines, objects, and persons are connected with each other and gather data to perform different tasks. This has revolutionized many sectors in improving and reaching new heights to provide a comfortable social and personal life. One such is smart agriculture, where devices are employed to monitor the soil moisture, temperature, turbidity, and so on. Aquaculture has more sensitive parameters to monitor as water plays an important role in the process. As we know water quality has deteriorated in recent years due to climate change, pollution, and global warming. According to aquaculture management studies, any anomaly in the characteristics of water that affects the survival, reproduction, or growth of aquatic species has an impact on the environment as well as the quality of the production. Aquaculture water quality became a serious issue towards the end of the twentieth century as a result of its extensive production. By continuously evaluating the water quality parameters, the issue

DOI: 10.1201/9781003436461-6

can be resolved. This is done by using monitoring systems and the IoT. In IoT-enabled water quality monitoring (WQM) systems, the physiological, biochemical, and microbiological characteristics of water are monitored without the involvement of a human [1], whereas laboratory-based testing is a traditional method, in which water samples are directly collected from various watercourses for numerous analyses [2]. The conventional arrangement, which gathers, processes, and then transmits water quality parameters to the appropriate water stations, can be replaced by a wireless sensor network (WSN). Every day, over a million new IoT devices [3, 4] connect to the Internet, and the number is growing.

According to estimates [5], there will be up to 75 billion IoT devices in use by 2025. The huge number of these devices, such as sensor nodes and gateways, will result in more CO_2 generation and, as a result, more greenhouse gas emissions in the atmosphere. In the information and communications technology industry, network devices and network servers account for more than half of CO_2 emissions [6]. As a result, new ways for optimizing the energy usage of IoT devices need to be eco-friendly [7, 8]. New blockchain computing and other optimization techniques are evolving to make energy-efficient systems [9]. Likewise in WQM systems, energy efficient techniques are inevitable. However, with IoT-based WQM systems, energy constraint is a major concern.

The battery life of most WQM systems is limited, which is a major drawback. The WQM systems have power-hungry sensors which use energy scarce battery units. This frequently demands battery replacement, which becomes extremely inconvenient. These systems require a high level of service reliability, rapid sensor data delivery, increased range, and low power consumption [10]. The energy usage of the WSN system is heavily influenced by the physical layer chosen. After correct encoding, this layer sends the signal through a communication channel. The power with which the data is transmitted, data rate, modulation technique used, and time on air are all factors that contribute to energy depletion. Improving the energy efficiency of these factors will result in a significant reduction in the amount of energy required. Cellular, ZigBee, Bluetooth, Wi-Fi (Wireless Fidelity), and LoRa are all common communication technologies in IoT-based WSN systems. ZigBee and Bluetooth are less energy intensive technologies, but their communication range is only a few hundred meters. Wi-Fi provides a quicker data transfer rate for low-range applications, but at the expense of more battery life. Cellular technology can support high data rates, but it also uses the most energy of any current communication technology. Because LoRa has a transmission range of up to 10 kilometers, it is an appropriate technology for IoT applications. When compared to competing technologies with similar transmission ranges, LoRa technology consumes fewer resources [11–13]. The present study deals with energy saving in the WQM devices used for aquaculture monitoring. This chapter starts with an energy-aware algorithm for dynamic nodes that can adjust the transmission power based on the distance between the transmitter and receiver. The variability and distribution of the path loss exponent are examined in the study to understand the dynamics of environmental factors impacting it. The implementation of the suggested algorithm in a real-world setting is advised to measure the performance. The major findings, such as the node's energy consumption, power saving, and sensor data collected via ThingSpeak, are summarized before the conclusions are reached. The rest of this chapter is organized as follows. Section 6.2 reviews the state of the art of LoRa and its applications, Section 6.3 describes the related work, and Section 6.4 describes the research significance. Section 6.5 presents the energy-aware power control algorithm. Section 6.6 discusses the experimental setup and Section 6.7 discusses the results from the experimental implementation to evaluate the performance of the solutions. Finally, Section 6.8 concludes this chapter.

6.2 LoRa AND LoRaWAN APPLICATIONS

LoRaWAN is a popular low-power wide area network (LPWAN) technology that has recently received a lot of attention. A large number of static and mobile end devices are envisaged to be connected using LoRaWAN. LoRaWAN takes advantage of the limited radio bandwidth of the industrial, scientific, and medical (ISM) band, which is also shared by numerous co-existing technologies. Apart from the issue of limited radio resources, end nodes are frequently placed in remote locations having harsh radio environments, resulting in substantial variability in the quality of connection due to a variety of factors such as blockages, device mobility, and environmental factors. LoRa is a promising technology that is capable of communicating across a distance of approximately 10 km in a line-of-sight environment [14]. The architecture of LoRaWAN is shown in Figure 6.1. The endpoints, which can be sensors, connect to the gateway through a radio frequency (RF) interface for LoRa/LoRaWAN. Through a non-LoRaWAN network, such as Ethernet, 3G/4G, or Wi-Fi, the gateway sends frames to the server.

The LoRaWAN communication stack is shown in Figure 6.2. The LoRaWAN specification defines three classes that are accessible for various power usage techniques in consideration of application requirements. Figure 6.3 depicts these classes, which can be concisely defined as follows. In class A, based on the requirements of sensors, it can start an uplink transmission. With this class, communication is possible in both directions; each uplink transmission is followed by two quick downlink messages. The power usage of class A is the lowest. In class B, ping slots are transmitted by the gateway to start the connection so the end device can receive more windows at predetermined fixed time intervals. For synchronization, the gateway must periodically send out a beacon. This class uses a medium amount of power. In class C, the receive windows on end devices are essentially always open and can only be closed during transmission. Class C endpoints

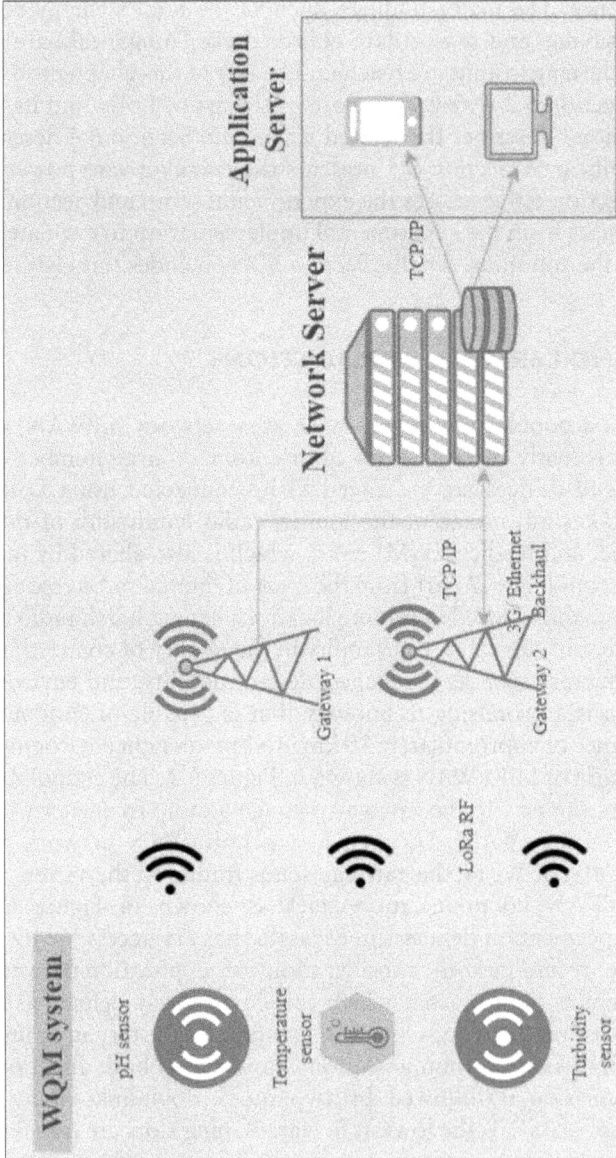

Figure 6.1 The architecture of LoRaWAN.

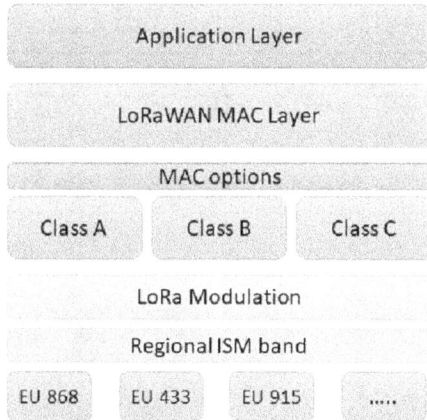

Figure 6.2 LoRaWAN communication stack.

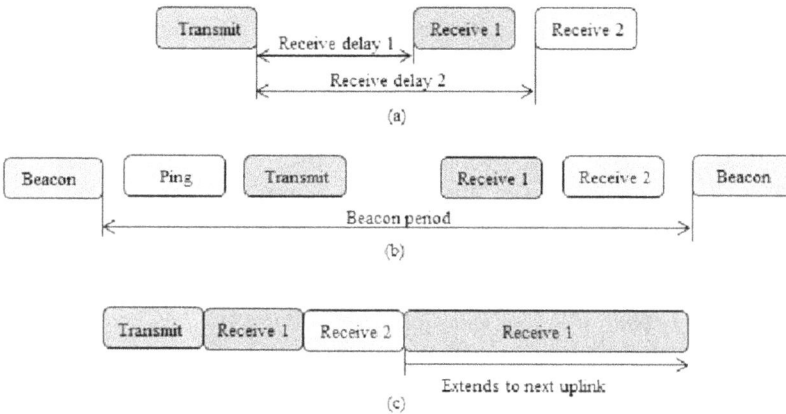

Figure 6.3 Different LoRaWAN classes: (a) class A; (b) class B; (c) class C.

have lower communication latency between servers and endpoints than class A or B, but they require more power to operate.

The carrier frequency, spreading factor (SF), bandwidth (BW), and coding rate (CR) are some of the configuration parameters for the LoRa radio. These parameters work together to produce various energy values and transmission ranges. The carrier frequency is the center frequency used for the transmission which can take values of 433, 868, and 915 MHz depending upon the region of operation. The number of chips per symbol is the spreading factor.

Its value is an integer between 6 and 12. The receiver's capacity to separate noise from signal increases with the spreading factor value. The transmission band's frequency range is represented by the bandwidth. There are

Table 6.1 The LoRa parameters influencing energy

	Parameter	Values
1	SF	6, 7, 8, 9, 10, 11, 12
2	CR	4/5, 4/6, 4/7, 4/8
3	Frequency of operation	433/868/915 MHz
4	BW	125/250/500 kHz
5	Transmission power	2–20 dBm

just three possibilities available: 125, 250, or 500 kHz. A higher value is preferable if a quick transmission is required. Table 6.1 displays the parameters and their values.

6.3 RELATED WORK

A floating device is used in an application [15] to monitor environmental data including the pH of water, environmental temperature, and humidity. As the data changes slowly, the system may function at a modest data rate, making LoRa a better fit for such applications. The sensor nodes collect data from the environment and deliver it to a LoRa gateway located far away. The gateway is made up of a radio module with an antenna and a data processor module that is connected to the cloud for real-time monitoring [16]. The industrial application of LoRa nodes for detecting the anomaly in production, battery replacement, and damage penalty is significant. The low power consumption of the modules and sleep modes helps to reduce the overall consumption and thereby increases the lifetime of the battery by the selection of an optimal sensing interval in various settings. The battery life significantly improves when the sensing period increases from 1 to 5 minutes. The application of LoRa in the health care system helps people [17] with dementia, where they are tracked using energy-efficient LoRa nodes linked with GPS to prevent them from roaming. This allows dementia patients to be more flexible and safe, while also giving carers ease of mind. LoRa-based monitoring systems are used to track boats in maritime scenarios. This system is more useful for small sailboats, recreational boats, or radio control ships where energy restriction is a concern [18]. The safety of laborers in dangerous working environments is critical as measures are required to safeguard and avoid accidents in working sites such as mines and gas plants. LoRa-based systems are widely used for safety applications and water quality monitoring systems as they can provide reliable and real-time data [19]. The LoRa-based WQM devices have a wide range of applications in irrigation, aquaculture, and water distribution systems. An energy-efficient monitoring system for river water contamination utilizing deep neural networks and LoRa connectivity is suggested in some literature

[20] where game theory is employed to choose the spreading factor and minimize energy usage. In [21], a nitrate monitoring device with an energy harvesting module is installed in a water body for WQM. A comparison of LoRa to other common communication technologies indicates that it extends battery life by switching to sleep mode when there is no transmission. The majority of past research in this field is only applicable to static nodes. Dynamic nodes have received very little attention [22]. For optimizing the energy of the battery, some studies use methods based on path loss estimation, and transmission parameters are selected accordingly. In the case of a cellular network, it is necessary to exchange channel state information, user traffic, and so on. However, LoRa has duty cycle constraints that will not permit frequent exchange of data [14]. Taking into account all these challenges, an energy-aware algorithm is proposed in this work which ensures a minimum energy consumption.

6.4 RESEARCH SIGNIFICANCE

Besides LoRa's built-in low-power advancement and sleep modes, energy-saving strategies are required for an extended battery life. Several factors contribute to the drain of energy such as data processing, transmitting power, and receiving windows. Amongst transmitting power is the key energy spender of the end node, according to some previous literature [12, 23, 24].

Considering the state of the art, the adaptive data rate (ADR) is a technique used in static nodes where the LoRaWAN network controls the data rate and transmission power with respect to the successive signal-to-noise ratio (SNR) of the received signal packets. ADR is not ideal for dynamic nodes as the signal parameters change as the node moves. In this study, the node is a dynamic, GPS-enabled, floating system. The GPS helps to track the location of the floating buoy and reduces the power consumption of the system by finding the distance between the node and gateway.

The energy-aware algorithm provides a node-centric strategy for lowering the transmitting power by utilizing a log-normal shadowing model. The end node's transmitting power adjusts itself in response to its distance from the gateway. Consequently, as the nodes get closer together, the transmission power will decrease. In this study, we introduce a novel technique for configuring the best transmission power settings in mobile nodes to achieve the required communication reliability. Based on the distance between the moving node's present location and the closest gateway, the suggested technique chooses the ideal configuration for each transmission dynamically. Table 6.2 shows the different applications of WQM in static and mobile devices. Only two applications are dynamic, according to the table, and the majority of the research is focused on static nodes. The energy efficiency of these mobile nodes is not addressed in these works, which shows the importance of this area.

Table 6.2 The application of WQM in static and dynamic nodes

Technology	Static node	Mobile node	Energy efficient	Reference
WiFi	✓	✗	✗	[25]
Bluetooth	✓	✗	✗	[26]
Cellular	✓	✗	✗	[27]
LoRa	✓	✗	✗	[21]
ZigBee	✓	✗	✗	[28]
Satellite	✗	✓	✗	[29]
GSM and GPS	✗	✓	✗	[30]
Present study	✗	✓	✓	

6.5 ENERGY AWARE POWER CONTROL ALGORITHM (EAPC)

The requisites of the algorithm include a suitable radio propagation model, path loss exponent, and various other channel characteristics which may affect the data transmission. Environmental clutter causes random variations in path loss as a result of object obstruction in the propagation path or changes in reflecting and scattering items. As a result, path loss can vary significantly between places with the same transmitter and receiver distance. In such situations, statistical models should be used to reflect the non-deterministic properties of path loss [31]. The log-normal shadowing model is most commonly used for describing non-deterministic effects. The path loss in dB is expressed as:

$$PL = -K(dB) + 10\gamma \log_{10}\left[\frac{D}{D_0}\right] + X_\sigma \qquad (6.1)$$

D refers to the distance between the transmitter and the gateway, and D_0 is used as the reference distance. PL stands for path loss. The path loss exponent γ, which changes depending on the environment, the average channel attenuation K, and the zero mean Gaussian distributed random variable X_σ are all defined.

$$T_p = P_r - K(dB) + 10\gamma \log_{10}\left[\frac{d}{d_0}\right] + X_\sigma \qquad (6.2)$$

where T_p and P_r are the transmitting and receiving power.

Maximum coupling loss, or *MCL* in wireless communication, is defined as the maximum total channel loss between the transmitter and receiver antenna ports at which data service may still be delivered. *MCL* should be greater than *PL* for a successful transmission [32].

MCL is expressed as:

$$MCL(dB) = T_p(dBm) - \left(-174(dBm) + BW(dBHz) + CNIR(dB)\right) \quad (6.3)$$

The transmitting power is T_p, while the carrier-to-noise and interference ratio is $CNIR$.

6.5.1 Variability analysis of path loss exponent

The variability in the path loss exponent of the environment under consideration is another factor that determines the validity of the system. The assumptions about the real path loss mechanisms have a major impact on the accuracy of an empirical model. Variability in environmental conditions affects the path loss exponent, which varies the distance over which the minimum power can be delivered, as with the EAPC algorithm's shadowing model. The approximate path loss exponent is determined using a sample size of 35 observations, as shown in Table 6.3. The histogram of the observation is shown in Figure 6.4. The average of the observations of the path loss exponent is 3.2. The major goal of variability analysis is to find

Table 6.3 The statistics of the path loss exponent

Statistics	Data
Sample size	35
Mean	3.2
Variance	0.05
Standard deviation	0.2
Coefficient of variance	0.06

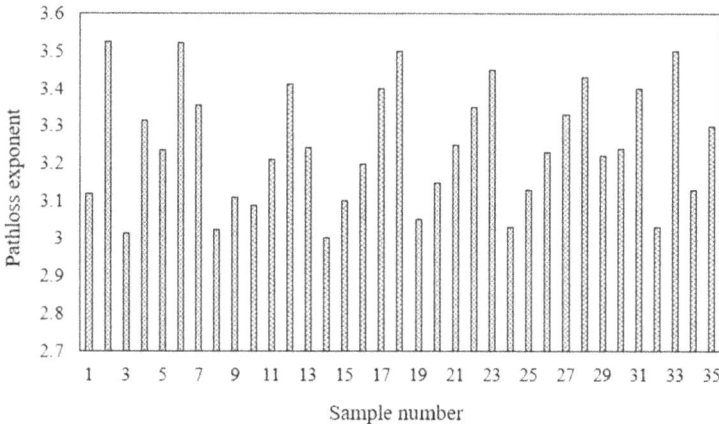

Figure 6.4 The path loss exponent for a sample size of 35 observations.

a probabilistic model that can account for the fluctuation in the path loss exponent. The two popular goodness of fit tests used for the dataset are Kolmogorov-Smirnov (K-S) [33] and Anderson-Darling (A-D) [34]. These tests are used to determine whether a sample belongs to a population with a particular distribution. A measure of the difference between the observed and expected cumulative distribution functions is determined by comparison to a threshold value in the K-S test. The sample size and significance level are used in determining the threshold value. The two goodness-of-fit tests provide the distribution parameter, rank, and statistical values of each path loss exponent distribution. Johnson SB distribution ranks first in both K-S and A-D tests and is the best fit for simulating path loss exponent variability.

The probability distribution function (PDF) and the cumulative distribution function (CDF) of the Johnson SB distribution are described as:

$$f(x) = \frac{\delta\lambda}{\sqrt{2\pi}(x-\eta)(\eta+\lambda-x)} \times \exp\left\{-\frac{1}{2}\left[\alpha + \delta\ln\left(\frac{x-\eta}{\eta+\lambda-x}\right)\right]^2\right\} \quad (6.4)$$

$$F(x) = \Delta\left(\alpha + \delta\ln\left(\frac{x-\eta}{\eta+\lambda-x}\right)\right) \quad (6.5)$$

The continuous shape parameters are α and δ, the continuous scale parameter is λ, and the continuous location parameter is η. The Laplace integral is denoted as Δ. In the present study x is the path loss exponent, γ. The effectiveness of the Johnson SB distribution to represent the path loss exponent probability distribution can be demonstrated in a variety of ways. The PDF plot is the most typical representation of variability. Figure 6.5 shows a PDF plot of the path loss exponent versus the Johnson SB distribution. The fit of the Johnson SB distribution to the observed data is illustrated in this figure. The cumulative distribution function, $F(\gamma)$ verses the path loss exponent, γ, is

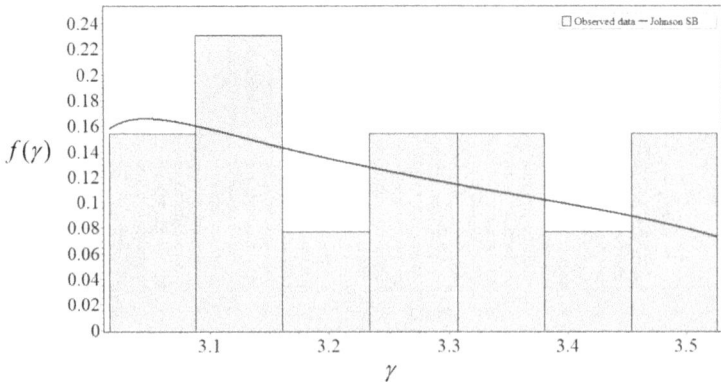

Figure 6.5 The probability distribution function of the path loss exponent.

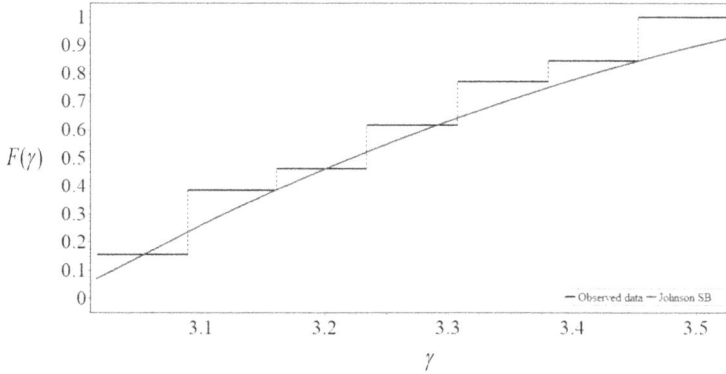

Figure 6.6 The cumulative distribution function of the path loss exponent within the sample size.

shown in Figure 6.6. While deploying in a real environment, the variability analysis gives an idea of its unpredictability and corresponding distribution. The value of γ obtained in the calculation is 3.2 which is used in the algorithm for further analysis.

6.5.2 Algorithm

The EAPC algorithm has a parameter that affects the energy consumption of the system, the transmitting power (T_p). The objective of the algorithm is to vary the transmitting power according to the distance of the end node from the gateway. To establish whether the connection has been successfully established, the measured path loss is compared to the maximum coupling loss. The SF is set to 12 and the transmitting power varies from 2 to 20 dBm. The minimum transmission power is represented as $T_{p\,min}$ and the maximum transmission power as $T_{p\,max}$. The proposed algorithm is described as:

ALGORITHM: EAPC

```
Input: γ, T_p min, T_p max
Output: T_p
Start: Retrieve 'D' from the transmitter node
Calculate MCL and P L
if MCL > P L
Connection ←True
then Find T_p from distance-transmit power relationship
if T_p ≤ T_p min
do T_p = T_p min
else if T_p min < T_p < T_p max
do T_p = T_p
```

```
else if T_p ≥ T_p max
do  T_p = T_p max
else Connection ←False
return NULL
Stop
```

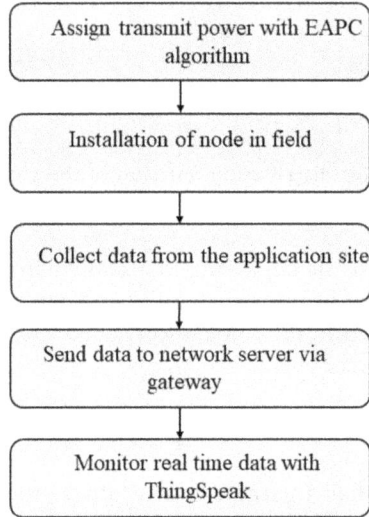

Figure 6.7 The flowchart of aquaculture field application.

The system before implementation in the field is first assigned with transmission power according to the EAPC algorithm. The node is then installed in the field to collect data. The data collected from the aquaculture site is then transmitted to the network server through a gateway. The real-time data collected from the sensors are monitored through the ThingSpeak application server. The flowchart of the system is shown in Figure 6.7.

6.6 EXPERIMENTAL SETUP

WQM in aquaculture is critical because any anomaly in the properties of water in the production line can affect aquaculture species survival, procreation, growth, and development, causing an environmental impact or compromising product quality and safety. Effective WQM strategies help in the production of high-quality aquaculture products on a large scale. The increased use of fertilizers and manures on agricultural lands and waste deposition of industries result in water pollution which greatly affects the health of aquatic bodies [35]. Some of the problems are the acidic nature of water, increased ion concentration, turbidity, and depletion of dissolved oxygen. Treatments such as aeration, water circulation, and exchange can be used to combat these issues on a timely basis. The real-time monitoring of the water bodies is made possible by an energy-efficient WQM system with the help of a set of sensors. The floating WQM system proposed in the study helps to check the anomalies in the characteristics of water which might risk the health of aquatic beings. Deployment of WSNs in the field has another challenge: limited battery life. An energy-efficient algorithm can minimize energy consumption and improve the lifetime of the battery.

The algorithm is implemented in the aquaculture WQM system to check the quality of water in an energy-efficient way. It is significant to check the pH, turbidity, temperature, and other parameters of water for the safety of aquatic bodies. The experimental setup was carried out in the reservoir of the National Institute of Technology Rourkela. The gateway is placed at the water station as shown in Figure 6.8 which is connected to the back-haul network to transmit the data. The terminal node with sensors is mobile and floats on the reservoir as shown in Figure 6.9. The sensors at the end node collect timely information from the water body and transmit data to the gateway. The spreading factor of the LoRa end node is specifically set to 12 because the reservoir spans over 1 km. As a result, the node may reliably transmit data across long distances with minimal losses. A GPS hat is mounted

Figure 6.8 Field test: (a) location of the reservoir; (b) location of the gateway.

Figure 6.9 LoRa node: (a) end node with sensors; (b) dynamic end node deployed in field.

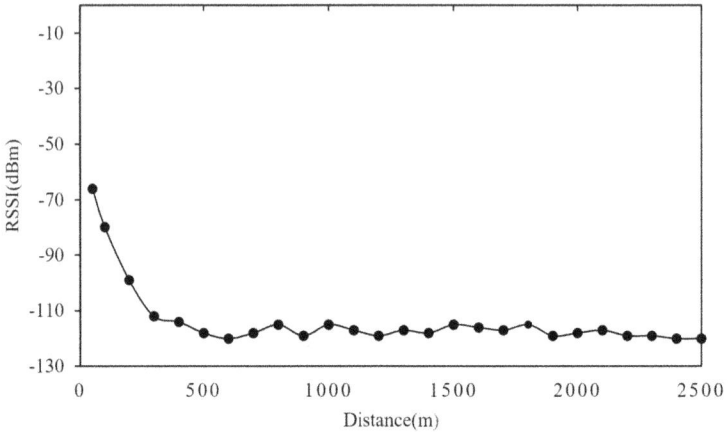

Figure 6.10 The variation of RSSI as the distance increases from 500 to 2,500 m.

over the LoRa module for finding the location of the end node. The gateway is fixed such that the GPS coordinates of the gateway will be known. The location coordinate of the gateway is fed to the end node such that the distance can be calculated using the current location. The system consists of an analog pH, turbidity, and temperature sensor which are initially tested and calibrated in the laboratory. This is then deployed in the field for taking real-time values of water parameters. The received signal strength indicator (RSSI) and SNR values of the received signal are taken to analyze the signal strength at various locations. When the node is closer to the gateway, the signal is strong and the RSSI value becomes –65 dBm as shown in Figure 6.10. The RSSI value decreases to a minimum of 120 dBm as the distance increases.

6.7 RESULTS AND DISCUSSION

This section discusses the variation of transmission power after implementing the EAPC algorithm. The variation of power allocation of the LoRa node with varying distances from the gateway is studied in the experiment. As shown in Figure 6.11 the transmission power increases linearly with distance. In practical applications of LoRa devices, the default power is taken as either 14 or 17 dBm. The present study considers 17 dBm as the default which is compared with the EAPC algorithm to find the improvement in energy consumption. The purpose of the EAPC algorithm is to minimize energy wastage while transmitting to a nearby area that takes less energy than to the farthest location. In the experiment, three locations, P, Q, and R, are considered for power consumption and analysis. The transmitting power at the location changes adaptively according to the estimated path loss relationship. At location P, the transmitting power sets to 2 dBm with an instantaneous power consumption of 565 mW. In location Q, the transmitting

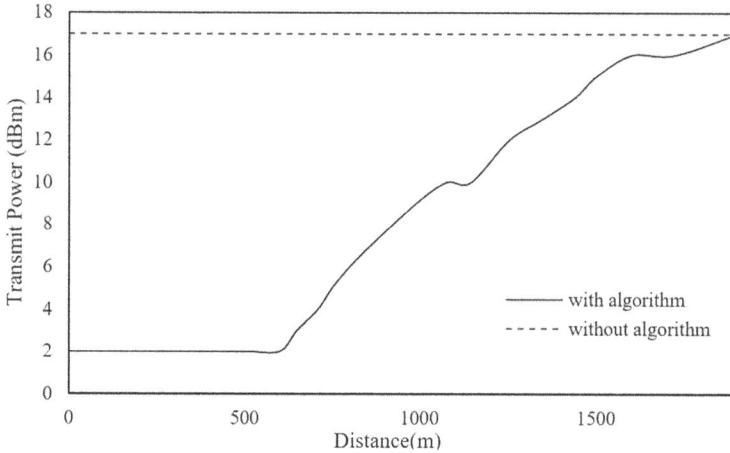

Figure 6.11 The variation of transmitting power with distance.

power changes adaptively to the distance as 10 dBm for a distance of 1,250 m. The power consumption changes to 700 mW, saving power compared to the default setting (930 mW). At location R, the transmitting power increases to 14 dBm with a power consumption of 785 mW. The data analysis shows that the EAPC algorithm reduces energy while sending to a closer area as compared to the farthest. The power consumption of the end node on increasing distance is shown in Figure 6.11.

When compared to the default power mode, it can be seen that site P, which is close to the gateway, uses less power, while site Q consumes 700 mW, and site R consumes more compared to others. As illustrated in Figure 6.12, the percentage of power savings is calculated at each location.

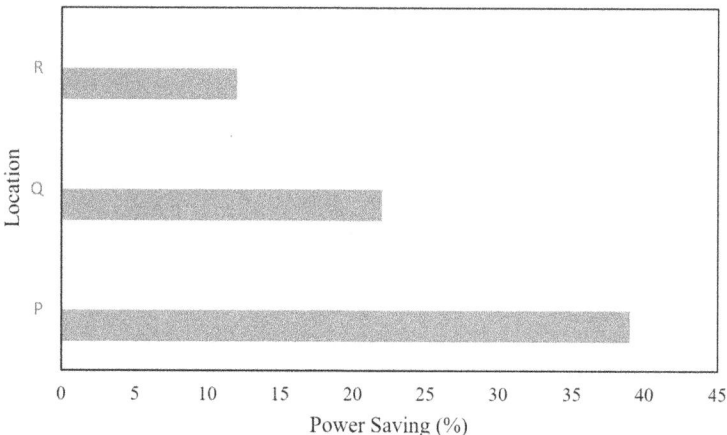

Figure 6.12 The percentage saving of energy at locations P, Q, and R.

(a)

(b)

(c)

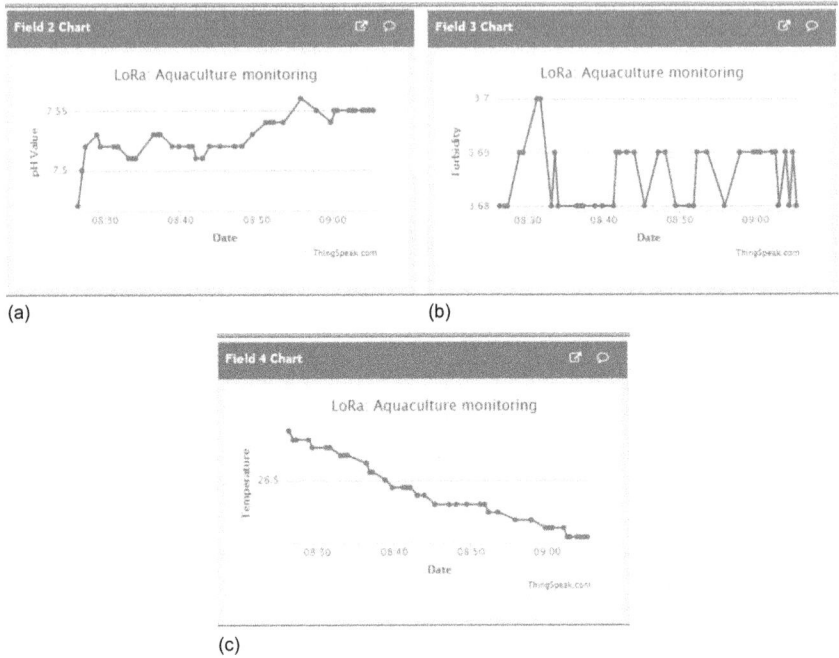

Figure 6.13 Sensor data on ThingSpeak: (a) pH; (b) turbidity; (c) temperature.

The power savings for sites P, Q, and R is 39, 22, and 11% respectively. The power consumption at different locations shows that power increases linearly after location P. The consumption will gradually reach a maximum of 930 mW as the transmitting power reaches 17 dBm such that the power saving decreases with distance. To assess the suitability of water for aquaculture, an energy-efficient WQM system is deployed in the reservoir. The main purpose of the algorithm is to check the quality of water on a real-time basis and ensure that it is fit for aquaculture. The sensors used to serve the purpose are pH, turbidity, and temperature. To improve the system, more sensors can be incorporated according to the application. In the study, the sensors are deployed in the field in a floating system that senses the pH, turbidity, and temperature of the water body. The pH value of the water is sensed as 7.5, the temperature as 28°C, and turbidity as 3.66 NTU as shown in Figure 6.13. All these water quality parameters are within the limits such that the water is suitable for aquaculture.

6.8 CONCLUSION

In wireless sensor systems, energy-efficient techniques are crucial. An EAPC algorithm has been proposed to reduce the energy consumption of the dynamic end nodes for WQM systems. This system enables the user to monitor the characteristics of water such as pH, turbidity, and temperature to assess water quality suitable for aquaculture applications. The proposed algorithm uses a log-normal shadowing model and selects appropriate transmitting power based on the distance of the end node from the gateway. A statistical analysis of the variability of the path loss exponent was performed in this approach, and the most appropriate distribution was determined to be Johnson SB. The EAPC algorithm when implemented on a dynamic node closer to the gateway has an energy saving of 39%. This node-centric approach is very beneficial for dynamic end nodes like floating devices or buoys in the water body, commodity tracking, and electric vehicle fleet management.

REFERENCES

1. O. Postolache, J. D. Pereira, P. S. Girão, Wireless sensor network-based solution for environmental monitoring: Water quality assessment case study, *IET Science, Measurement & Technology* 8 (6) (2014) 610–616.
2. S. O. Olatinwo, T.-H. Joubert, Enabling communication networks for water quality monitoring applications: A survey, *IEEE Access* 7 (2019) 100332–100362.
3. S. T. John, B. K. Roy, P. Sarkar, R. Davis, IoT enabled real-time monitoring system for early-age compressive strength of concrete, *Journal of Construction Engineering and Management* 146 (2) (2020) 05019020.
4. S. T. John, A. Mohan, M. S. Philip, P. Sarkar, R. Davis, An IoT device for striking of vertical concrete formwork, *Engineering, Construction and Architectural Management* (ahead-of-print) (2021).
5. F. K. Shaikh, S. Zeadally, E. Exposito, Enabling technologies for green internet of things, *IEEE Systems Journal* 11 (2) (2015) 983–994.
6. C. Zhu, V. C. Leung, L. Shu, E. C.-H. Ngai, Green internet of things for smart world, *IEEE Access* 3 (2015) 2151–2162.
7. S. T. John, P. Sarkar, R. Davis, Energy-efficient long range wide area network for construction industry applications, *Automation in Construction* 136 (2022) 104150.
8. S. T. John, P. Sarkar, R. Davis, A long-range wide-area network system for monitoring early-age concrete compressive strength, *Journal of Construction Engineering and Management* 149 (1) (2023) 04022148.
9. S. Namasudra, G. C. Deka, P. Johri, M. Hosseinpour, A. H. Gandomi, The revolution of blockchain: State-of-the-art and research challenges, *Archives of Computational Methods in Engineering* 28 (3) (2021) 1497–1515.
10. S. O. Olatinwo, T.-H. Joubert, Energy efficient solutions in wireless sensor systems for water quality monitoring: A review, *IEEE Sensors Journal* 19 (5) (2018) 1596–1625.
11. M. Aref, A. Sikora, Free space range measurements with semtech LoRa™ technology, in: 2014 *2nd International Symposium on Wireless Systems within the Conferences on Intelligent Data Acquisition and Advanced Computing Systems*, IEEE, 2014, pp. 19–23.
12. M. S. Philip, P. Singh, Energy consumption evaluation of LoRa sensor nodes in wireless sensor network, in: *2021 Advanced Communication Technologies and Signal Processing (ACTS)*, IEEE, 2021, pp. 1–4.

13. T. Bouguera, J.-F. Diouris, J.-J. Chaillout, R. Jaouadi, G. Andrieux, Energy consumption model for sensor nodes based on LoRa and LoRaWAN, *Sensors* 18 (7) (2018) 2104.

14. J. C. Liando, A. Gamage, A. W. Tengourtius, M. Li, Known and unknown facts of LoRa: Experiences from a large-scale measurement study, *ACM Transactions on Sensor Networks (TOSN)* 15 (2) (2019) 1–35.

15. W.-K. Lee, M. J. Schubert, B.-Y. Ooi, S. J.-Q. Ho, Multi-source energy harvesting and storage for floating wireless sensor network nodes with long range communication capability, *IEEE Transactions on Industry Applications* 54 (3) (2018) 2606–2615.

16. H. H. R. Sherazi, M. A. Imran, G. Boggia, L. A. Grieco, Energy harvesting in LoRaWAN: A cost analysis for the industry 4.0, *IEEE Communications Letters* 22 (11) (2018) 2358–2361.

17. T. Hadwen, V. Smallbon, Q. Zhang, M. D'Souza, Energy efficient LoRa GPS tracker for dementia patients, in: *2017 39th annual international conference of the IEEE engineering in medicine and biology society (EMBC)*, IEEE, 2017, pp. 771–774.

18. R. Sanchez-Iborra, I. G. Liaño, C. Simoes, E. Couñago, A. F. Skarmeta, Tracking and monitoring system based on LoRa technology for lightweight boats, *Electronics* 8 (1) (2019) 15.

19. M. S. Philip, P. Singh, Adaptive transmit power control algorithm for dynamic LoRa nodes in water quality monitoring system, *Sustainable Computing: Informatics and Systems* 32 (2021) 100613.

20. S. Chopade, H. P. Gupta, R. Mishra, P. Kumari, T. Dutta, An energy efficient river water pollution monitoring system in internet of things, *IEEE Transactions on Green Communications and Networking* 5 (2) (2021) 693–702.

21. M. E. E. Alahi, N. Pereira-Ishak, S. C. Mukhopadhyay, L. Burkitt, An internet-of-things enabled smart sensing system for nitrate monitoring, *IEEE Internet of Things Journal* 5 (6) (2018) 4409–4417.

22. A. Gupta, M. Fujinami, Battery optimal configuration of transmission settings in LoRa moving nodes, in: *2019 16th IEEE Annual Consumer Communications & Networking Conference (CCNC)*, IEEE, 2019, pp. 1–6.

23. G. Callebaut, G. Leenders, et al., Long range iot connections: Experimental confirmation of the energy drain and exploration of known escape routes, in: *Proceedings of the 2018 Symposium on Information Theory and Signal Processing in the Benelux, Werkgemeenschap voor Informatie-en Communicatietheorie (WIC)*, 2018.

24. M. S. Philip, P. Singh, An energy efficient algorithm for sustainable monitoring of water quality in smart cities, *Sustainable Computing: Informatics and Systems* (2022) 100768.

25. D. Nguyen, P. H. Phung, A reliable and efficient wireless sensor network system for water quality monitoring, in: *2017 International Conference on Intelligent Environments (IE)*, IEEE, 2017, pp. 84–91.

26. A. Faustine, A. N. Mvuma, et al., Ubiquitous mobile sensing for water quality monitoring and reporting within lake victoria basin, *Wireless Sensor Network* 6 (12) (2014) 257.

27. M. Carminati, A. Turolla, L. Mezzera, M. Di Mauro, M. Tizzoni, G. Pani, F. Zanetto, J. Foschi, M. Antonelli, A self-powered wireless water quality sensing network enabling smart monitoring of biological and chemical stability in supply systems, *Sensors* 20 (4) (2020) 1125.
28. A. M. Ilie, C. Vaccaro, J. Rogeiro, T. E. Leitao, T. Martins, Configuration, and implementation of 3 smart water network wireless sensor nodes for assessing the water quality, in: *2017 IEEE SmartWorld, Ubiquitous Intelligence & Computing, Advanced & Trusted Computed, Scalable Computing & Communications, Cloud & Big Data Computing, Internet of People and Smart City Innovation*, IEEE, 2017, pp. 1–8.
29. C. Hu, B. B. Barnes, B. Murch, P. R. Carlson, Satellite-based virtual buoy system to monitor coastal water quality, *Optical Engineering* 53 (5) (2013) 051402.
30. A. M. A. Helmi, M. M. Hafiz, M. S. Rizam, Mobile buoy for real time monitoring and assessment of water quality, in: *2014 IEEE Conference on Systems, Process and Control (ICSPC 2014)*, IEEE, 2014, pp. 19–23.
31. S. Kurt, B. Tavli, Path-loss modeling for wireless sensor networks: A review of models and comparative evaluations., *IEEE Antennas and Propagation Magazine* 59 (1) (2017) 18–37.
32. K. Staniec, M. Kucharzak, Z. Joskiewicz, B. Chowanski, Measurement based investigations of the nb-iot uplink performance at boundary propagation conditions, *Electronics* 9 (11) (2020) 1947.
33. X. Jin, T. W. Chow, Y. Sun, J. Shan, B. C. Lau, Kuiper test and autoregressive model-based approach for wireless sensor network fault diagnosis, *Wireless Networks* 21 (3) (2015) 829–839.
34. R. D'Agostino, M. Stephens, *Goodness-of-fit techniques*. New York: Marcel Dekker (1986).
35. R. D. Zweig, J. D. Morton, M. M. Stewart, *Source water quality for aquaculture: a guide for assessment*, The World Bank, 1999.

Chapter 7

Harmonic distortions in smart devices

A comprehensive survey from conventional to future smart IoT devices

Uttam Kumar Gupta and Dinesh Sethi
JECRC University, Jaipur, India

Pankaj Kumar Goswami
Teerthanker Mahaveer University, Moradabad, India

Garima Goswami
Teerthanker Mahaveer University, Moradabad, India

Kumar Gautam
Gwangju Institute of Science and Technology, Republic of Korea

CONTENTS

7.1 INTRODUCTION

The harmonic distortions in the mains supply due to the nonlinear performance dynamics of electronic components have been a global challenge for all commercial and noncommercial applications. The generation of higher-order harmonics is caused by many factors such as heterogeneous load characteristics, solid-state electronic devices, switching modules, variable drives, and many advanced smart sensor networks. The presence of higher-order harmonics is often observed in the power converter units of most electronic component-based devices [1]. Even household applications such as televisions, mobile phones, air conditioners, refrigerators, personal computers, and microwave ovens are responsible for the generation of harmonics in the mains supply.

The main cause is the use of a power converter unit and associative electronic onboard components. The power converters are complex circuits of hybrid components such as diodes, transistors, resistors, and capacitors. The devices connected over the internet are the biggest source generation of harmonic distortion in the mains supply. The newly added technological feature of the Internet of Things (IoT) is popular in commercial and industrial applications. Figure 7.1 shows conventional commercial and noncommercial industry applications which are in high demand for a mitigation of harmonic current. Active power filter (APF) compensation is the most popular scheme and various evolutions of it are the main focus of the researcher in various application environments.

Table 7.1 shows the major effects of serious harmonic distortion in smart devices used under the technical operation of the IoT, which is directly affected by harmonics at the device end as well as in the network signaling section. The main motive of harmonic distortion analysis is to identify the diversified causes of the depreciation of power quality and improvement techniques under different load conditions and application environments. Due to the nonlinear characteristics of the active electronic components, the higher-order harmonics of fundamental current are generated. This causes

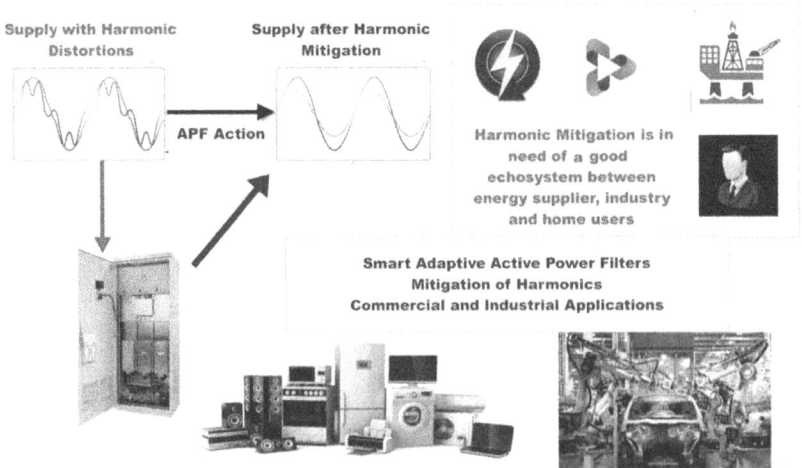

Figure 7.1 APF compensation utility for commercial and industrial applications.

Table 7.1 Effect of harmonic hazards

Issue	Smart systems	Effect of harmonic distortion
Harmonic generation	Devices	• Biomedical instrument malfunctioning • Recording instrument takes false measures • Networked device interference
	Signaling	• Signaling interference • High voltage flicking • Electromagnetic interference and glitches

interference to the fundamental frequency of the main current component and generates higher-order frequencies of fundamental current components. Nowadays, the evolution of smart devices comprises advanced features related to their monitoring and control from remote locations. The device advancement is the outcome of system embedding as an auxiliary interface to the main process. This will lead to the conclusion that in the next two decades 80% of systems and equipment will be connected to the network for remote sensing and controlling. Additionally, machine-to-machine (M2M) or a machine-to-network (M2N) communication involves the soft computing technologies of the main system. The IoT devices are the best examples among smart systems that elicit the application of all M2M and M2N smart systems. As the IoT devices comprise the soft computing mechanism in smart switching components, this advancement in technologies causes the harmonic current generation. Thus, the components-based load of power electronics exhibits highly non-linear voltage-current characteristics. The nonlinear V-I characteristics of smart electronic devices are a major source of the generation of higher-order harmonics. The presence of higher-order harmonic current components causes malfunctioning of the systems connected to the mains supply. This results in several severe issues including false readings, malfunctioning of devices, flickers, RF interference, and low efficiency. Hysteresis current compensation (HCC) was one of the major contributions to the mitigation of harmonics. The scheme deals with harmonic distortion caused by the nonlinear characteristics of IoT devices. The novel contribution of this survey is to analyze the most effective techniques to reduce harmonic generation in power management units of smart IoT devices. It also explores the new machine learning (ML) approaches to deal with the uncertainty of load variation on a real-time basis. The survey helps to identify, analyze, and minimize the harmonics generation issues in smart devices. This survey is organized into three major sections: (1) the recent causes of harmonic distortions in the power management unit of smart devices; (2) a state of art review of traditional to smart compensation techniques; and (3) process effectiveness and the existing scope of research, followed by the conclusion.

7.2 CAUSES AND EFFECTS OF TOTAL HARMONIC DISTORTION IN THE POWER MANAGEMENT UNIT (PMU) OF SMART DEVICES

The National Institute of Standards and Technology (NIST) for IoT devices releases information on the attributes of connected devices over the internet. The basic features in IoT systems are sensor-based devices, aggregator action, a networking and communication channel, and application-based e-utility [1]. The technology and device resources are the elements of the IoT types that are self-made. The IoT has different classes of connections between the heterogeneous hybrids for controlling and monitoring. Electronic devices are the main components of an embedded system in most IoT applications. According to studies, the active components in electric power tools are sensitive to harmonic interference in electrical systems. Finally, electrical power possesses many unwanted quality issues in electrical distribution systems [2]. The nonlinear V-I graph of semiconductor electronics components was the main source of complexity in major IoT devices [3, 4]. The strategy behind switching to that was shunted active power filters (SAPFs) which are very popular for eliminating harmonics in the current [5]. SAPF performance was observed and developed by the smart balance of the current reference and harmonic current components. Various current reference ideas have been reported, such as applicable functional theory, production-based templates and symmetrical framework, and concordance of the reference compensation violation [6–8]. The quality of power and reliability of the supply are the crucial elements for the consistent operation of industry and home automation systems. The rapid reaction time of the maneuvers creates a call for a proficient and intelligent control system to foil unpleasant PQ issues [9]. The basic internal structure of IoT devices and the corresponding requirement of the PMU is presented in Figure 7.2.

This is an illustrative approach that defines that the advanced PMUs include soft computing as an essential component. This deals with fault protection, voltage conversion, monitoring, serial interfacing, and controlling operations smoothly. A defective power supply significantly affects the system and results in insufficient mobility for the efficiency of smart power converters. As per IEC 61000-4-7 [10], static reactions and their variations such as harmonic deviation are allowed to exceed the allowable limit [11–13]. The PQ risk analysis in smart devices is shown in Figure 7.3 for various risk attributes. The main constituent entities of a smart IoT system, such as aggregators, sensors, transports, and user interface, experience, aristocratic risk, security risk, reliability risk, and the risk of layer interconnection due to the presence of unwanted current harmonics.

This section has revealed the impact and effect of harmonic distortion in traditional as well as smart devices. The study shows that not only the PQ is a major concern but the security at the sensor interface or reliability in the network connection is equally crucial in smart devices. In one such case, the

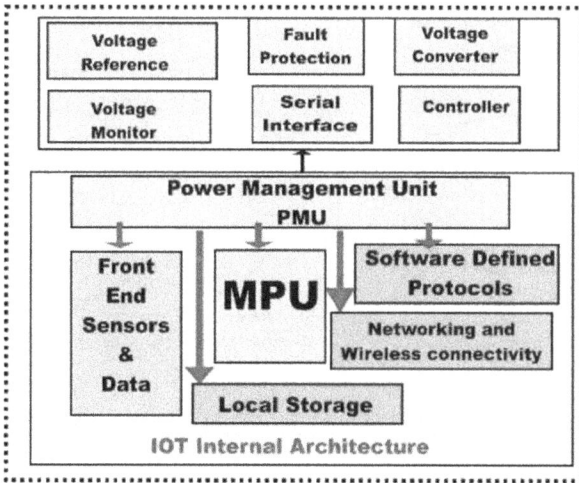

Figure 7.2 PMU architecture in smart devices [16].

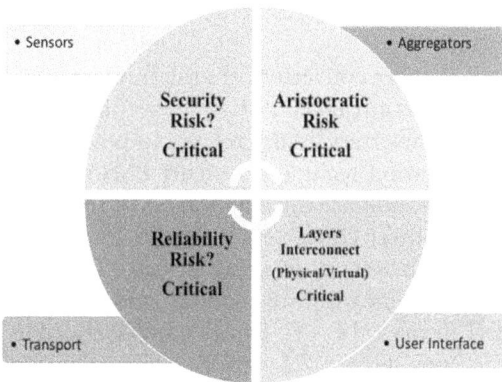

Figure 7.3 PQ risk analysis in smart devices.

smart IoT devices used for industrial control may be deviated from their usual precise operation of sensitive equipment due to the presence of harmonics at the layer of wireless network interconnections.

7.3 COMPREHENSIVE CASE STUDIES ON TOTAL HARMONIC DISTORTION (THD) ANALYSIS IN TRADITIONAL AND SMART IoT SYSTEMS

This section includes state-of-the-art analysis to determine the causes of PQ and effective compensation techniques. Conceptually, the power converter units are the main and essential circuitry associated with the PMU of smart

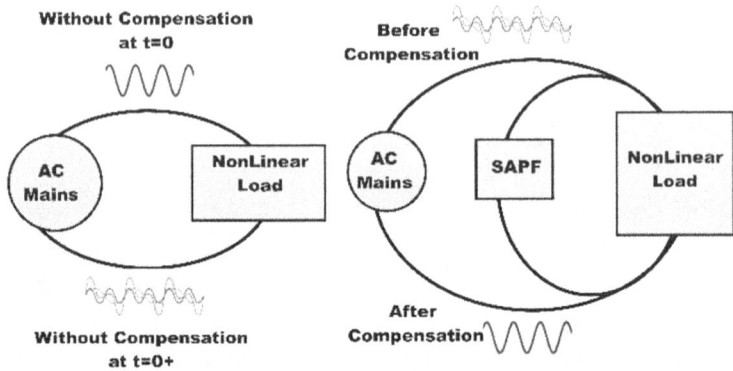

Figure 7.4 APF association and compensation model.

IoT devices. From the fundamental architecture of smart IoT devices, power converters are the main operational unit for a suitable energy feed to the constituent units of the system. Figure 7.4 shows the effect of the insertion of harmonics at $t = 0+$ and the effect of the basic compensation scheme after settling time.

The conventional power converters are the fundamental building blocks for soft computing-based advanced PMUs. Therefore, the analytical study of conventional power converters, issues related to power converters, conventional harmonic mitigation techniques, smart PMUs, and soft computing-based current compensation are the major areas of research. The generation of harmonics due to power converters produces several reasons to analyze the performance of such systems in smart devices. Therefore, any deviation in quality output can significantly affect device performance and the network connectivity of IoT devices [14, 15]. Before moving ahead in the direction of state-of-the-art analysis, Figure 7.5 depicts a summary of the evolution of techniques and impact projection up to the present.

This summary shows the comprehensive research done, the gaps, and existing problem domains of PQ issues. It was observed in [16] that the %THD in line currents before compensation was 78% with a load power factor of 0.783, and 125% with a power factor of 0.592 for load A and for load B respectively. After compensation/filtration by the Active Power Line Capacitor/Peripheral Interface Controller the %THD was reduced to 25% with the load power factor improved to 0.98, and 30% with the load power factor improved to 0.93 for load A and for load B respectively. In [17] the authors proposed the use of Takagi–Sugeno (TS), a controller to improve energy quality and compensation for workloads that require a nonlinear load. The use of the Mamdani controller in a three-phase functional filter was previously investigated but had a limited number of incomprehensible sets and had to increase the coefficient degree for improving the performance of the controller. The current hysteresis operating mode was used for

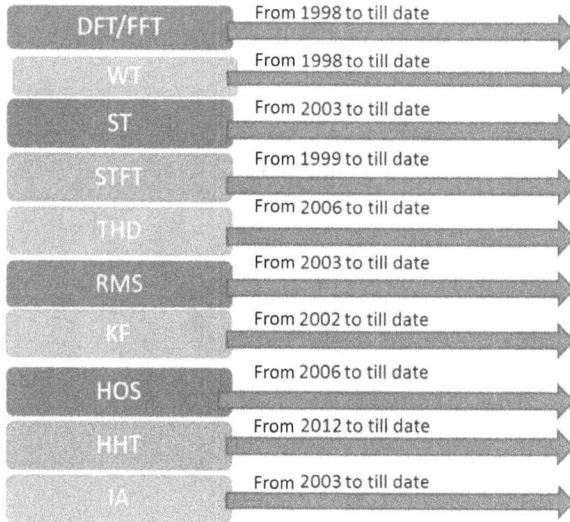

DFT/FFT	From 1998 to till date
WT	From 1998 to till date
ST	From 2003 to till date
STFT	From 1999 to till date
THD	From 2006 to till date
RMS	From 2003 to till date
KF	From 2002 to till date
HOS	From 2006 to till date
HHT	From 2012 to till date
IA	From 2003 to till date

Figure 7.5 Technology evolution and areas of improvement [38].

the production of pulse-wide signal transition. The computer simulation results show that the performance of the TS controller appeared more effective than the standard proportional integral (PI). If there are any variations found in our system, the critical TS controller appears as a strong harmonic current compensator. Complete harmonic distortion included in the current line after filtration was obtained below the acceptable limit of 5%. It was better to use TS fuzzy logic control instead of Mamdani for the non-compliant control type as the former requires only 4 rules, 2 sets, and 5 coefficients for doing well as compared to 49 rules, 7 sets not included, and 17 coefficients prepared for Mamdani's mindless mind control [17]. A shunt active power filter based on P-I, P-I-D, and FLC was proposed to be implemented for conditioning the power line and to improving the harmonic generation of the distribution system. The main current or line current which needs compensation for harmonics and reactive power was sensed by appropriate sensors and used for the extraction of the reference current extraction. Figure 7.6 shows the work of researchers in three major domains of harmonic mitigation.

The experimental real-time signal leads the survey with 47.3% over another two classes of theoretical review and simulation-based modeling. Real-time signaling and synthetic signals both are the preferred choice of researchers to deal with harmonic mitigation under various nonlinear load conditions. The three controllers proportional-integral, proportional-integral-derivative, and FLC were used to assess the peak value of the reference current. This is performed through error signal control, generated via a comparison of the DC voltages of shunt-connected converters. The gate

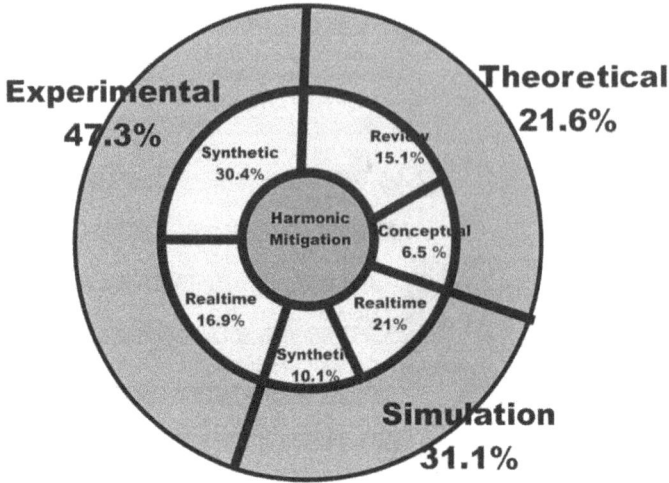

Figure 7.6 Research focus on three major domains [38].

pulse for the shunt-connected multilevel inverter was generated with the help of the hysteresis current controller. This provided a satisfactory dynamic response under all conditions, whether steady-state or transient operations. The system evolved with a shunt-connected active power filter, used for various nonlinear loads under steady-state and transient conditions. The results obtained from simulation models depict the perfect performance of the SAPF in conjunction with PID, PI, and FLC. The source current without APF has a THD in the steady-state of 25.38% and in the transient condition of 25.32% including a power factor of 0.8772. The source current with an APF and PI controller has a THD of 2.61% in the steady-state and 2.59% in the transient condition with a 0.9733 power factor. The line current with an APF and PID controller consisted of a THD of 2.58% in the steady-state and 2.59% in the transient condition with a 0.9721 power factor. The line current with an APF and fuzzy logic controller experienced a THD of 2.53% in the steady-state, and 2.48% in the transient condition with a 0.9829 power factor [18]. In 2014, the authors of [19] suggested a boost converter-based approach for current control which was utilized for integrating the ability to mitigate the system harmonics with the help of the power generation function of the primary direct current (DC) set. In this method of harmonic compensation, not only the harmonic voltage of the distribution system or the harmonic current of the nonlinear load as the fundamental components, but also the harmonic components of the DCcurrent are controlled independently through the proposed current controller which has two decoupled control branches [19]. Additionally, a power control scheme having closed-loop control was employed to derive the fundamental value of the reference current directly. The proposed scheme for power control successfully abolished the influence of the errors generated in the current tracking of DG units. In such a manner, precise power control has been recognized even when the functions for harmonic compensation were actuated. Moreover, this work also discussed briefly how the proposed method performs satisfactorily when the DG unit has been coupled to a grid, even with variations in system frequency. The correctness of the proposed methodology was validated by results obtained from the simulation and experiment of a 1-phase DG unit. A distributed generation unit (DG unit) was used to supply a nonlinear load and reference current extraction was executed by using a digital controller based on a phase-locked loop and a conventional integral controller. It was observed that initially without compensation the DG unit current was polluted with 201.5% THD while after compensation it became 5.99% THD with the grid current having 3.64% THD. The THD of node1, node2, and the power over coaxial (POC) voltage are 11.46, 14.09, and 17.03% [19].

The distributed power generating systems be used to control the operation of biomass/biogas diesel engine derived synchronous reluctance generators (SYRGs) [20]. The SYRG was referred to as a primary source to supply various nonlinear and linear loads. An adaptive control-scheme-based neural network controller was used here to mitigate the harmonic distortions and to improve the voltage profile of the SYRG, which was provided with a battery to store energy, by generating an appropriate gate pulse for the switching device used in a DSTATCOM (distribution static compensator). To get the actual values of the real and reactive power components of the nonsinusoidal or distorted line current the control algorithm was used. The reference value of the source current was extracted by using the active, reactive, and harmonic components of line current to generate the gate trigger for a bridge converter implemented in VSI used as a shunt compensator (DSTATCOM) for active power filtration. It was observed that the FACT device DSTATCOM performed satisfactorily for unbalanced and balanced load conditions of the generating unit. The performance of SYRG with linear loads was observed and it was found that 6.7 and 1.7% that the generator current and power control centre The reactive power demand by the nonlinear load was provided by DSTATCOM by which the THD in the line current and load current were reduced to on an average 4.5 and 22.5% respectively [4, 21]. A flexible power conversion control system that acts as an effective compensator for power supply connectors was a revolutionary move. Each current component was measured with compensation constants, adjusted immediately and independently at any percentage using load-alignment devices, thus providing online flexibility about compensation purposes and active force injections. The results of the simulation and testing were submitted to validate the efficiency of the process. In the first case, the regulatory body was designed to compensate for all the disruptions; the power disturbance was 9.5%. The current wavelength of the grid was maintained as before the harmonic step. The idea was proposed of designing a single compensator to reduce the THD and suggested one more compensator to balance the reactive power demand by the nonlinear load. A transformerless hybrid multilevel for the harmonic generation improvement of a 1-phase domestic home appliance was also proposed. The proposed methodology reflects the new movements of consumers in the direction of polluting electronic loads and a combination of renewable energy sources by which a sustainable and reliable power supply can be achieved. This work contributed to harmonic generation improvement for a contemporary single-phase distribution system. A separate energy source provided a supply to the compensating converter which reduces the need for a bulky transformer. This gives a good solution to the problem of harmonics, voltage variation, power factors, and harmonic generation without significant losses in bulky transformers [22, 23].

A harmonic reduction system was reported with a star-delta multifunction balance transformer and a 3-phase bridge converter. This work completely explored the natural −ve current thus reducing the capability of star-delta MFBT, because of which the flow of power distorted by the conventional compensating system (transformer-based) was much more than that of the proposed passive filter aided by shunt compensators based on the conservative power theory system. Additionally, because the YD-MFBT consists of 3-phase comparatively low voltage level output ports, it was possible to connect the full-bridge converter directly to the main transformer through these out-put ports exclusive of the need for an auxiliary transformer. In this work, the transforming relationship between the current and the multifunction balance transformer (MFBT)-power quality compensating system (PQCS) was implemented and the recognition and control methods have been given also. The efficacy of the proposed work has been verified by both the experimental and simulation results. MFBT-PQCS started operation; the primary currents of YD-MFBT were changed from unbalanced and distorted 3-phase waveforms to the sinusoidal ones (because some harmonic currents from other adjacent nonlinear loads flow into the primary side in our experimental period. A 3-phase power filter driven by a single-step non-iterative algorithm that operates under unbalanced supply conditions is also very popular. Unlike previous methods, this algorithm was simpler and faster as it does not include more complex methods of use (such as Newton–Raphson and quadratic sequence systems), which make them work better under heavy load conditions. The performance of the algorithm was evaluated in comparison to a control algorithm based on efficiency and verified using a real-time system. The APF system achieves a new state of stability within a single cycle without affecting the APF compensation between the two states of up and down of the load. Also discussed in the literature is the fact that the IoT system enables greater rapidity, direct control, and distant access for social and business uses. Easy access to distant controls reverses the problem of system ambiguity due to the existence of integrated loading signals. Converters based on high electronic performance and nonlinear loads produced a few unpleasant harmonic generation issues [22, 24–28].

The various operational detections and harmonic distortion examinations were particularly important for smart devices. This work analyzes the merits of a hybrid load system on devices based on smart IoT to identify energy excellence issues and the effective application of a flexible independent learning method to reduce harmonic distortion. The combination of FLC and APF has solved the problems related to power quality to some extent. A switched-mode APLC can be implemented effectively which utilizes fuzzy logic to gate triggering switching devices. The results obtained from simulations and experiments were matched at a perfect scale which shows that the fuzzy-proportional-integral controller (FPIC) can improve harmonic generation in a more intended manner than the conventional proportional-integral controller by mitigating the harmonics in a line current within the range of 5% and improving the power factor in transient and steady-state conditions [29–32]. This survey deals with the flexible computer system called the adaptive neural fuzzy interface system (ANFIS) which has both ANN and FLC benefits. It was an integrated conversion system for measuring the IO functions of the current reference. The ambiguous IO logic sections introducing dynamic relationships between NGOs and ANN help to create and refine unambiguous rules depending on the assets of the training algorithm. The model suggests ANFIS has learned to control offline loading interference on smart IoT devices. The current hysteresis controls are used in the production of the current reference to create an SAPF to reduce harmonic distortions in the distribution. The test setup confirms the abnormality data set for NN training and a flexible control strategy. The neuron analysis weight was adopted using different study layers and reduced the THD to a significant assessment from 72.58 to 0.81%. This ensures maximum performance of the control plan according to IEEE 519 standards [33–36].

This APF model was highly compatible with improving the performance and protection of device modules based on the non-linear component of various applications. As per [37, 38], Table 7.2 indicates harmonic flickers in the test devices of power converters. Figure 7.7 shows the analytical surveys on various parameters for the specific load conditions; the impact on the current waveform is observed. It is highly indicative that various nonlinear loads severely affect the mains supply and need to be mitigated for the malfunctioning of devices. The issues were dealt with and adequate APF has been suggested by many of the researchers and the wave shaping is presented for the widespread integration of electronic devices into modern systems, which enhances the scale of small-scale work with automation in AC

Table 7.2 Harmonic flickers and disturbances [37, 38]

Disturbance	Class of disturbance		Duration	Range	
				Minimum	Maximum
Voltage	Voltage (average)		10 min	0.85 unit	1.1 unit
parameters	Flickers		–	–	7%
	Sag	Duration short	10 msec^{-1} sec	0.1u	0.9u
		Duration long	1 sec^{-1} min	0.1u	0.9u
		Duration long	>1 min	0.1u	0.9u
	Under-voltage	Short	<3 min	0.99u	0.99u
		Long	> 3 min	0.99u	0.99u
		Temp short	10 ms^{-1} s	1.1u	1.5 kv
		Temp long	1 s^{-1} min	1.1u	1.5 kv
		Over voltage	<10 ms	1.1u	6 kv
Frequency of operation	Deviation: slight		10 sec	49.5 Hz	50.5 Hz
	Deviation: critical		10 s	47.0 Hz	52.0 Hz
Harmonics	Harmonics			THD > 8%	

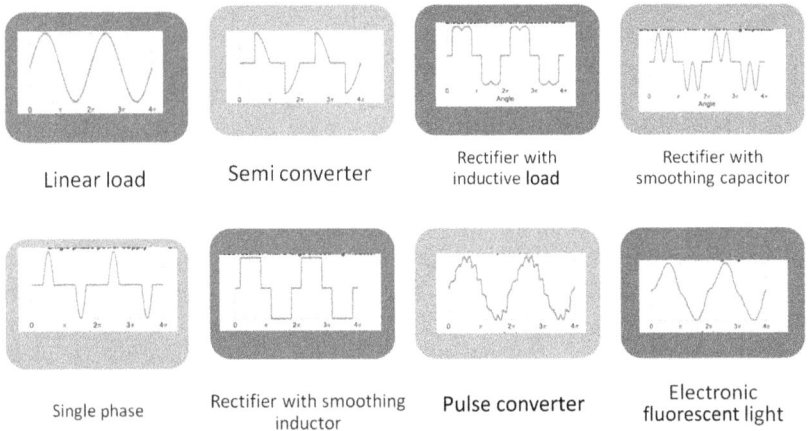

Figure 7.7 Effect analysis of a few common existing nonlinear loads [37].

and electric-powered devices. Smart homes and industries form offline load components on a large scale. In addition, the inconsistency of such materials results in significant harmonic distortions, operational deviations, and a high risk of device failure [22].

The compensation signal switching and control are monitored by the ANFIS to supervise the PID action and optimize the standard shunt APF action. A state-of-the-art consistent response of soft computing algorithms was considered to determine the feasibility of a real-time intervention program. This offers an algorithm to control THD in an offline loading system with the importance of reducing THD without diverting the system's complexity. The filtering process reduced the THD significantly from 92.23 to 0.49% with a 0.99 power factor for line-of-line load parameters [25]. Goswami and Goswami [17] noted that with the great advancement of technology in electrical equipment, there was a significant increase in nonlinear loads, such as switch mode power supply, devices used in telecommunications systems, home appliances, fast-moving drives, and so on. These types of loads cause a decrease in efficiency and energy quality. Repeated results are the generation of harmonics of many orders. An efficient coupling converter was designed for a significant reduction in harmonics and thus the overall quality of the rotating power. The main control strategy was implemented through the artificial neural network improvement process and the specific PID controller compensation strategies. To avoid system losses, an unused low-power filter was designed. By modeling a nonlinear system below normal hazard levels, the reduction of harmonics was reached at a level of 0.21% with the adjustment of the saturating power to 0.99. Results are validated according to IEEE519 standards.

7.4 PROCESS EFFECTIVENESS AND THD REDUCTION TECHNIQUE UTILITY

Based on a comprehensive survey and analytical studies of THD issues in various commercial and noncommercial applications, the process effectiveness and the utility of the methodology adopted for the reduction of harmonics are shown in Table 7.3. Interestingly, a hybrid combination of fault effects and control signaling produces a large set permutation of observation analysis. Pulse width modulation to various filters performs adequately well with the listed limitations which define the process effectiveness. The parallel processes challenge event analysis; and synthetic or real-time signal stimuli are assessed on a hybrid or a single-mode fault operation.

The real vs synthetic signals are the two main classes of the harmonic mitigation process in a conventional single event or advanced complex event problems. This makes a sequence of combinations and the degree of effectiveness to be based on the complexity of the event. Figure 7.8 shows analytics from 1998 to the present, where the focus on hybrid events has been soaring continuously and smart system complexities are the prominent cause of the parallel events [38]. This exhibits the end probability of occurrence of a single event as harmonics and associative parallel events sag, flicker, swell, notch, spike, and so on as combining factors. For a better

Table 7.3 Harmonic flickers and disturbances

Technique	Effect analysis
Use of PWM	The use of PWM to control harmonic generation performs gate triggering, but its effectiveness in THD reduction has limitations.
DSTATCOM	Harmonic reduction by DSTATCOM appears effective but total power quality improvement is a challenge.
Functional filters	High-end utility in distribution networks for harmonic reduction but increase the system complexities and expense.
Optimization algorithm	The optimization devises the strategies to be used for the iterative complex stimuli of harmonic distortion. This includes soft computing techniques.
LCL filters	The high volume noise is effectively reduced with the use of the LCL filters.
Power factor improvement	A major domain is emphasized; and effective improvement is a crucial associative attribute.
Air loss reductions	Very less emphasis is observed, but a few researchers have focussed on the sequential effects of air loss over harmonic generation for PQ improvement.

understanding and validation of the research area, an active harmonic filter (AHF) of the YIY company has been taken into consideration.

Figure 7.9 shows the AHF model with specifications for four classes of voltage 220/400/590/600 for a wide range of harmonic mitigation from 2nd to 50th at switching frequency 12.8 kHz. This represents the real-time application of harmonic filters; and technical specification reveals the gap under the occurrence of parallel events in advanced operations of smart devices. Figures 7.10 and 7.11 are included for understanding real-time signal synthesis from the power grid to smart home applications. The YIY database for AHF performance exhibits THD mitigation up to the level of <5%. The adequate compensation harmonic currents in the PMU of smart devices are underestimated and eventually stimuli uncertainty needs a large focus on AHF. More attention is required for complex operations in smart devices, where not only the device performance is a major concern but also the device interconnects are equally crucial.

The survey analysis reveals that synthetic single signals with hybrid event attributes are used more frequently. However, many faults related to harmonic distortion are a single event, but the combination of disturbances is equally anticipated. The event complexities are increasing continuously, therefore synthetic signal analysis and combination of fault events are the major areas of research work. More than 50% of research work carries the effect of harmonics as a signal disturbance over the other derivative issues (Table 7.4).

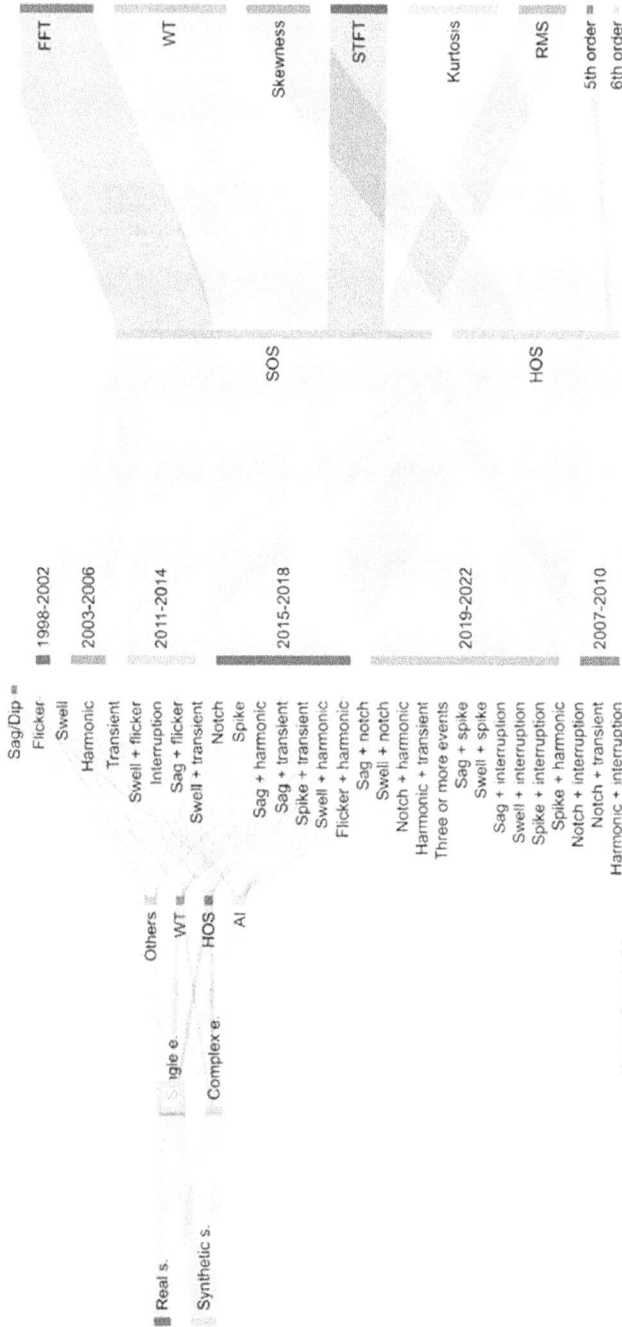

Figure 7.8 Signal vs fault occurrences for the period 1998–2022 [38].

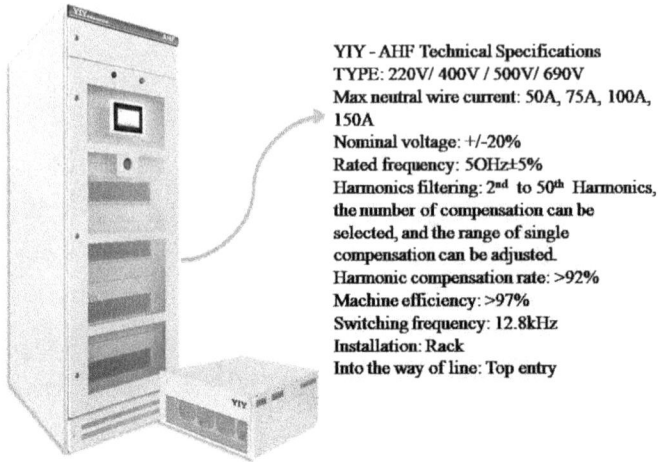

YIY - AHF Technical Specifications
TYPE: 220V/ 400V / 500V/ 690V
Max neutral wire current: 50A, 75A, 100A, 150A
Nominal voltage: +/-20%
Rated frequency: 5OHz±5%
Harmonics filtering: 2ⁿᵈ to 50ᵗʰ Harmonics, the number of compensation can be selected, and the range of single compensation can be adjusted.
Harmonic compensation rate: >92%
Machine efficiency: >97%
Switching frequency: 12.8kHz
Installation: Rack
Into the way of line: Top entry

Figure 7.9 YIY-active harmonic filter (AHF) with technical specification [37].

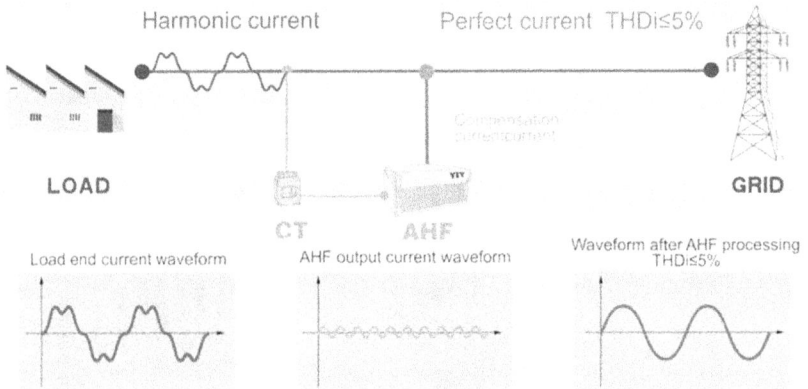

Figure 7.10 AHF implementation for harmonic current compensation [37].

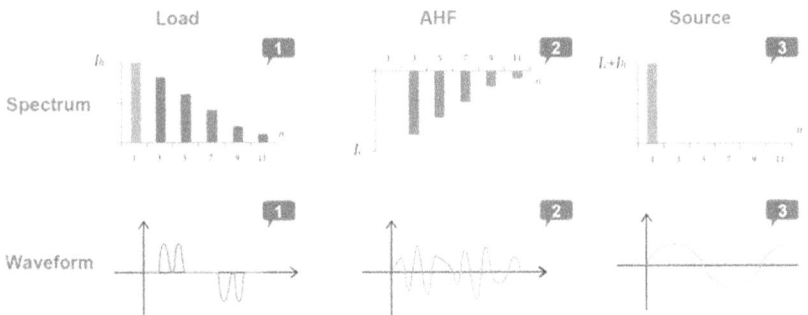

Figure 7.11 AHF current compensation [37].

Table 7.4 Harmonic flickers and disturbances

Reference	Major focus	Techniques used	Outcomes
Goswami et al. [16]	Harmonic issues in smart IoT	Self-adaptive converters	Smart IoT systems identify energy excellence issues and the effective application of a flexible independent learning method to reduce harmonic distortion.
Goswami et al. [7]	Commercial and noncommercial nonlinear loads	ANFIS supervised SAPF	THD is an offline loading system for reducing THD without diverting the system's complexity.
Goswami et al. [17]	Nonlinear industrial application	Artificial neural network	The main control strategy is implemented through the artificial neural network improvement process and the specific PID controller compensation strategies.
Han et al. [34]	Smart IoT device challenges	Self-adaptive learning-based controller	RMSE and MSE 0.15 and 3.77 for residential and commercial datasets respectively.
Mayer et al. [33]	Experimental results	ANFIS supervised PID controlled SAPF	Good efficiency above 85% and adequate control.
Ababneh et al. [32, 36]	Conventional PMU designs	ANFIS supervised PID controlled SAPF	The PMU performance optimization is executed using a buck-boost DC-DC converter; PSO implementation.
Badoni et al. [24]	3-phase distribution system	DSTATCOM	The line current by the nonlinear load in the absence of DSTATCOM was 26.95% which was reduced to 2.44% after using the proposed methodology of shunt active power filter.

7.5 CONCLUSION

This survey has shown that several articles were published by the year 2021 to discuss and mitigate higher-order current harmonics due to nonlinear loads. The descriptive details have provided a complete perspective of harmonic distortion issues in future technologies and existing practices. It has been observed that much less emphasis has been placed on the detection and mitigation of harmonics in smart devices. An adequate review analysis of harmonic generation issues to predict the future hazards of IoT devices was made. A statistical overview of harmonic hazards on sensitive components of IoT devices like sensors and actuators was carried out, as was an overview of the design and implementation of the adequate adaptive controlled APF for the HCC in the PMU of IoT devices. This survey concluded with the identification of PQ challenges in traditional and smart electrical and electronics systems. The major attributes are the nonlinear load systems and smart techniques used for machine-to-machine communication. Therefore, smart devices would be a great cause of the depreciation of PQ and the soaring use of smart systems would be a matter of great concern for PQ measures. Total harmonic distortion threshold minimization and soft computing compensation techniques are the next level of research. Machine learning and adaptive training models for APF control need the attention of PQ improvement analyzers.

ABBREVIATIONS

AHF   Active harmonic filter
AI   Artificial intelligence
ANFIS   Adaptive neural fuzzy interface system
ANN   Artificial neural network
APF   Active power filter
DFT   Discrete Fourier transform
FFT   Fast Fourier transform
FLC   Fuzzy logic controller
HCC   Hysteresis current control
HHT   Hilbert-Huang transform
HOS   Higher-order statistics
IoT   Internet of Things
KF   Kalman filter
M2M   Machine to machine
ML   Machine learning
MPU   Micro processor unit
PMU   Power management unit
PQ   Power quality
RMS   Root mean square
SAPF   Shunt active power filter
SSA   Singular spectrum analysis
ST   Transform
STFT   Short time Fourier transform
THD   Total harmonic distortion
VSI   Voltage source inverter
WT   Wavelet transform

DATA AVAILABILITY STATEMENT

Data sharing is not applicable to this article as no new data were created or analyzed in this study.

REFERENCES

[1] Voas J, "NIST Special Publication 800-183.," *Networks of 'Things'*, 2016. Available online: https://csrc.nist.gov/publications/detail/sp/800-183/final

[2] Kewat S, Singh B, "Modified amplitude adaptive control algorithm for power quality improvement in multiple distributed generation system," *IET Power Electronics*, vol. 12 wass. 9, pp. 2321–2329, 2019.

[3] Samuel VJ, Keerthi G, Prabhakar M, "Ultra-high gain DC-DC converter based on interleaved quadratic boost converter with ripple-free input current" *International Transactions on Electrical Energy Systems*, vol. 30, wassue 11, September 2020, https://doi.org/10.1002/2050-7038.12622

[4] Bonaldo JP, Paredes HKM, Pomilio JA, "Control of single-phase power converters connected to low-voltage distorted power systems with variable compensation objectives," *IEEE Transactions on Power Electronics*, vol. 31, no. 3, pp. 2039–2052, 2016.

[5] Busarello TDC, Pomilio JA, Simões MG, "Passive filter aided by shunt compensators based on the conservative power theory," *IEEE Transactions on Industry Applications*, vol. 52, no. 4, pp. 3340–3347, 2016.

[6] Sijia H, Li Y, Xie B, Chen M, Zhang Z, Luo L, Cao Y, Kubis A, Rehtanz C, "A Y-D multi-function balance transformer based harmonic reduction system for single-phase power supply system," *IEEE Transactions on Industry Applications*, vol. 52, no. 2, pp. 1270–1279, 2015.

[7] Goswami G, Goswami P. K "ANFIS Supervised PID Controlled SAPF for Harmonic Current Compensation at Nonlinear Loads", *IETE Journal of Research*, June 2020. https://doi.org/10.1080/03772063.2020.1770134

[8] Singh Y, Hussain I, Mishra S, Singh B, "Adaptive neuron detection-based control of single-phase SPV grid integrated system with active filtering," *IET Power Electronics*, Vol. 10wass. 6, pp. 657–666, 2017.

[9] Agrawal S, Palwalia DK, Kumar M, "Performance analysis of ANN based three-phase four-wire shunt active power filter for harmonic mitigation under distorted supply voltage conditions," *IETE Journal of Research*, 2019, https://doi.org/10.1080/03772063.2019.1617198

[10] International Electrotechnical Commission, IEC 61000-4-7: 2002+A1: 2008, Electromagnetic Compatibility (EMC). Part 4-7: Testing and Measurement Techniques—General Guide on Harmonics and Inter Harmonics Measurements and Instrumentation, for Power Supply Systems and Equipment Connected Thereto; IEC: Geneva, Switzerland, 2018.

[11] Blair SM, Booth CD, Williamson G, Poralis A, Turnham V, "Automatically detecting and correcting errors in harmonic generation monitoring data," *IEEE Transactions on Power Delivery*, vol. 32, no. 2, pp. 1005–1013, 2017.

[12] Dehnavi E, Afsharnia S, Gholami K, "Optimal allocation of unified power quality conditioner in the smart distribution grids," *Electrical Engineering (springer)*, November 2019, https://doi.org/10.1007/s00202-019-00861-2

[13] Balamurugan R, Nithya R, "FC/PV Fed SAF with Fuzzy Logic Control for Harmonic generation Enhancement," *International Journal of Power Electronics and Drive System (IJPEDS)*, vol. 5, no. 4, April 2015, pp. 470–476.

[14] Singh B, Pal S, Shrivastava A, "A universal input PFC CSC converter in low power consumer lighting applications", *IETE Technical Review*, 7 Aug 2019, https://doi.org/10.1080/02564602.2019.1645621.G.

[15] Karuppanan P, Mahapatra K, "PI, PID and fuzzy logic controlled cascaded voltage source inverter based active filter for power line conditioners," *WSEAS Transactions on Power Systems*, vol. 6, no. 4, pp. 100–109, 2011.

[16] Goswami G, Goswami PK, "Self-adaptive learning based controller to mitigate PQ issues in internet of things devices," *International Transactions on Electrical Energy Systems*, e12888, 2021, https://doi.org/10.1002/2050-7038.

[17] Goswami G, Goswami PK, "Power quality improvement at nonlinear loads using transformer-less shunt APF with ANFIS supervised PID controllers", *International Transactions on Electrical Energy Systems*, 2020, https://doi.org/10.1002/2050-7038.12415.

[18] Kanjiya P, Khadkikar V, Zeineldin HH, "A non iterative optimized algorithm for shunt active power filter under distorted and unbalanced supply voltages," *IEEE Transactions on Industrial Electronics*, vol. 60, no. 12, pp. 5376–5390, 2013.

[19] Bhende CN, Mishra S, "TS-fuzzy controlled active power filter for load compensation," *IEEE Transactions on Power Delivery*, vol. 21, no. 3, pp. 1459–1465, 2006.

[20] Jinwei HX, Li YW, "Active harmonic filtering using current-controlled, power control," *IEEE Transactions on Power Electronics*, vol. 29, no. 2, pp. 642–653, 2014.

[21] Arya SR, Niwas R, Bhalla KK, Singh B, Chandra A, Al-Haddad K, "Power quality improvement inwasolated distributed power generating system using DSTATCOM," *IEEE Transactions on Industry Applications*, vol. 51, no. 6, pp. 4766–4774, 2015.

[22] Alireza Javadi A, Hamadi A, Ndtoungou A, Al-Haddad K, "Power quality enhancement of smart households using a multilevel-THSeAF with a PR controller," *IEEE Transactions on Smart Grid*, vol. 8, no. 1, pp. 465–474, 2016.

[23] Nikum K, Saxena R, Wagh A, "Effect on Harmonic generation by Large Penetration of Household Non Linear Load" *IEEE International conference on Power Electronics, Intelligent control and Energy Systems*, 2016.

[24] Badoni M, Singh A, Singh B, "Control Algorithm for DSTATCOM," *IEEE Transactions on Power Electronics*, vol. 30, no. 5, pp. 2353–2361, 2015.

[25] Kumar R, Bansal HO, "Real-time implementation of adaptive PV-integrated SAPF to enhance power quality," *International Transactions on Electrical Energy Systems-Wiley*, 2019, https://doi.org/10.1002/2050-7038.12004.

[26] Jayasankar LVN, Vinatha U, "Advanced control approach for shunt active power filter interfacing wind-solar hybrid renewable system to distribution grid," *Journal of Electrical Systems* vol. 14, no. 2, pp. 88–102, 2018.

[27] Nelson S, Helder R, Flavio O, Wanderley C, Daniel C, Esequie P, "A proposal to Harmonic generation improvement based on SHAPF with P-SSI controller parameters optimized by multi-objective optimization and applied through gain scheduling technique," *Electric Power Systems Research*, vol. 177, December, 2019, https://doi.org/10.1016/j.epsr.2019.105939.

[28] Patel A, Mathur HD, Bhanot S, "An improved control method for unified Harmonic generation conditioner with unbalanced load," *Electrical Power and Energy Systems*, vol. 100, pp. 129–138, 2018, https://doi.org/10.1016/j.ijepes.2018.02.035.

[29] Mishra R, Saha TK, "Modelling and analysis of distributed power generation schemes supplying unbalanced and non-linear load," *Electrical Power and Energy Systems*, vol. 119, 2020, https://doi.org/10.1016/j.ijepes.2020.105878.

[30] Zang H, Cheng L, Kwok TD, Cheung W, Wei Z, Sun G, "Day-ahead photovoltaic power forecasting approach based on deep convolutional neural networks and meta learning," *Electrical Power and Energy Systems*, June 2020, https://doi.org/10.1016/j.ijepes.2019.105790.

[31] Kumar A, Srungavarapu G, "Algorithm-based direct power control of active front-end rectifiers," *IET Power Electronics*, vol. 12wass. 4, pp. 712–718, 2019.

[32] Ababneh MM, Ugweje O, Jaesim A, Optimized Power Management Unit for IoT Applications. *2019 15th International Conference on Electronics, Computer and Computation (ICECCO)*, 2019, https://doi.org/10.1109/icecco 48375.2019.9043189.

[33] Mayer P, Magno M, Benini L, "Smart Power Unit - mW-to-nW Power Management and Control for Self-Sustainable IoT Devices," *IEEE Transactions on Power Electronics*, 1–1, 2020, https://doi.org/10.1109/tpel.2020.3031697.

[34] Han T, Muhammad K, Hussain T, Lloret J, Baik SW, "An efficient deep learning framework for intelligent energy management in IoT networks," *IEEE Internet of Things Journal*, 1–1, 2020, https://doi.org/10.1109/jiot.2020.3013306.

[35] Monedero I, Leon C, Ropero J, Garcia A, Elena JM, Montano JC, "Classification of electrical disturbances in real time using neural networks," *IEEE Transactions on Power Delivery*, vol. 22, no. 3, pp. 1288–1296, 2007.

[36] Khetarpal P, Tripathi MM, "A critical and comprehensive review on power quality disturbance detection and classification," *Sustainable Computing: Informatics and Systems*, vol. 28,100417, 2020.

[37] www.powerquality.yiyen.com

[38] Carmona PR, José J, Rosa G, Oliveros OF, Fernández JMS, et al., "Current status and future trends of power quality analysis," *Energies*, vol. 15, 2328, 2022, https://doi.org/10.3390/en15072328

Chapter 8

Design of nano- and microgrids using the HetNet switching strategy

D. Shruthi
PSN College of Engineering & Technology, Tirunelveli, India

R. Rajesh Kanna
Vellore Institute of Technology, Vellore, India

R. Rengaraj
Sri Sivasubramaniya Nadar College of Engineering, Chennai, India

R. Raja Singh
Vellore Institute of Technology, Vellore, India

CONTENTS

DOI: 10.1201/9781003436461-8

8.1 INTRODUCTION

Renewable energy resources (RERs) have become the strongest choice to satisfy the growing energy demand, resulting in a gradual decrease in fossil fuels [1]. As renewable green energy technology such as solar/wind power is becoming more prevalent, end-users may also generate some power for their own uses. This is called a distributed generating system, and it is made possible by two-way power flow technology [2]. A microgrid is a small network of electrically isolated distributed energy resources (DERs) and loads. A microgrid can plug in to or plug up from the main grid, depending on the needs of the environment and/or the economy. Both the grid-connected mode (synchronized with the grid) and the island mode are functional. Figure 8.1 depicts an overview of the typical microgrid. Within the larger smart grid's three principles, microgrids are a "new distribution network architecture, capable of more fully harnessing the full spectrum of benefits arising from the integration of several small-scale DERs into relatively low-voltage energy distribution networks", according to one description of Lawrence Berkeley National Laboratory [LBNL]) [3].

In the modern environment, microgrids have been considered to be a leading tool for managing and coordinating distributed generation units [4]. Microgrids afford different profits to the consumers and utilities, which include enhanced energy efficiency, minimized overall energy ingestion, lesser impact on the environment, and improved supply consistency. Other network operational benefits include voltage and frequency management [5], as well as providing the most cost-effective power environment. On the other hand, the nanogrid concept is gaining attention. A nanogrid follows many of the principles of microgrids with some key exceptions. A nanogrid

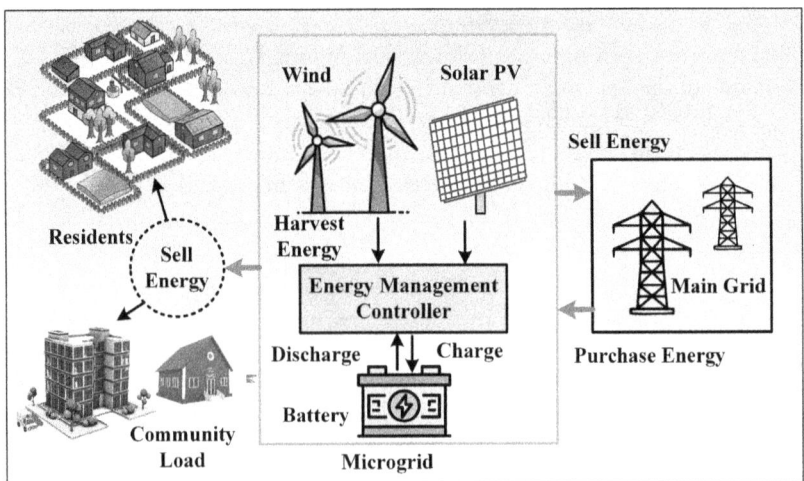

Figure 8.1 Overview of typical microgrid.

is a basic building block of microgrid architecture which integrates its energy into the community in an accessible manner [6]. In the smart grid, gateways can be used to link nano- and microgrids [7]. Each gateway controls the nano- or microgrid network, which may contain certain primary electricity sources [8].

8.1.1 Communication topology of microgrids

The interaction between microgrid nodes and the behavior of the loads and dispersed generation add to the complexity of the power system's operation and control. Hence, communications become increasingly important in the implementation of microgrids to ensure optimal operation with stability and reliability. The connectivity between the microgrid resources must be bidirectional and interoperable for the communication network of the microgrid to provide a lofty degree of dependability. A microgrid's communication network may be thought of as a connection between its physical hardware and its control and monitoring systems. Grid-connected microgrids are not always independent; instead, they interact and communicate with a variety of end users, including sub-grids in commercial, institutional, and residential contexts, as well as with the traditional electric grid. A home area network (HAN), field area network (FAN), body area network (BAN), industrial area network (IAN), wide area network (WAN), and neighbourhood area network (NAN) are the main categories used to describe microgrids. Figure 8.2 displays the communications protocols and technologies employed in these applications.

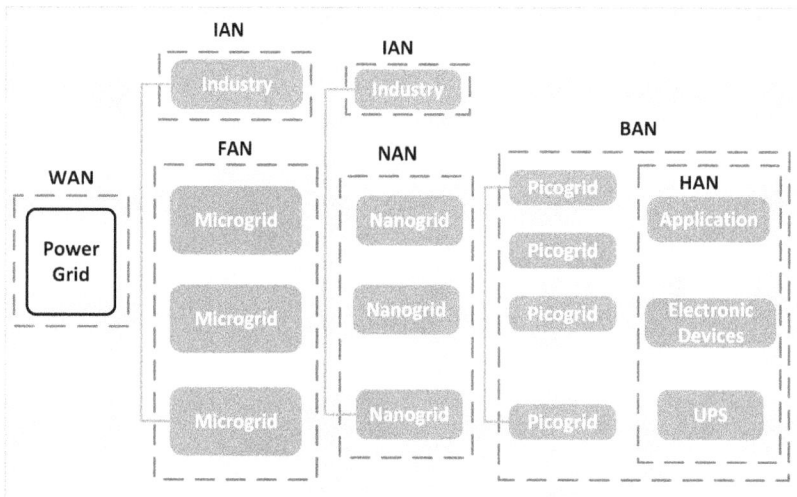

Figure 8.2 High-level communication topology of microgrids and their interfacing networks.

8.1.2 Heterogeneous network (HetNet) strategy

The HetNet strategy is about the fragmentation of sources and loads to enhance the performance, efficiency, and reliability of the grid system [9]. HetNet-based RER architecture consists of three types of layers, namely macro, metro, and femto [10]. The interactions between consecutive layers of "transfer of power" are considered during demand time. As a result, the femto layer responds to femto users' indoor needs, whereas the metro-layer responds to outside events as well as the femto layer's excess needs. The macro-layer of HetNet is solely for macro users, and the discovered metro is overspilled from the lower layer [11]. The macro-layer covers an extensive area and serves multiple users. Macro-layer power management is in the order of a few kilowatts. After careful preparation and definition of the site, a macro-distribution framework deploys the macro-layer. Because of the heavy energy handling, both capital and operating expenses would be high. The metro-layer has a lower power range than the macro. They are also installed directly by the customer in this situation, but the smart energy production system (SEPS) supplying the metro-layer has lower computational capabilities and power consumption than a macro-layer, resulting in lower capital and operating costs. The femto-layer includes low power and low demand. They have enhanced coverage and efficiency in public spaces in the central, metropolitan, and rural areas. The femto-layer can be implemented by operators with loose planning or even with absolutely no planning [12].

These cutting-edge technologies are used by the HetNet to get around the coverage issues seen in Figure 8.3. In addition, HetNet may dynamically take use of the structure of current cellular networks to close the distance between the access network and subscribers. The emergence of HetNets marks a turning point in the feasibility of the confined area usage

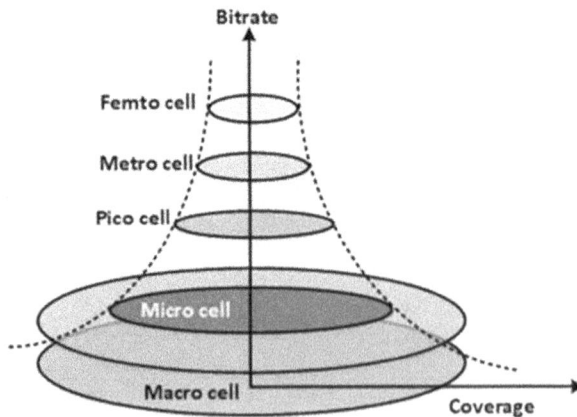

Figure 8.3 Approaching a heterogeneous network.

of renewable energy. Each layer consumes little electricity, so feeding on RES with its capability to reduce CO_2 production using solar energy and long-term cost savings, reduced site capital costs, and operational costs is relatively cheap [13]. SEPS generates electricity using RER including solar, wind, and batteries. The SEPS of the macro-layer requires a high amount of power, therefore larger solar panels and high storage batteries are needed during off-power generation. These are highly costly and will raise the cost of acquiring/renting the site because of the extra room needed. Combining multiple hybrid energy sources such as solar, wind, and diesel generators at the same location helps reduce the size of solar panels and batteries. Cost optimization by minimizing a site's initial and operational costs is always the prime objective for setting up a system [14]. A heterogeneous network consists of several interconnected nodes, as shown in Figure 8.4. The intelligence that governs the network and its components may be distributed (the devices come to their own conclusions), centralized on a controller, or housed in the cloud. These nodes are also linked together utilizing wired or wireless technologies. Significant issues still exist, particularly in the management of wireless technology and ecosystems. As the diversity of nodes and technologies grows, so does the management workload. Particular applications are operating on various pieces of hardware at the node level, each with a different set of requirements and capabilities (e.g., high throughput or low power consumption, supported technologies, and functionalities). Similarly, each technology has its distinct characteristics as a result of the diversification of technologies (e.g., capacity, range, power consumption).

The chapter is organized as follows. The detailed mathematical modelling of the microgrid is presented in Section 8.2, followed by the analytical approach of the heterogeneous network system and its sub-layers in Section 8.3. In Section 8.4, challenges in the self-organizing electric system are discussed. Finally, the simulation results are presented in Section 8.5 and Section 8.6 concludes.

8.2 MATHEMATICAL MODELLING OF MICROGRIDS

The models for individual systems are used to form the aggregated grid models. This technique is known as the component-wise system modelling technique. Therefore, the modelling of microgrids [15] explained through small individual system models like solar, wind, other RES, storage devices, and loads are presented.

8.2.1 Solar PV

Figure 8.5 depicts an analogous arrangement of a PV cell. The cell's photocurrent is represented by I_{ph}. The series and shunt resistances of the

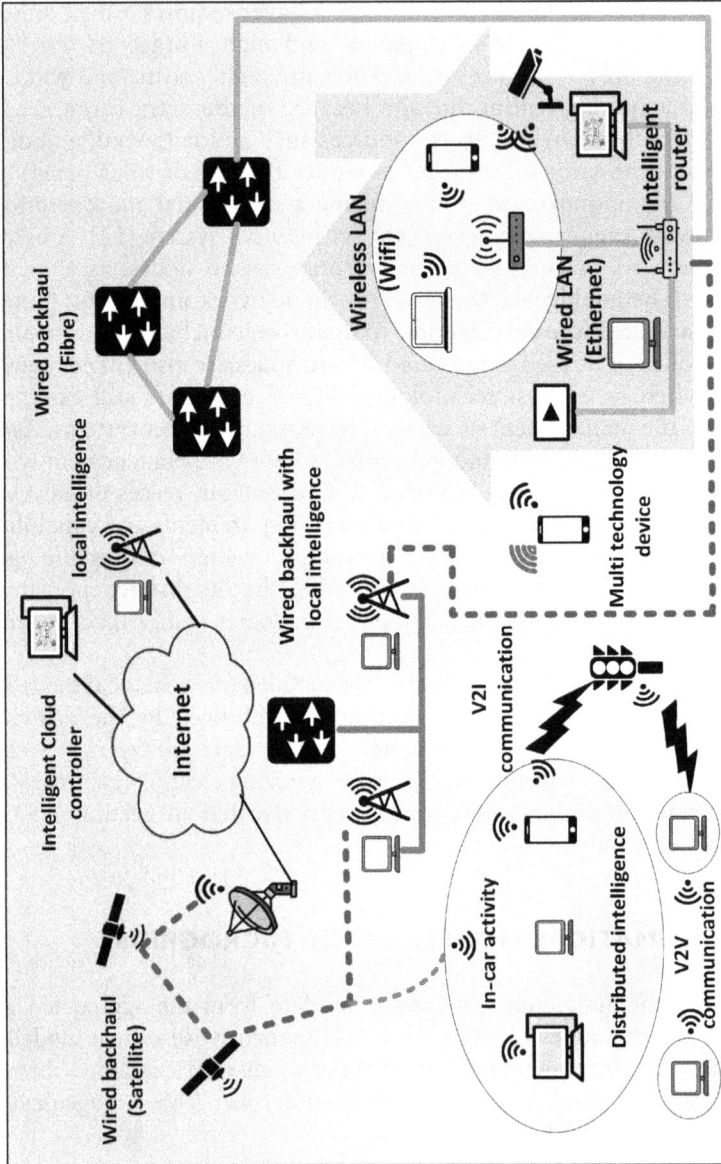

Figure 8.4 Representation of a heterogeneous network and its interconnections.

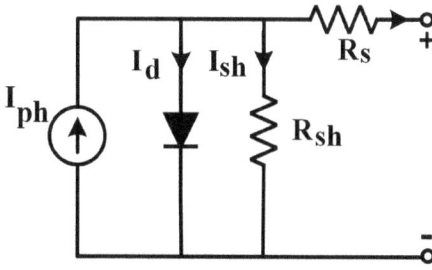

Figure 8.5 Equivalent circuit of PV cell.

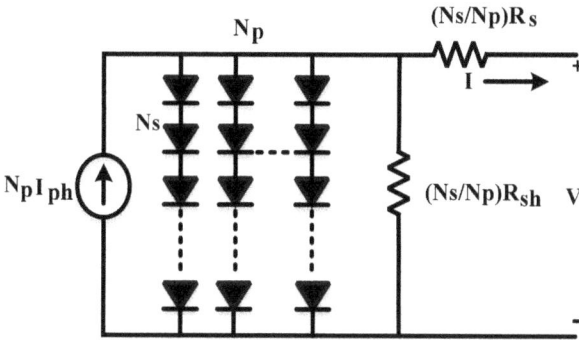

Figure 8.6 A solar array's circuit model.

photovoltaic (PV) cell are denoted by R_s and R_{sh} respectively. In general, R_s is quite small while R_{sh} is much larger. PV cells are grouped into superior units known as PV modules [16]. As illustrated in Figure 8.6, these modules are attached in parallel or series to form PV arrays that are being used to generate energy in solar power plants [17].

Module photo current $\left(I_{ph}\right)$

The module's photocurrent could be calculated as:

$$I_{ph} = \left[I_{sc} + P_i\left(T - 298\right)\right] * \frac{I_{sr}}{1000} \tag{8.1}$$

where I_{ph} represents the current of PV cell (A), I_{sc} denotes a short circuit current flowing through shunt resistance (A), P_i is equivalent to I_{sc} at 25°C, T_k denotes the temperature of the PV panel (K), and I_{sr} signifies solar irradiation

(W/m²). The reverse saturation current of the PV module (I_{rs-pv}) can be calculated by:

$$I_{rs-pv} = \frac{I_{sc}}{\left[\exp\left(\dfrac{q V_{oc}}{N_s k n T} \right) - 1 \right]}$$ (8.2)

where q represents the electron charge, the open circuit voltage is denoted by V_{oc}, N_s is the series connected cell, the diode's ideality factor is denoted by n, and k is Boltzmann's constant.

Module saturation current (I_o)

This differs from the temperature of a PV cell, which is given by:

$$I_o = I_{rs-pv} \left[\frac{T_k}{T_r} \right]^3 \exp\left[\frac{q * E_g}{nk} \left(\frac{1}{T_k} - \frac{1}{T_r} \right) \right]$$ (8.3)

Here, T_r is the nominal temperature and E_g is the semiconductor's band gap energy.

The solar PV module's current output can be expressed as:

$$I = \left\{ \left(I_{ph} N_p \right) - I_d - I_{sh} \right\}$$ (8.4)

where I_d is the diode current and I_{sh} is the shunt current.

From Equations (8.3) and (8.4) the diode current can be derived as:

$$I_d = N_p * I_o * \left[\exp\left(\frac{\dfrac{V}{N_s} + I * \dfrac{R_s}{R_p}}{n * V_t} \right) - 1 \right]$$ (8.5)

where V_t is the thermal voltage and the shunt current is given by:

$$V_t = \frac{kT}{q}$$ (8.6)

$$I_{sh} = \frac{V * \dfrac{N_p}{N_s} + I * R_s}{R_{sh}}$$ (8.7)

where N_p is the parallel connected PV modules.

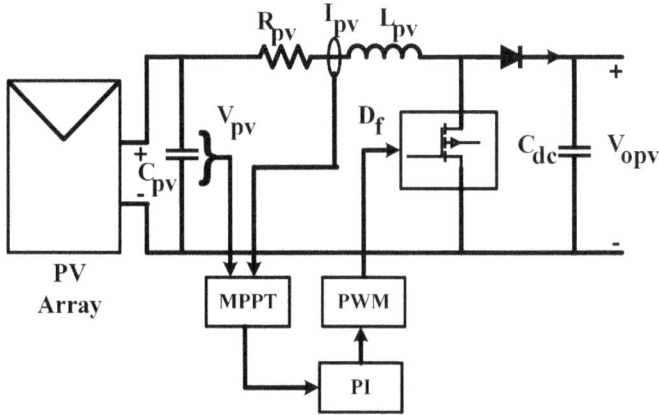

Figure 8.7 PV array with MPPT algorithm.

The maximum power point tracking (MPPT) algorithm is used to extract the most power from PV panels and transmit the supreme power from PV modules to loads [18]. DC to DC converters are used, which is shown in Figure 8.7. An MPPT perturb and observation approach is established to run a stationary PV module at the maximum power point (MPP).

8.2.2 Wind turbines

The wind's power is produced in the turbine blades, as represented by [19]:

$$P_w = \frac{1}{2} * \rho \pi r^2 V_w^3 \tag{8.8}$$

where ρ is the constant value according to air density, V_w signifies wind velocity in m/s, and r represents the blade radius m.

The power collected by a wind generator is determined by the Bentz constraint $c_p(\lambda, \beta)$:

$$P_t = \frac{1}{2} * \rho \pi r^2 V_w^3 c_p(\lambda, \beta) \tag{8.9}$$

where λ is the tip speed ratio and β denotes the pitch angle.

$$\lambda = \frac{(\omega_m \cdot R)}{V_w} \tag{8.10}$$

where ω_m denotes the speed of the generator (in rad/sec) and c_p is the power coefficient.

Various power coefficient models have been established, mainly in [20], which incorporate C_p by:

$$C_p(\lambda, \beta) = 0.517\left(\frac{116}{\lambda_i} - 0.4\beta - 5\right)e^{\wedge}\left(-\frac{21}{\lambda_i}\right) + 0.006795 * \lambda_i \qquad (8.11)$$

where the λ_i value can be calculated as:

$$\lambda_i = \frac{1}{\dfrac{1}{\lambda + 0.008 * \beta} - \dfrac{0.03}{\beta^3 + 1}} \qquad (8.12)$$

The graph in Figure 8.8 illustrates the relationship between produced output power (W) and wind velocity V_w (m/s). The wind energy conversion system runs across cut-in to cut-off speeds, which are 3.1 and 25 m/s respectively [22]. To extract the most electricity from the wind, the wind turbine should operate at the optimal all of the time. The rotor speed of a permanent magnetic synchronous generator can be derived as:

$$\omega_{opt} = \frac{\lambda_{opt} * V_w}{R} \qquad (8.13)$$

where λ_{opt} is the optimum tip speed ratio.

The turbine's torque equation is derived as:

$$T_t = \frac{P_t}{\omega_m} \qquad (8.14)$$

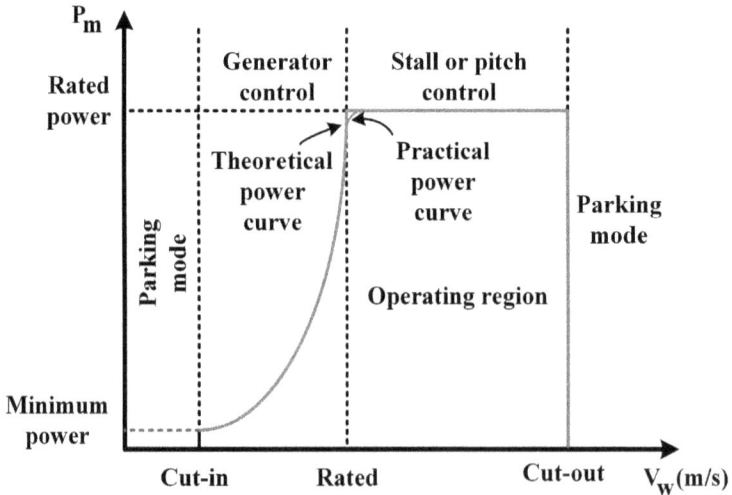

Figure 8.8 Output power vs wind speed relationship curve.

Therefore, the torque equation will be:

$$T_t = \frac{0.5 \cdot \rho \cdot \pi \cdot r^2 \cdot V_w^3 \cdot C_p(\lambda, \beta)}{\omega_m} \tag{8.15}$$

8.2.3 Battery storage system

The total daily charge/discharge transfers make up the battery's state of charge (SOC). The battery's condition at each given time t is dependent on its prior state of charge as well as the system's energy consumption and production patterns from time t_1 to time t. When the combined productivity of all generators exceeds the load demand throughout the charging process, the accessible battery bank capacity at hour t may be defined as:

$$E_{BATT}(t) = E_{BATT}(t-1) + \left[E_{SUR-DC}(t) \right] * \eta_{CHG} \tag{8.16}$$

$$SOC(t) = \left[1 - \frac{E_{Req}(t)}{E_{Batt}(t)} \right] * 100 \tag{8.17}$$

Meanwhile, the charged quantity of the battery is subject to the following constraints:

$$SOC_{min} \leq SOC(t) \leq SOC_{max} \tag{8.18}$$

The lowest SOC value is defined by the maximum depth of discharge (DOD); the maximum value of SOC is 1.

$$SOC_{min} = 1 - DOD \tag{8.19}$$

8.3 HETEROGENEOUS NETWORK (HetNet) SWITCHING STRATEGY AND SUB-LAYERS

A microgrid usually spans a physical region ranging from a few hundred metres to several kilometres in length. As a result, deploying a committed network infrastructure at such a location is generally not viable, both technically and financially [22]. Furthermore, microgrids are generally deployed across existing infrastructures, linking existing resources and connecting new ones to provide grid autonomy. Similarly, the network expected to control and monitor the microgrid can take advantage of existing communication infrastructures (such as field buses, the ethernet local area network, and fibre optic connections), combining them with low-cost technologies to create a HetNet. The utilization of a HetNet infrastructure to enable grid

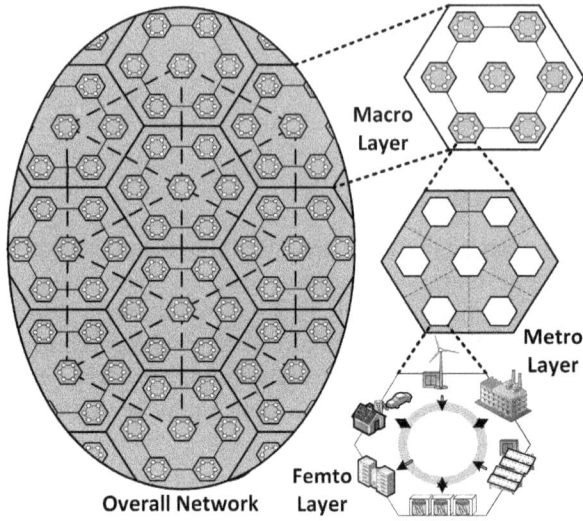

Figure 8.9 Functional layers of heterogeneous network.

automation isn't new, but it does necessitate careful planning to meet IEC 61850 standards [23].

With three multilayers, Figure 8.9 is an illustration of HetNet architecture in a large city. There are three multilayers in HetNet. Femtocells are thought to be the lowest layer, whereas macrocells, which pass through metrocells, are thought to be the highest layer. If every metrocell is followed by a mega-cell, then several femtocells will be superimposed. For the femto, metro, and macrocells, we are assuming a circular form. Small cells, also known as femtocells or metrocells, are low-power, short-range access points designed to increase capacity for homes, companies, and events in both rural and densely populated regions. They can be placed adjacent to difficult high-data traffic regions both indoors and outdoors. As a result, metrocells assist femto users' external traffic operations as well as femtocell excess traffic, whereas femtocells service femto users' call activity in interior settings. But only traffic that has backed up from either the metrocell layer or traffic from macro users is handled by HetNet's top layer, the macrocell. A limited number of subscribers depending on close groups of subscribers have access to femto and metrocells.

When renewable energy is found to be inadequate to meet the demand, many dispatch systems have been devised to regulate the functioning of generators and storage batteries. The HOMER optimization algorithm seems to employ a variety of scheduling patterns, including load following (LF), cycle charging (CC), generator order, and combination dispatch, depending on the requirement, the accessibility of the sources of power, and the circumstances of the environment. The battery and generator operations are managed using the LF and CC dispatch techniques. Generators turn on to charge the battery pack When CC employs dispatch strategies and turns off during off-peak hours, the overall number of hours that generators are in operation is decreased. However, due to the order of both charging and discharging operations used in this technique, the battery replacement expense is considerable. The LF method specifies that a generator should only have the configuration changes to meet the load demand when one is required. To maintain the system's sustainability and stability, the total load should be met by the generation of RERs. Because the generator is only used to satisfy load requirements in the network under the LF method and cannot charge the battery storage, RERs are utilized to their fullest potential. Typically, just the surplus electricity from the renewable system is used to recharge the battery. Additionally, because of its decreased consumption, the battery lives longer. The load's power source comes first in this dispatch strategy. This approach has a very high level of system dependability as a result. In this study, simulation was performed using the load flow strategy, and the suggested load following the method's main flowchart is illustrated in Figure 8.10.

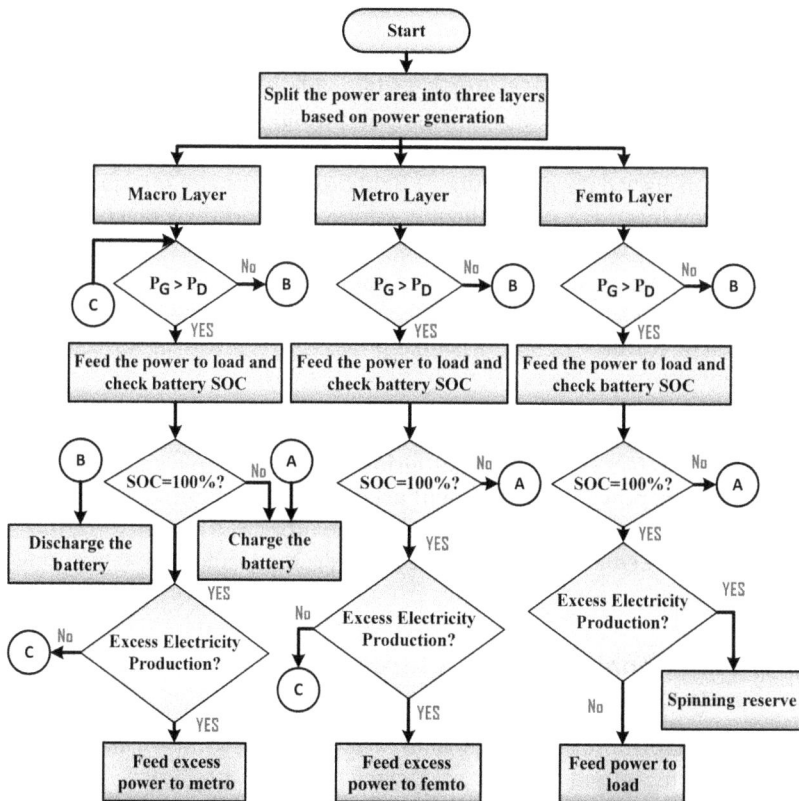

Figure 8.10 Diagrammatic representation of load-following dispatch strategies.

8.3.1 Self-organizing heterogeneous microgrid network technology

A self-organizing network is an automation technique that simplifies and speeds up the administration, configuration, optimization, planning, and healing of HetNets, and which is shown in Figure 8.10. All operational grids, including pico, nano, and micro, should self-optimize parameters and algorithmic behaviours regularly in reaction to observed network functionality and radio conditions. Newly added base stations should be self-configured in a "plug-and-play" manner. While awaiting a more long-term fix, self-healing procedures can also be utilized to make up for a recognized equipment loss [24]. Self-organizing networks have grown in importance in wireless communication technologies in recent years due to their characteristics as ad hoc networks, including self-optimization, dynamic routing, adaptive environment awareness, self-healing, and load balancing. A computer communication network that swaps packets through wireless communication channels is known as a wireless self-organizing network. The network as a whole operates with a dispersed self-organizing management system and is independent of the base or core node. It is possible for the retirement of any node or the insertion of a new node to be handled automatically. These features make the diverse self-organizing network advantageous in several ways, such as quick and flexible networking, simple access, powerful anti-capability, and substantial topological variation. The self-organizing network's networking strategy generally fits the Energy Internet's standards for user data interaction. Highly dispersed user interface devices in the Energy Internet may be flexibly networked and their reaction can be modified according to the environment's changes to enable decentralized end-to-end transactions. This is made possible by self-organizing network technology.

The salient features of self-organizing heterogeneous microgrid network technology are:

- Unless there is a chance of trading power at a price lower than the utility, connectivity between microgrids is usually simple and minimal;
- Exchange power – the grid will need to be reconfigured such that one microgrid serves as a generator and the other as a load;
- The total demand for microgrids will decrease since some portions of the power grid will get power from renewable sources.

8.3.2 Functional layers of heterogeneous microgrid networks

The IEC 61850 standard, which is already widely used in substation automation systems, is used to automate the microgrid. IEC 61850 is largely concerned with the architecture of communication networks that transport any kind of data, including sensor readings, security warnings, and monitoring signals. Sadly, for economical or opportunistic reasons, the microgrid is typically established using existing facilities and technology rather than being created from scratch. A microgrid is thus compelled to rely on a wide range of communication networks with diverse transport layers, including fibre optics, the ethernet, bespoke field buses, wireless communications, and power line communication, to mention a few. The generators, storage, and loads that make up a microgrid each have their independent automation/control system. This is typically the case even if the proposed block has a conventional IEC 61850 interface to the (outside) microgrid control system. Additionally, it is common practice to install distributed renewable energy producers and storage in urbanized environments (such as within buildings or along busy thoroughfares) which may already contain building auto-mation technology. Such circumstances can be improved from a microgrid automation standpoint if the complete infrastructure is combined. To put it differently, beyond the physical layer, the idea of a HetNet is expanded to include coexisting technologies with networking devices other than IEC 61850 [22, 24].

8.3.3 Physical layer

The physical parts of a microgrid (or smart grid in general) are all included in this layer, including the loads, distributed sources (such as solar panels or wind turbines), distributed storage (such as batteries or electric vehicles), and energy communications devices. To ensure plug-and-play functional-ity of all entry points to the various levels of the microgrid, power elec-tronics solutions are necessary. To be linked without affecting the system, new distributed energy resources must have this plug-and-play capability, and the microgrid must be able to flip between grid-connected and island modes. Electronic power systems are therefore used in the connection points of DERs to the nanogrid, loads to the picogrid, and the communication links between the microgrid and the electric grid (using a "back-to-back voltage source converter," or VSC, or an "electronic control unit," or ECU). For instance, inverters are necessary to supply the correct frequency and amplitude to electric batteries and solar panels. These inverters are two-stage bidirectional inverters with buck or buck-boost choppers (typically single-phase or three-phase) and a VSC (AC to DC stage). Many appliances and household gadgets need DC, which is generated by some DERs, like PV or DS. Because of this, employing DC equipment in the house may be an alternative, leading to a study into control systems and appropriate DC voltage ranges in residences.

8.3.4 Communication layer

The communication layer is responsible for supplying all the details and knowledge on the state of the communication protocol description of existing applications, features, and accessible network interconnection rates to the intelligence layer. The lowest stages in the design of correspondence are the building area network (BAN) and the home area network (HAN). The former controls the correspondence within each building (nanogrid) and is responsible for gathering each apartment's consumption (picogrid) and RES output within the house. The home energy management systems (HEMS) and systems for home automation which carry out these tasks are in charge of gathering data on consumption and managing the smart appliances within each apartment. For a picogrid in a HetNet, the pico base station must coordinate interference for the control and traffic streams with the predominant macro-path loss, and the user terminals should support advanced transmitters for interference cancellation for a login entry point to obtain infrastructure from a low starting station in the presence of macro-base stations with stronger downlink signal strength. HEMS also includes graphical user interfaces that enable customers to see their usage and supervisory control and data acquisition (SCADA) systems to save the details. Wi-Fi, Home Plug, LAN, ZigBee, ZigBee Smart Grid, the Internet Protocol (IPv6) and Low-power Wireless Personal Area Networks (6LoWPAN), Dash7, or programmable logic controller (PLC) are the commonly used protocols in such networks. ZigBee, SEP, and 6lowPan are built on the IEEE 802.15.4 specification, but they use the IP protocol in their network layer to accommodate the more common interfaces, such as TCP/IP. A NAN may be used for interactions inside a microgrid at the upper stage. A field area network (FAN) which is responsible for the interactions between the substations and the microgrid may also be located at the distribution point. Such regions include substation and control requirements, such as IEC 61850. Once each nanogrid's usage and output data are submitted to the distribution company (DISCO), each microgrid has a NAN with a central data centre responsible for processing and storing the data packets. The networking motors used today are Wi-Fi for homes, PLC, DSL, telephone (EDGE, GPRS, GSM), or an ethernet. Lastly, energy management systems have the task of balancing the usage of electricity and gathering data on the power use and output of different microgrids. In each microgrid, the usage of buildings and industry-specific data is passed from the smart meters to a gateway, and then the gateway sends the notification to the WAN. There would be a central controller in the WAN which may be the DSO or some other operator which optimizes the transmission parameters of the electrical power system to the microgrid operator (MGO).

8.3.5 Intelligence layer

This layer contains all control and decision-making mechanisms, local or centralized, which process information from physical layer elements (loads, DG, picogrids, nanogrids, and microgrids). This uses the contact layer to link in the physical layer with sensors, metering devices, and actuators. The picogrid will attempt load-balancing and load-shifting to reduce the expense of the energy consumed. To control its loads, the energy boxes (EBs) in the picogrid should take into account the needs of the end user and the demand signals. The nanogrid objectives will be to acquire the bulk of the renewable sources and the energy shortages by peak shaving, load lowering, and load shifting algorithms. Building management systems (BMSs) will strive to optimally handle the generation and storage. Based on their style of service, microgrids will have two primary objectives. If the microgrid operates in a grid-connected environment, an aggregate management system (AMS) will combine customers to communicate with wholesale markets. However, as the microgrid works in enclosed mode, AMS will try to control the microgrid resources and economic movements, utilizing a local sector; for example, AMSs will include facilities related to energy quality.

8.3.6 Business model

Many real-world factors, such as the possibility of lower energy costs, better reliability, and maybe extra factors like the amenity value of self-supply, form the basis of microgrid business models. Since they are at the core of any commercial offer, business model layers frequently focus on the financial benefits and drawbacks of self-supply and point to more work that may be done to increase the analysis's reliability. The business case for local energy provision, which we define as when a utility customer utilizes a microgrid to self-generate (completely or partially) electricity and maybe thermal energy (heating and cooling) loads, should be the model. The sole basis for this business case is the microgrid's ability to provide these loads at a lower total cost than a traditional utility service. Due to these factors, there will be two structures in this network: an AMS run by an email service provider (ESP) or aggregator/retailer, who is in charge of the microgrid's electricity and financial transactions as well as providing energy-efficient facilities, and a microgrid management system (MGMS) from a DISCO, whose main duty is to ensure network reliability and protection.

8.4 CHALLENGES IN SELF-ORGANIZING ELECTRIC GRIDS

A network of self-organizing microgrids confronts both technological and legal difficulties. It is crucial to make sure that the commercial power source is isolated before the microgrid begins supplying power to loads since the electric grid continually transfers electricity from the utility. If either source is not separated, it might result in back currents that harm the machinery. At the point of common connection, one supply is separated from another point of common coupling (PCC). A switch that opens and shuts at the common coupling point enables either the utility or the microgrid to connect to the load. However, PCC is limited to one microgrid. It will not be possible to swap power across microgrids by opening and shutting a PCC.

A substation's design would need to be changed to allow electricity to be routed between microgrids, or extra lines between microgrids would need to be added, and the logical connections would need to be changed each time a different microgrid had excess electricity, to ensure that electricity was exchanged between microgrids. It is expensive to connect more lines (nearly $5 per foot), and utilities are unlikely to alter the substation's design for privately operated microgrids. Because of this, exchanging power across microgrids is difficult [24]. Since building transmission lines is prohibitively expensive, exchanging electricity between microgrids that are close to one another and have a reasonable chance of doing so can be useful and feasible. The interchange and self-organization of the grid between microgrids located at various substations, however, becomes quite challenging. This is because there isn't a point of common connection to cut off energy from the utility and create a way for electricity to be exchanged across microgrids. Additionally, building a mile-long independent network is not practical since the costs associated with doing so would be far higher than any potential gains. Finally, the utility won't be particularly cooperative in allowing locally generated electricity to pass via their system. Due to these factors, creating a self-organizing microgrid network that is more complicated than one that is contained inside a single substation is not now possible. A substation typically houses no more than seven or eight microgrids, and we'll suppose that a different transmission line is built between microgrids that are no more than 5 km apart from one another. Natural renewable energy sources (RES) and diesel generators (DG) are used to produce electrical energy for microgrids using traditional backup generators. The DG electricity is heavily polluting and expensive, leading to CO_2 pollution and global warming. Moreover, the fuel is particularly costly in rural areas owing to shipping costs and conditions for secure storage. Diesel engines (DEs) are noisy and, when run at low loads, experience premature fatigue, while increasing maintenance costs. Compared to DG, a squirrel-cage induction machine (SCIM) powered by a wind turbine with a condenser bank is scalable and needs additive power for fixed frequency and voltage at the point of coupling (PCC). Microgrid configurations are graded as two, three, and four energy sources depending on the sources used. These microgrids are selected according to their capacity, the energy sources accessible in distant places, and the generator unit utilized for distributed generation units (DGUs). With the exception of DGUs with a double fed induction generator (DFIG), many DGUs in these microgrids are linked to a DC bus using proper converters to avoid issues with AC bus synchronization and reduce the number of power converters. The rotor of the DFIG is connected to the DC bus using VSC to drain energy while the stator is directly attached to the AC bus. The DC bus is linked to the various synchronous generators using VSC: a synchronous generator, synchronous reluctance generator, permanent magnet brushless DC generator, and squirrel cage induction generator. The permanent magnetic synchronous generator and permanent magnet brushless DC generator are additionally linked to the DC bus through a boost converter and diode rectifier to lessen control sophistication, the requirement for extra sensors, and related expenses. A transformer and a VSC link both power sources to the AC bus. The transformer is employed in conjunction with the remainder of the network to galvanically isolate DGUs, resulting in a four-wire distribution system with a delta-star architecture. All the loads are stored on an AC bus termed PCC. To accommodate for the shifts in production strength from DGUs, such microgrids are frequently strengthened with a battery energy storage system (BESS). In order to prevent overflowing the batteries and having the dump load impact the AC bus's power production, a dump load is linked to the DC bus. DG is linked as an ES for emergencies in each of these microgrids. It only functions if the total load power demanded by all DGUs exceeds the total power generated in that microgrid. It is used to prepare and charge the BESS simultaneously. The VSC is maintained for constant AC bus voltage, consistent frequency, and improved power efficiency under all load situations. The boost converters, or VSCs, connected to PV and WT are controlled by MPPT.

8.5 RESULT AND DISCUSSION

HOMER simulates all possible system arrangements that meet a certain area's load requirement while utilizing its energy resources. It eliminates unfeasible configurations, ranks the viable ones based on total net present cost, and then offers the feasible system configuration with the least total net present cost as the ideal system. The simulation of the metro and femto-layers is presented in Figure 8.11.

The simulation results for the required load of metro and femto-layer power capacity of the various system configurations are listed in Table 8.1. The results presented in this section are mainly focused on two parts, first the metro-layer output for electrification is presented, including annual load analysis, grid extension, and RER details, followed by the femto-layer for

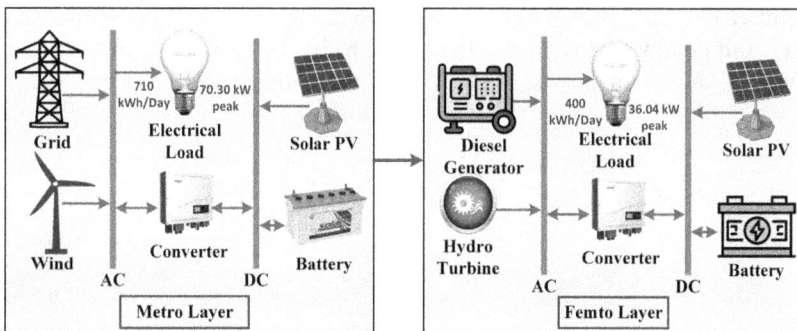

Figure 8.11 Simulation model of metro-layer and femto-layer.

Table 8.1 Power sharing of various components in layers

						Architecture			System	
Serial number	System	PV (kW)	Hydro (kW)	Wind (kW)	Number of batteries	Converter (kW)	Excess electricity (%)	Unmet load
1	Metro	126	–	111	2219	62.9	43.7	0
2	Femto	84.8	11	–	295	38.9	10.02	0

the same. Here, this metro-layer is powered by a grid, solar PV, and wind. Moreover, the actual load requirement on this metro-layer was estimated as 710 kWh/day and was then synthesized by HOMER software by adding randomness for different days and months. The annual load profile of the metro and femto-layers is shown in Figure 8.12.

In addition, there will be more loads soon, which will increase the need for electricity.

In the metro layer, grid/solar PV/wind/battery supply metro consumers. At the same time, it can receive the power from the macro-layer in case the excess power is required for outer needs in the metro-layer. On the other hand, the femto-layer is powered by an off-grid microgrid, such as hydro, solar, or diesel generator, for providing power to the inner circle of femto users. It should be noted that the load requirement of the femto-layer was estimated at 400.95 kWh/day, which is shown in Figure 8.12. The hydro/solar/diesel/battery systems of the femto-layer contribute their full capacity with their inner needs. Each year, the metro system produces 4,72,475 kWh of energy. More than 77.3% of the renewable energy output in the metro-layer is generated by wind and 22.7% by solar PV, according to research; 72.72% of the total power produced (710 kWh/day) is used directly by end customers, while the remaining 10% is sent to battery storage. However, over 18% of the total output (127.8 kWh/day) is surplus to demand and must be moved to another microgrid or femto-layer where power supply needs exist. From Figure 8.12, it can be seen that the power production in the metro-layer is higher than the required load. So, this excess power from the metro-layer is considered to be the spinning reserve of the femto-layer. The wind energy contribution was found to be the maximum over the year, which is shown in Figure 8.13c. Solar and wind power generation over

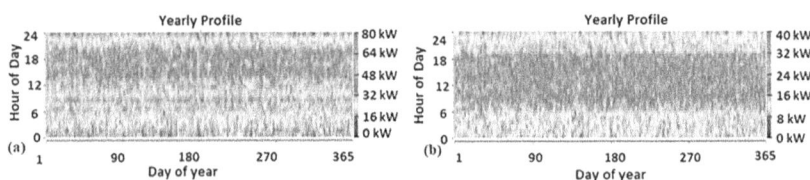

Figure 8.12 Yearly load profile: (a) metro-layer; (b) femto-layer.

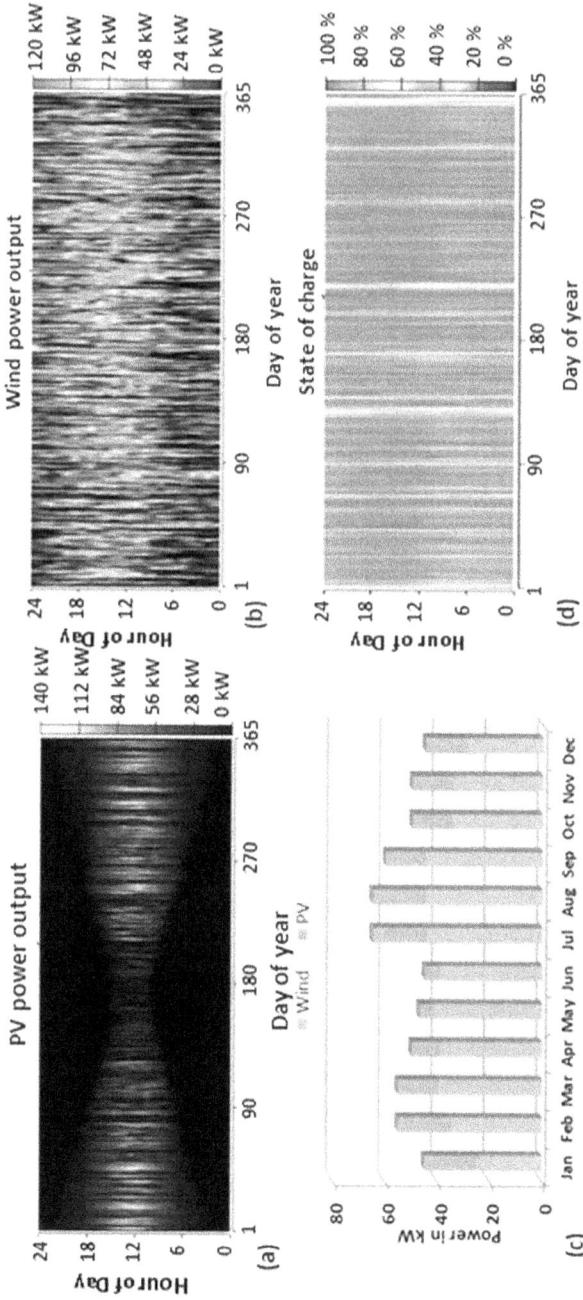

Figure 8.13 Annual profile of metro-layer: (a) annual solar power production; (b) annual micro-hydro production; (c) power production by hybrid system over year; (d) rainbow profile of battery's state of charge (SOC).

various months of the year are shown in Figure 8.13a and b. The findings in Figure 8.13d show the charging and discharging of a battery bank according to the renewable energy system generation. The metro system also generates excess electricity of 43.7% over the year, which passes to the femto system.

The femto system generates 2, 79,861 kWh of power annually; 62.5% comes from solar PV and 37.5% from micro-hydro turbines. Additionally, as seen in Figure 8.14c, PV power is extremely large in the summertime. Therefore, a higher proportion of this hybrid system is found in solar plants than in hydroelectric plants. The hydropower generation contribution, on the other hand, was determined to be at its peak from June to August. Figure 8.14a and b display the solar and micro-hydro power generation across various periods of time of the year. The findings in Figure 8.14d show how energy can be stored in a battery system when a renewable energy system offers extra power. The findings in Figure 8.14d show how energy can be stored in a battery system when a renewable energy system offers extra power. The rainbow profile indicated that the SOC was comparably modest. For the yearly transmission, values between 40 and 80% occur around 70% of the time, showing that the batteries have been fully or almost charged up for a significant amount of time for regular distribution. The battery storage system generally charges at noon and delivers power between mid-evening and early morning. Because of the higher agricultural load in January through May, there is a higher flow rate. Hence, there should be sufficient room for a battery storage system throughout those five months to provide a consistent and long-term supply of power. In the case where the load requirement is increasing in the femto layer, then reserve power from the metro-layer will be fed into that femto to compensate.

Figure 8.15a displays an example of the results of the hourly computation for 24 hours on a particular day of a month when analysing the energy balance of the suggested model. It is evident that in order to fulfil local energy demands, the available grid, solar, and wind production are used first. Energy storage is then charged once RES production outpaces the requirement for the load. In the event of a power outage, the battery storage system is drained to produce energy and thereby satisfy the net load. When RES is not available, this energy storage battery system can create up to 420 kWh. When massive amounts of electricity are produced by renewable energy sources, the surplus energy from the metro-layer is transferred to the femto-layer, as seen in Figure 8.15b.

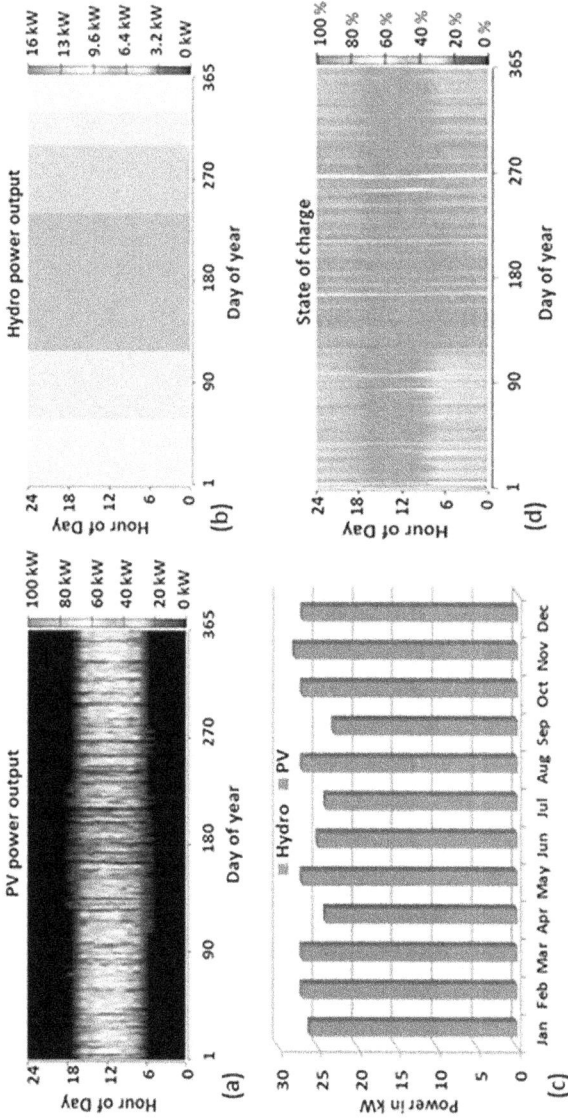

Figure 8.14 Annual profile of the femto layer:: (a) annual solar power production; (b) annual micro-hydro production; (c) power production by hybrid system over year; (d) rainbow profile of battery's SOC.

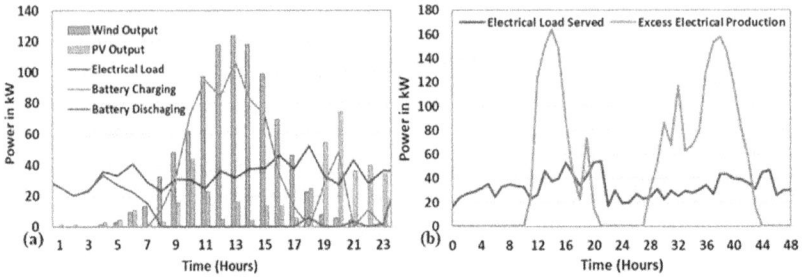

Figure 8.15 (a) Energy equilibrium of metro-layer; (b) excess energy production in metro-layer.

8.6 CONCLUSION

The development of the smart grid of tomorrow requires the modelling of microgrids. The heterogeneous network and its self-organizing strategy based on its functional layer have been thoroughly surveyed in this study. The important characteristics of self-organizing networks were also covered. The functional layers and sub-layers interact with all microgrids and loads. Along with many other research challenges, the results of lower-order decreased models produced by using model order reduction approaches and dynamic equivalence concepts have also been examined. Lastly, the challenges of self-organizing electric grids were summarized to obtain efficient controller design techniques as well as feature research on heterogeneous networks.

NOMENCLATURE

E_g	Band gap energy of the semiconductor
β	Pitch angle (θ)
I_d	Diode current (A)
I_{ph}	Current in photovoltaic cell (A)
I_{rs-pv}	PV module reverse saturation current
I_{sc}	Short circuit current (A)
I_{sh}	Shunt current (A)
I_{sr}	Solar irradiation (W/m²)
K	Boltzmann's constant, J/K
K_i	Equivalent of I_{sc} when cell at 25°C
n	Ideality factor of the diode
N_p	Ssummation of cell joined in parallel
N_s	Summation of cell joined in series
ρ	Air density
PD	Power demand (kW)

PG   Generated power from RES (kW)
Q Charge of electron (C)©
RES   Renewable energy sources
R_p Resistance connected in parallel (Ω)
R_s Resistance connected in Series (Ω)
T Operating temperature (K)
T_r Nominal temperature (°C)
V_t Diode thermal voltage (V)
V_w Wind velocity
ω_m Generator speed in rad/sec
λ Dip speed ratio

REFERENCES

[1] K. Cleary and K. Palmer. Renewables 101: Integrating Renewable Energy Resources into the Grid How Is Renewable Energy Integrated into the Grid. 2020. https://media.rff.org/documents/Renewables_101.pdf

[2] D. Gielen, F. Boshell, D. Saygin, M. D. Bazilian, N. Wagner, and R. Gorini. The role of renewable energy in the global energy transformation. *Energ. Strat.* 2019. 38–50. doi: 10.1016/j.esr.2019.01.006

[3] R. Work. Site Environmental Report for 2019 September 2020 Lawrence Berkeley National Laboratory. pp. 301–302, 2019.

[4] A. Muraleedharan Pillai, R. Rajesh Kanna, and R. Raja Singh. Advancement of Inventive Solar Power Based Frameworks for Rural India. *IOP Conf. Ser. Mater. Sci. Eng.* 2020. 906. doi: 10.1088/1757-899X/906/1/012001

[5] J. Kumar, A. Agarwal, and N. Singh. Design, operation and control of a vast DC microgrid for integration of renewable energy sources. *Renew. Energy Focus* 2020. 17–36. doi: 10.1016/j.ref.2020.05.001.

[6] S. Micro, M. Latifi, S. Member, and A. Rastegarnia. A Self-Governed Online Energy Management 2020. vol. 67. 7484–7498.

[7] S. Yousaf, A. Mughees, M. G. Khan, A. A. Amin, and M. Adnan. A comparative analysis of various controller techniques for optimal control of smart nano-grid using GA and PSO algorithms. *IEEE Access* 2020, 205696–205711. 2020. 10.1109/ACCESS.2020.3038021.

[8] I. Worighi, A. Maach, A. Hafid, O. Hegazy, and J. Van Mierlo. Integrating renewable energy in smart grid system: Architecture, virtualization and analysis. *Sustain. Energy, Grids Networks* 2019. 18. 100226. doi: 10.1016/j.segan.2019.100226.

[9] S. Nesmachnow, H. Cancela, and E. Alba. A parallel micro evolutionary algorithm for heterogeneous computing and grid scheduling. *Appl. Soft Comput. J* 2012. 12.626–639, doi: 10.1016/j.asoc.2011.09.022.

[10] Y. Xie, B. Yu, S. Lv, C. Zhang, G. Wang, and M. Gong. A survey on heterogeneous network representation learning. *Pattern Recognit.* 2021. 116, 107936. doi: 10.1016/j.patcog.2021.107936.

[11] A. N. Montanari, E. I. Moreira, and L. A. Aguirre. Effects of network heterogeneity and tripping time on the basin stability of power systems. *Commun. Nonlinear Sci. Numer. Simul.* 2020, 89, 105296. doi: 10.1016/j.cnsns.2020.105296.

[12] L. U. Meilian and Y. E. Danna. HIN_DRL: A random walk-based dynamic network representation learning method for heterogeneous information networks. *Expert Syst. Appl* 2020, 158, 113427. doi: 10.1016/j.eswa.2020.113427.

[13] Q. Han, B. Yang, N. Song, Y. Li, and P. Wei. Green resource allocation and energy management in heterogeneous small cell networks powered by hybrid energy. *Comput. Commun.* 2021. 160. 204–214. doi: 10.1016/j.comcom.2020.06.002.

[14] H. Golpîra, S. A. Rehman Khan, and Y. Zhang. Robust Smart Energy Efficient Production Planning for a general Job-Shop Manufacturing System under combined demand and supply uncertainty in the presence of grid-connected microgrid. *J. Clean. Prod.* 2018. 649–665. doi: 10.1016/j.jclepro.2018.08.151.

[15] S. Sen and V. Kumar. Microgrid modelling: A comprehensive survey. *Annu. Rev. Control.* 2018. 46. 216–250. doi: 10.1016/j.arcontrol.2018.10.010.

[16] A. K. Singh and R. R. Singh. An Overview of Factors Influencing Solar Power Efficiency and Strategies for Enhancing. 1–6, 2022. doi: 10.1109/i-pact 52855.2021.9696845.

[17] R. R. Kanna, M. Baranidharan, R. Raja Singh, and V. Indragandhi. Solar Energy Application in Indian Irrigation System. *IOP Conf. Ser. Mater. Sci. Eng.* 2020. 937. doi: 10.1088/1757-899X/937/1/012016.

[18] K. Patel, S. Borole, K. Ramaneti, A. Hejib, and R. Raja Singh. Design and implementation of Sun Tracking Solar Panel and Smart Wiping Mechanism using Tinkercad. *IOP Conf. Ser. Mater. Sci. Eng.* 2020. 906. doi: 10.1088/1757-899X/906/1/012030.

[19] Y. Nie, F. Li, L. Wang, J. Li, and M. Sun. A mathematical model of vibration signal for multistage wind turbine gearboxes with transmission path effect analysis. *Mech. Mach. Theory* 2021. 167, 104428. doi: 10.1016/j.mechmachtheory.2021.104428.

[20] B. Benyachou, B. Bahrar, A. Moufakkir, K. Gueraoui, and M. S. Hassani. Materials Today: Proceedings Optimization & control strategy for offshore wind turbine based on a dual fed induction generator. *Mater. Today Proc.* 2022. 12. doi: 10.1016/j.matpr.2022.02.546.

[21] W. Torki, F. Grouz, and L. Sbita. Vector control of a PMSG direct-drive wind turbine. *Int. Conf. Green Energy Convers. Syst. GECS* 2017. doi: 10.1109/GECS.2017.8066247.

[22] F. Martin-Martínez, A. Sánchez-Miralles, and M. Rivier. A literature review of Microgrids: A functional layer based classification. *Renew. Sustain. Energy Rev.* 2016. 62. 1133–1153. doi: 10.1016/j.rser.2016.05.025.

[23] S. Wei, L. Qi, L. Yiqun, B. Meng, C. Guoli, and Z. Hongke. Small cell deployment and smart cooperation scheme in dual-layer wireless networks. *Int. J. Distrib. Sens. Network* 2014. doi: 10.1155/2014/929805.

[24] G. Piro et al. HetNets powered by renewable energy sources: Sustainable next-generation cellular networks. *IEEE Internet Comput.* 2013. 17. 32–39. doi: 10.1109/MIC.2012.124.

[25] R. H. Lasseter et al. CERTS microgrid laboratory test bed. *IEEE Trans. Power Deliv.* 2011. 26. 325–332. doi: 10.1109/TPWRD.2010.2051819.

Chapter 9

Adaptive Smart Power Saving Techniques for Machine-to-Machine Communication-Enabled Wireless Sensor Networks

J. Paruvathavardhini

Jai Shriram Engineering College, Avinashipalayam, India

R. Sudarmani

Avinashi Lingam Institute for Home Science and Higher Education, Coimbatore, India

CONTENTS

DOI: 10.1201/9781003436461-9

9.1 INTRODUCTION

In today's fast-moving world, technology has shown its dominance on people where they cannot match their needs without the mastery of technology, especially in information sharing. Technology is the use of skills, knowledge, process, and technique to accomplish a specific task and to achieve higher production capacity, scientific investigation, and efficiency, among other things. Simply put, technology is the advancement in utilizing machines to perform tasks efficiently and work less. Unlike in the past, people aren't involved in work physically; rather they started using machines in all kinds of work, and later, with new inventions, people started to control these machines remotely. With a new era of inventions, machines started to communicate with each other with null or limited human intervention. With the enhancement of technology, the usage of multiple devices has increased and which are connected to each other. At the same time, end-users also use multitudinal devices like digital smart TVs, smart phones, Kindles, and i-pads. On such occasions, the concept of machine-to-machine (M2M) enters the scenario. An M2M system enables networked devices to communicate information and coordinate actions with each other automatically, without requiring human intervention. M2M devices use point-to-point communication systems between machines, sensors, and hardware [1].

Communication between the systems first happened in industries with the help of the IoT, which includes a wired or wireless network. Today artificial intelligence and machine learning are used to interface across systems, allowing them to make decisions on their own [2]. Sensors and communication modules are entrenched within M2M devices, permitting data to be conveyed from one device to another through wired and wireless communications networks. As a general rule, the IoT operates in the same way as a brain, storing real-world data (in cloud-based services or databases) and monitoring, interpreting, and even making decisions based on it. When seen from a more vertical and closed perspective, M2M with Internet protocols could be viewed as a subset of the IoT. However, the IoT takes a more broad and meaningful approach, combining vertical applications to serve the demands of a wide range of users.

A wireless sensor network (WSN) serves as the eyes and ears of the IoT. It serves as a link between the physical and digital worlds. Additionally, it is in charge of communicating to the Internet the values perceived in the real world. These sensor nodes are battery powered but, due to different activities like sensing, computation, and communication, with analysis the battery gets depleted. Researchers have analyzed and developed many methods to avoid unnecessary battery usage and to extend the lifetime of a sensor node. Even then, it remains a challenge when the sensors lose all energy. However, the recent technique of energy harvesting would support prolonging the lifetime of a node [3]. Energy harvesting (EH) is a method of obtaining/scavenging energy from external or internal sources and subsequently using it to power the sensor nodes to extend their operational lifetime and capacity. The rest of the chapter is organized as follows. First discussed are M2M communication, the IoT, and their applications; the following sections explain WSN, its characteristics and types, battery requirements, different battery materials, and battery modeling. This is followed with energy harvesting techniques.

9.2 MACHINE-TO-MACHINE COMMUNICATION

M2M communication occurs automatically between devices without human intervention. It includes machine monitoring and control, as well as crawling for results on search engines. Intelligent machines are capable of exchanging information without human intervention, as well as coordinating and carrying out tasks. The exchange of data or information, which is called "communication", started in the early 20th century, and grew through telemetry, caller ID, and automation. Then, in the late 20th century and early 21st century M2M became involved in the combination of sensor networks and the IoT. Zero human intervention is expected but it does not happen at all times as in some places there is a need for it. Examples of M2M communication include a vending machine reporting stock levels to the supply chain software, an ATM requesting permission from bank servers to issue cash, and a credit card reader approving a transaction [4]. M2M is intended for cross-platform integration, whereas supervisory control and data acquisition (SCADA) is designed for isolated systems employing proprietary technologies. It resembles industrial SCADA. In M2M systems, end users can get information from assets, such as temperature or inventory levels, about events [5]. This consists of three components [6]:

1. Data end point (DEP). This is the device or system which has the data to be transmitted, e.g. sensors that send weather data or a patient's medical data, ATM machines, vending machines, microcomputer systems, or transmitters connected to receivers. Numerous connected devices and data endpoints might make up an M2M communication

network. The desired data is transported from the data endpoints to the network, where it is then sent to the data integration point. The network is also used by individual data endpoints to communicate with one another.

2. A wireless or wired connection to the Internet can be used to transfer data over communication networks. Some basic techniques used include RFID and Bluetooth.

3. Data integration point (DIP). This is the data integration point which can be the server or control center which collects the data.

Data transmission in the Industrial Internet of Things (IIoT) is emerging as the most crucial component of the jigsaw as essential industries move more and more toward automated solutions. Industries of all stripes are using sensor technology to traverse the constant stream of data in order to make better decisions, increase operational effectiveness, and boost production [5]. Digitally enhanced business models are being driven by the needs of the real-time corporate network that effortlessly coexists inside the industrial sectors as WSNs continue to grow. As a result, proactive IT departments and cutting-edge networking technology are given more attention.

M2M communications require some basic requirements for smooth operation, and the European Telecommunications Standards Institute (ETSI) is responsible for developing international standards for information and communications technology, which may affect anyone wishing to conduct business in the EU or elsewhere using M2M systems. The following are the requirements mentioned by ETSI [7]:

1. System scalability: As more devices are linked, the system must continue to function effectively.

2. System anonymity: Device identities must be hidden from the system.

3. Logs: M2M systems must be able to capture inaccurate data, failed installations, and other errors and store these records.

4. A system must follow the rules described above regarding M2M communication.

5. Transmission techniques: To lessen the burdens during M2M data transmission, the system must support a variety of transmission techniques, including Unicast, Anycast, Multicast, and Broadcast, and be able to switch between them.

6. Message transmission scheduling: The system should be able to schedule data transfer periods as well as manage or delay communications according to priority levels.

7. Communication channel selection: An optimization strategy should be used for communication channels in an M2M system to account for transmission failures, delays, and network costs.

9.2.1 Applications of M2M

Some of the applications of M2M are (see Figure 9.1):

- Remote monitoring frequently makes use of M2M. In the event of low product levels, a vending machine can alert the distributor or machine. As a result of M2M, warehouse management systems (WMSs) and supply chain management (SCM) are possible tracking and monitoring assets.
- M2M communication is used by energy firms to read meters and bill clients. Additionally, it can monitor every device and guarantee service continuity by using sensor data. Even when a patient is not physically present, clinicians can utilize M2M technology to track the patient's heartbeat or other vital indicators in real time.
- Services for mobile payments combine machine learning, artificial intelligence, and the IoT. Customers are embracing this technology and using e-wallets like Apple Pay or Google Wallet. In smart homes, there are many ways to use M2M communication, such as controlling lighting, turning electrical equipment on and off, or automatically generating shopping lists.

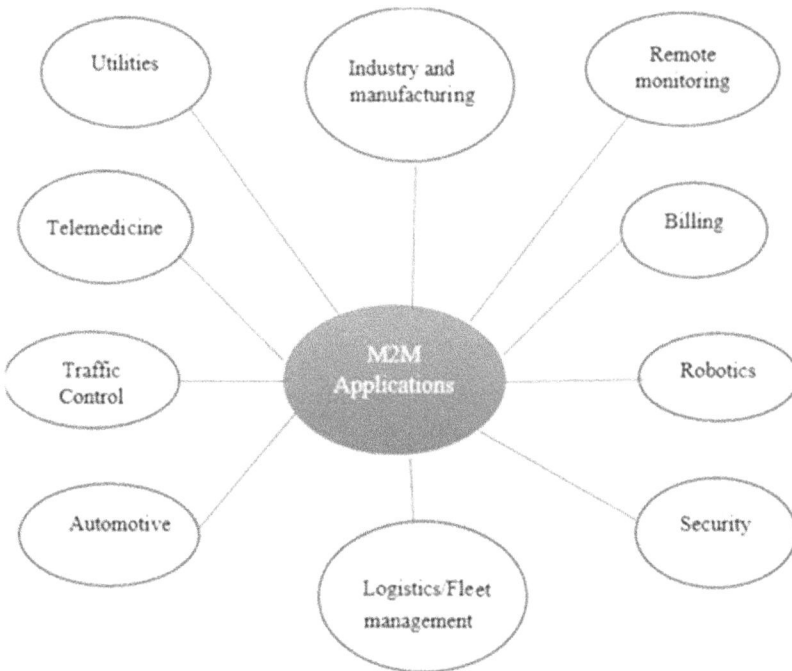

Figure 9.1 Applications of M2M.

- M2M communication is directly related to online marketing. As already mentioned, one instance of M2M communication is the process of search engines scanning the Internet. This technique is crucial for paid advertisements on social networking platforms or search engines, though.
- While performing this procedure, the algorithms interact with the machines that provide them with data. There might be sources of these problems such as the server on which your website is hosted or the advertisements you place in your Google Ads account.
- Additionally, when we utilize web services to schedule appointments, stream movies, or shop online, M2M communication is at work. The client and server exchange requests and responses throughout these operations.
- M2M technology has also been used in smart home systems. Home appliances and other technologies are given real-time operational control and remote communication capabilities thanks to the employment of M2M in this embedded system.
- M2M is considered an important element in different domains in addition to robotics, traffic control, security, logistics, fleet management, and automotive.

The demand for more intense monitoring, measurement, and automation via M2M communication and networking technology has inevitably followed as industrial markets continue to place greater emphasis on safety, automation, and increasing operating efficiencies. The focus of these businesses today is on utilizing and maximizing the sensor technologies that can meet this wide range of needs. For businesses striving to use information in real time, understanding the developing advancements of WSNs via M2M wireless communications and examining the requisite security and reliability issues reveals an exciting but challenging future.

9.3 M2M TO IoT

Both M2M and IoT are outcomes of the advancements in technology over the past few decades, including not only a decline in the price of semiconductor components but also a remarkable uptake of the Internet Protocol (IP) and wide-scale Internet adoption. There are numerous imaginable applications for such solutions; however, the role M2M and IoT will play in business and society has only now begun to become clear for a number of interrelated factors [1].

9.3.1 What is the IoT?

The IoT refers to a network of real-world objects connected via software, sensors, and other means. It is a setting where gadgets exchange data over the Internet, a networking medium. By utilizing the IoT to connect a wide range of devices via technology, users can create quick, scalable, high-performance networks. Sensor data is shared by Internet-connected devices to an IoT gateway, where it is forwarded to be processed either locally or in the cloud. Additionally, these devices converse with one another and act on the knowledge they have gained from one another. However, in most cases, the smart device will handle all the work in the background. Configuration or retrieval of acquired data can be performed by humans on these devices; the future of human productivity at home and in the workplace will purely depend on these devices. Data-driven IoT devices produce the highest level of ease since they automatically respond and learn from the data they receive. The IoT has applications in various fields, namely healthcare, industry automation, manufacturing industries, agriculture, home automation, traffic control, remote monitoring, smart cities, air quality monitoring, and so on. When industry automation is considered, initially machines were to communicate with each other, which does not require much Internet connection, but as the technology developed, people expected the sophistication which leads to an improved M2M. Hence it is easy to give guidance in emergency situations. This requires a set of protocols for communication [1].

The IoT ecosystem may develop similarly to the current web, where devices, networks, and application layers are all interconnected, as shown in Figure 9.2. The following technologies are used for M2M communication:

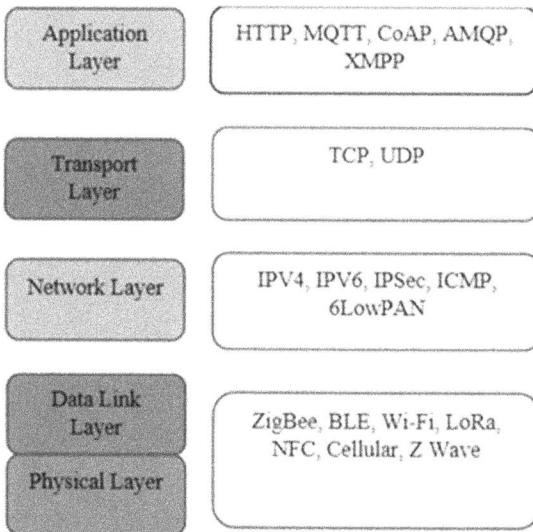

Figure 9.2 IoT network protocol stack.

sensor nodes with networks; radio-frequency identification (RFID); the mobile internet; wired and wireless communication networks; Bluetooth LE/Smart; IEEE 802.15.4 (low-rate wireless personal area networks; LR-WPAN), for example, ZigBee; internet engineering task force (IETF); IPv6-enabled low-power wireless personal area networks (6LoWPAN); routing protocols for low power and lossy networks (RPL); IPv4/IPv6; Wireless M-BUS; WirelessHART); M-BUS; power line communication (PLC); ISA100.11a; constrained application protocol (CoAP); and KNX. There are other low power wireless communication technologies available for connecting devices in personal area networks (PANs), home area networks (HANs), local area networks (LANs), and field area networks (FANs) to the M2M Gateway node, including Wi-Fi, Bluetooth low energy (BLE), ZigBee, 6LoWPAN, and Z-wave. The main server can be connected to a gateway node by using GSM, 3G, or 4G, as well as fixed bandwidth or FTTH [8].

9.3.2 Game changing

Game changers result from a series of social, economic, and environmental changes that motivate us to solve problems, but also present opportunities to rethink how we approach them. Technological and scientific advancements are teaming up with game changers to meet the extraordinarily high need for monitoring, regulating, and understanding the physical environment. One of the crucial aspects of the technological progress needed to meet these issues is the shift from M2M to the IoT, as shown in Figure 9.3.

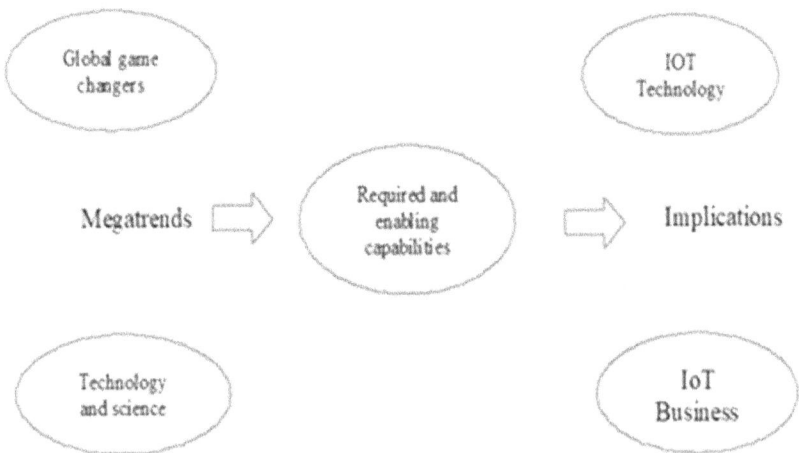

Figure 9.3 Game changing representation.

9.3.3 Differences between M2M and IoT

- A subset of M2M technology is IoT. The M2M communication system is a component of the IoT where two machines communicate without human intervention.
- The primary distinction between M2M and IoT technologies is point-to-point connectivity.
- An IoT system, on the other hand, typically places its devices within a global cloud network that enables more automation and more complex applications.
- Scalability is another significant distinction between IoT and M2M. Because devices may easily be added to the network and incorporated into already existing networks, the IoT is made extremely scalable.
- As new point-to-point connections must be created for each system, M2M network setup and maintenance may be more labor intensive.

9.4 WIRELESS SENSOR NETWORKS

WSNs are utilized in several real-time and practical applications due to their scalability and compact size. The sensing, computation, and communication capabilities of the huge number of sensor nodes connected to WSNs are constrained. Ad hoc wireless networks like WSNs use wireless sensors to remotely monitor system, physical, and environmental variables [9]. An integrated CPU is used in WSNs to control and monitor the environment at specific points in space. A WSN collects, processes, and analyzes all real-world data. To share data, a WSN system's base stations are linked to the Internet. The WSNs with IoT play a crucial role for applications that require continuous monitoring and automation. A WSN can be used for a wide variety of applications, including security monitoring, threat detection, environmental monitoring, noise measurement, health care applications, agricultural applications, and landslide detection. Since all the sensor nodes are battery-powered, their energy usage ought to be lower.

9.4.1 Elements of WSN

A WSN consists of the following elements (see also Figure 9.4):

Controller: Processes all pertinent data and has the ability to run any code.

Memory: Utilized to store data and programs; typically, several types of memory are used for this. The actual interface with the physical world is made up of sensors and actuators, which are tools for observing or modifying environmental physical properties.

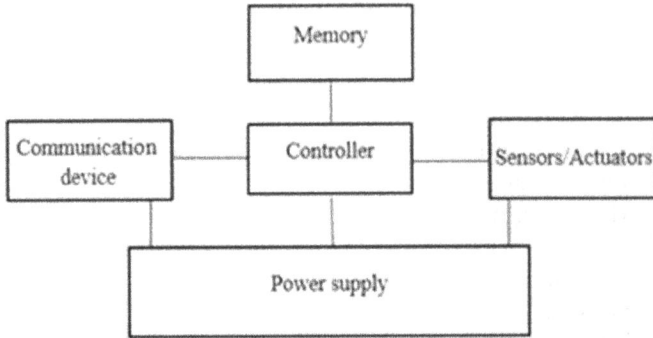

Figure 9.4 Sensor node's hardware components.

Communication: Creating a network of nodes requires a tool for transmitting and receiving data over a wireless channel.

Power supply: Since tethered power supplies are typically not accessible, energy must be provided by batteries of some kind. There are also occasionally options for recharging by getting energy from the surroundings (e.g., solar cells).

Radio nodes: These devices are used to gather data generated by sensors and which is transmitted to a WLAN access point. This device is composed of a microcontroller, a transceiver, an external memory, and a power source. An access point receives data that is wirelessly transmitted by radio nodes, usually over the Internet.

Evaluation software: A WLAN access point transmits data to the evaluation software for further processing, analysis, storage, and mining, as well as providing the users with a report for further processing.

WSNs have the following traits:

- Sensor nodes form a connection with them as a network called a WSN. Due to their compact size, sensor nodes have minimal power, little memory, and limited energy.
- Wireless networks may be vulnerable to hostile attacks and can be set up in harsh climatic circumstances.
- Despite their ad hoc deployment, they must be self-organized, self-healing, and capable of frequent reconfiguration [10].

9.4.2 Classification of WSNs

- Mobile and static;
- Deterministic and nondeterministic;
- With one or more base stations;
- With static and mobile base stations;
- Single-hop and multiple-hop;

- Self and non-self-reconfigurable;
- Homogeneous and heterogeneous.

Different WSNs are utilized depending on the application area where the nodes are to be put. WSNs are divided into categories based on the application.

The position of a sensor node is computed and fixed in a **deterministic** WSN. Only a few applications allow for the deployment of sensor nodes that have been designed in advance. Due to a number of reasons, such as hostile operating conditions or severe environments, it is generally not possible to determine the position of sensor nodes. Such networks require a sophisticated control mechanism since they are **non-deterministic**.

Sensor networks in a non-self-configurable WSN cannot organize themselves into a network on their own, so a control unit is required to gather data for them. In the majority of WSNs, sensor nodes can maintain a connection and work together to accomplish their missions.

The energy consumption, computing power, and storage of all sensor nodes are the same when using a homogeneous WSN. WSNs that use heterogeneous sensor nodes separate communication and processing activities based on their computational and energy demands [11].

Grid, mesh, star, and tree networks are preferred in **terrestrial** WSNs. A secondary power source would be solar cells.

WSNs that are placed in **underground** are more expensive in terms of deployment, upkeep, equipment costs, and so on. Due to the high amount of attenuation, wireless communication is a major issue in this situation, and battery recharge is also a major challenge.

In **underwater** WSN architecture, autonomous vehicles are utilized to collect data from underwater nodes, and challenging problems with long propagation delays, bandwidth, and sensor failure are present.

Multimedia WSN demands advanced data compression, processing, and high bandwidth for the transmission of multimedia data.

A **mobile** WSN is a network of sensor nodes that can move, reconfigure themselves, and sense their surroundings. Compared to static nodes, mobile sensor nodes can offer more efficient coverage and communication. They are used frequently for military surveillance, target tracking, and habitat monitoring [12].

9.4.3 Challenges for WSNs

Many challenges arise while designing a network (as shown in Figure 9.5):

- Level of service;
- Security concern;
- Energy savings;
- Internet throughput;
- Performance;
- Adaptability to node failure;
- Cross-layer enhancement;
- Scalability for extensive deployment [12].

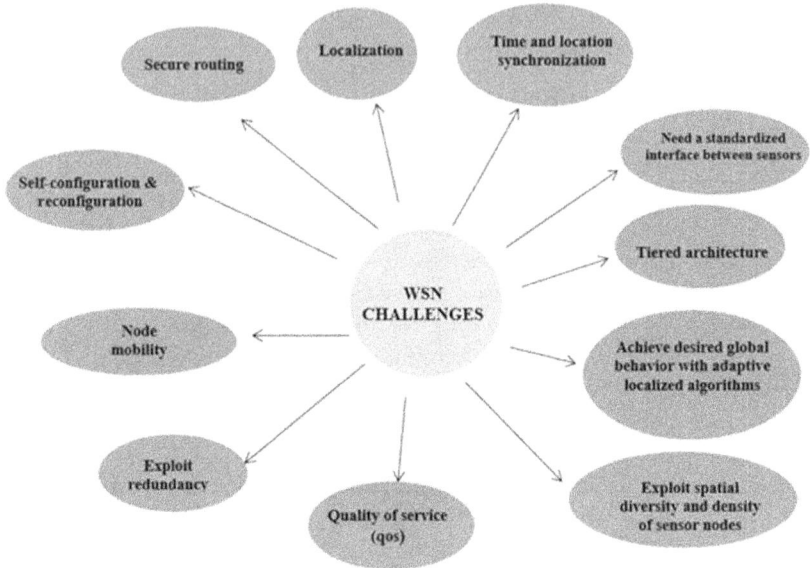

Figure 9.5 Challenges for WSNs.

9.5 ROLE OF A BATTERY

The sensor nodes in WSNs can only be powered by the primary batteries they are shipped with. Batteries will need to be replaced when they run out in addition to having a finite amount of energy, which raises the expense of their upkeep. For a vital application, it is crucial to stop a battery from emptying completely before it happens because the state of charge (SoC) of a battery is challenging to determine. The expense of changing a battery can also be unreasonably high or perhaps impossible for particular applications. To solve these issues and boost the battery's power density, suitable materials are required. This section discusses the various methods for harvesting energy and preserving battery energy in order to get around these issues. Batteries can be made of a variety of different materials [14].

Batteries must be made with a tiny size in mind due to the compact dimensions of the nodes of the sensors. Thus, their total operation is constrained by the energy sources that are accessible, whether from batteries, energy harvesting, or both [15]. Wireless sensors, on the other hand, must typically run for extended periods of time in order to continuously collect the necessary data. The energy of the battery needs to be utilized carefully to ensure long-term operation. For wireless sensor nodes, a variety of batteries are available. There should be several factors taken into account while contrasting various battery technologies. There are many factors to consider, such as energy density (the amount of energy stored per unit weight of the battery), cycle life (the number of discharge and charge cycles before disposal), safety, environmental impact, and charge/discharge characteristics. The past 20 years have seen the development of battery technologies aimed at meeting the growing demand for smaller, lighter, and higher capacity batteries. This section summarizes the characteristics of typical battery technologies [16].

9.5.1 Nickel-cadmium

For many years, nickel-cadmium has been utilized to power wireless sensors. Rechargeable nickel-cadmium batteries are frequently used in residential structures to power appliances. A cadmium anode, a nickel-hydroxide cathode, and an alkaline electrolyte are all components of a nickel-cadmium cell. They can withstand physical abuse and inefficient usage cycles, and the batteries they are used to make give large currents at a reasonably steady voltage [17]. High discharge rates can be sustained by a nickel-cadmium battery without negatively compromising battery capacity. Nickel-cadmium cells that produce 1.2 V cost a lot since cadmium is pricey. Cadmium is also poisonous and unfriendly to the environment.

Nickel metal hydride (NiMH) is the main source of nitric oxide in nickel rechargeable batteries. The major distinction between this and a nickel-cadmium battery is that the latter tends to lose its properties after numerous recharges. In AA-sized NiMH batteries, the voltage is usually 1.2 V, not 1.5 V as in alkaline batteries because they are heavier and less energy dense than alkaline batteries. Like the alkaline battery model, a second-order polynomial can be used to characterize its behavior. Low temperatures are beneficial for NiMH performance. As the current draw grows, it does not drain quickly. It can be used in high-temperature situations where lithium-based cells may explode because of its wide operating temperature range [45].

9.5.2 Alkaline

The most popular battery type in WSNs is alkaline. They have a set energy rating, which results in a finite lifespan. Batteries must be replaced every 17.4 months for a single sensor node powered by a common 3000 mAh AA battery with a 1% duty cycle; every week for a sensor node running at a 10% duty cycle [18]. When compared to zinc-carbon batteries, alkaline batteries offer a greater energy density and a longer shelf life. They have a five to six times longer lifespan than zinc-carbon batteries. A brand-new alkaline cell has a nominal voltage of 1.5 V. Cells can be used in series to provide different voltages. The zinc oxide and manganese dioxide content in the cathode and electrolyte determine the effective zero-load voltage of a non-discharged alkaline battery [19]. The discharge-dependent average voltage under load ranges from 1.1 to 1.3 V. The voltage of the fully discharged cell is between 0.8 and 1.0 V. A sloped discharge curve is typical for alkaline batteries. To account for this sloping discharge profile, the majority of devices are made to function within a specific voltage range (for instance, between 1.6 and 0.9 V per cell). In alkaline batteries, a sloped discharge occurs due to an increased internal resistance caused by chemicals building up on the electrode surfaces and a reduced availability of fuels.

Although reusable alkaline-manganese batteries have been created as a low-cost alternative to disposable alkaline batteries, their energy density and cycle life are inferior to those of nickel-cadmium batteries. Alkaline disposable batteries have been used for many years. In spite of the fact that reusable alkaline batteries have a higher starting energy density than nickel-cadmium batteries, this energy density rapidly diminishes after some years of usage.

9.5.3 Zinc-carbon

Zinc-carbon batteries are utilized a lot because of how inexpensive they are. They are not rechargeable and produce a working voltage of 1.5 V. In these batteries, the electrolyte is composed of a zinc anode, a carbon and manganese dioxide cathode, and zinc chloride. Their main shortcoming is the outside shell, which is made of zinc. The casing, which serves as the anode for the cell, may develop holes if the anode does not oxidize uniformly. Sometimes the pores that are made allow the mildly acidic electrolyte to leak, which could further damage the powered device. This sort of battery should therefore be used for non-critical applications [45].

9.5.4 Polymerized lithium

Because lithium has a high electropositive, it is a chemical element that can be used to construct ultra-thin batteries. Because of this, some lithium-based cells may have a specific energy that is five times higher than a lead-acid cell of equal size and three times more than an alkaline battery [20]. Compared to lead-acid and alkaline batteries, lithium polymer batteries are lighter, more cost-effective per use, produce a starting voltage of 3.0 V, and have a higher and more stable voltage profile. However, lithium polymer batteries are more dangerous than lead-acid and alkaline batteries because lithium can explode when it comes into contact with water. As a result, lithium and lithium-ion batteries are made in compact sizes. The 3.7 V output of some lithium-based batteries are rechargeable.

Each battery exhibits a different discharge behavior, according to [17]. Nickel-cadmium batteries are frequently used in battery-powered items such as phones, toys, and hand tools because of their high current, generally constant voltage, and tolerance to physical abuse. They do, however, undergo a memory effect since they lose capacity if they are recharged before being completely drained. Numerous cycles of drain and recharge are needed in order to restore the battery to virtually full memory. Small amounts of mercury are added to zinc-carbon batteries to prevent internal gas production that would cause leaks, potential ruptures, or a short shelf life without addressing the environmental danger. They also have a constrained capacity. Thus, alkaline batteries have largely replaced zinc-carbon batteries in these situations. Alkaline batteries have a high energy density, high-rate capability, and a very long shelf life. They perform well across a broad temperature range. They have been used to create a variety of home items, including clocks, calculators, cameras, fire and smoke detectors, and communications devices. Lithium batteries, a relatively new technology, provide great specific energy, a long cycle life, and no memory effect. They are intended to take the position of nickel-cadmium batteries in portable power tools and are frequently used in electronic devices including laptops, hearing aids, and cell phones. For applications requiring plug-in electric vehicles, they are frequently regarded as the best choice. Regular zinc-manganese dioxide batteries are an example of common batteries that are not recommended for WSN applications due to their low capacity and excessive self-discharge rates. On the other hand, alkaline batteries are preferred because of their great capacity, low self-discharge rate, and low cost.

9.6 POWER BANK MODELING

According to experimental results, battery qualities affect several collection-related design decisions, the processing of data, and the transmission of sensor data in WSNs as well as sensor node longevity. Numerous research projects have shown the necessity of approaching the issue from a quantitative perspective for a deeper comprehension of battery discharge properties and to forecast the lifespan of the battery. Different types of battery modeling work has been recently carried out. The nonlinear battery discharge characteristics are to blame for a battery technology's complexity. The material substances used in the battery electrodes (anode and cathode) as well as the rate of active element diffusion in the electrolyte are only two examples of the many variables that affect how well a battery works. The quantity of energy held in a battery is indicated by the battery capacity, which is represented as Ah (Ampere hours) or mAh (milli-Ampere hours). Battery models should accurately predict the battery lifetime without requiring complicated measurements, which can limit the applicability of these models in real-world scenarios. On the basis of precision, computational intricacy, and configuration attempt, current models should be contrasted.

Three key considerations when evaluating battery model performance are accuracy, computational complexity, and configuration effort. The degree to which experimental data and expected values of the relevant battery variables, such as lifetime and voltage, match gives the accuracy. Computational complexity depends on the time required to derive the results and configuration effort depends on the number of parameters in the model. People would like to employ a model with high accuracy but little computational complexity and configuration effort. This is not always the case, though. People must take into consideration the aforementioned aspects in order to produce reasonably close modeling results. An initial approximate prediction result using a model with low computational complexity may be useful for marketing personnel in setting up the budget. On the other hand, researchers require a sophisticated model to precisely predict the network's lifetime in order to analyze network performance more thoroughly. As a physical layer's energy consumption is decreased, advantages from shorter radio duty cycles could be realized, which is undoubtedly helpful for hardware designers. It is also important to highlight methods for lowering the indolent mode switching losses (leakage current) in CMOS based processors [14].

9.7 EVALUATION OF BATTERY MODELS IN WSNs

The overall charge (i.e., time + current) spent by a system or a device will often be measured or estimated to determine the system's energy efficiency. It is well known that the time and magnitude of the load being applied have an impact on the battery capacity [21].

Due to the practical challenges of directly measuring a battery over its full lifetime, battery modeling will be an essential addition to measurement studies for battery-aware evaluation in WSNs. We will focus on the rate capacities effect and charge retrieval, two aspects of battery discharge that are very significant for scheduling the load equivalently. Regarding their suitability for assessing WSN protocols and systems, three currently used battery modeling techniques are compared: the hybrid battery model [22], which combines the kinetic abstraction-based KiBaM [23] with an electrical circuit abstraction; and the commercial electrochemical simulator Battery Design Studio [24].

Because batteries are complex electrochemical systems, the amount of charge that may be removed before the cut-off potential is obtained relies on the intensity and timing characteristics of the load.

Two crucial battery discharge behaviors are the rate capacity effect and charge recovery. In the first case, this means that a low discharge rate (i.e., current) is more efficient than a higher one since it enables the battery to hold more charge before it hits a certain cut-off voltage. This latter statement alludes to an intermittent discharge's greater efficiency compared to a continuous discharge. These effects lead to different battery loads using the same total charge not producing the same device lifetime.

The behavior patterns which WSNs display are a mix of periodic and non-periodic trials, each with a unique signature, such as switching between the listening, transmitting, and backoff modes during the transmission of a frame. We essentially categorize battery simulation into two groups: while analytical simulations employ more straightforward abstractions to depict the battery's physical processes, the physical operations in the battery are modeled by electrochemical replications.

Reaction to load is actually based on the battery's precise chemical and structural characteristics; electrochemical modeling accurately simulates the physical processes that occur within the battery. Although these models are regarded as the "gold standard", these batteries have the following drawbacks: they have a vast amount of extremely specific variables or parameters that describe the physical makeup and chemistry of the battery. A simulation model's parameters are manufacturer-specific: they are not only designed for the single battery type. It can be difficult to integrate the mathematical framework of these simulators with popular discrete-event simulators in WSNs because they are computationally costly and of this type.

Analytical prototypes express the battery reaction to load considerably more simply than sophisticated physical models. They are actually built on abstract representations of non-electrochemical phenomena including kinetics, electrical circuits, stochastic processes, and diffusion. As a result, these concepts are more general and manageable than the models of actual procedures, but parameterization presents other difficult problems. Contrary to physical parameters that are well-defined characteristics of a specific battery, like electrolyte content or surface area, abstract parameters, like diffusion coefficients, have no physical counterparts. Therefore, these parameters must be constructed using experimental values collected.

To research the impact of load on surface temperatures, WSN applications require battery characterization. WSNs are flexible enough to be used both inside and outside. Since sensor nodes are frequently placed in inhospitable locations, battery life is frequently a significant issue. To enhance the lifetime of the sensor network, several strategies have been put forth in the literature. According to this perspective, the environment may significantly affect how the sensor nodes behave. In particular, with batteries, which are chiefly sensitive to temperature and load fluctuations, the implanted platforms in the sensor nodes can be significantly impacted by the thermal effect [25, 26]. It is generally known that the effective battery capacity of electrochemical cells fluctuates with temperature [27, 28].

Owing to the restricted availability of energy and the very long operational times, energy optimization for WSNs is a hot topic. The three primary tactics employed in WSN energy-efficient approaches are the harvesting of energy, transmission of energy, and conservation of energy. The first two methods call for the use of extra specialized hardware to produce power for recharging the on-board batteries. However, because of the more expensive, difficult, and limited energy transmission in these tactics, they will be not extensively utilized in massive WSN deployments.

As a result, energy-conscious algorithms are often designed at different layers in the architecture of WSN using energy preservation methodologies. In order to lessen high implementation prices and complexity, the majority of WSNs are developed and enhanced utilizing emulators that mimic real-world application behavior. Practically all WSN simulators incorporate the behavior of batteries using straightforward mathematical models. As a result, environmental variables like load, temperature, and humidity are ignored in faulty modeling, which results in serious errors. As highly nonlinear devices, batteries discharge in a variety of ways based on the chemistry of the battery, the present SoC, and the functioning conditions. In order to create energy-aware algorithms, WSN batteries must be characterized according to their intended use.

9.8 WHY ENERGY HARVESTING?

In WSNs, energy is a crucial resource since it affects how well sensors per-form in terms of data detection, processing, and transmission. The power source is one of the most crucial challenges that should be solved for long-term applications because a fixed supply of utility or swapping the battery manually is not technically or financially feasible. EH technologies, that can collect energy from renewable energy sources in the environment (such as solar power, thermal energy, wind energy, kinetic energy (acceleration), and electromagnetic energy (radio waves)), have been recently presented to charge (power) the sensor nodes and thus attain the goal of perpetual network system operation. It thus takes care of the energy supply issue in order to create permanent, unattended networks. Rechargeable sensor networks or EH sensor networks are two names for these WSNs. As men-tioned above, a node in a WSN has different units with three basic func-tions: sensing, processing, and transmitting. As the technical idea gets its new form along with an optimized algorithm, the battery usage increases. However, researchers have developed various less complex and low energy consuming protocols, but always the battery loses its entire energy. To avoid this and to make the sensor node continuously work, EH is the only solution [29].

The general concept of EH is depicted in Figure 9.6. There are two basic categories of EH sources: (1) ambient sources (direct energy conversion from ambient sources to electrical energy, which is subsequently utilized to charge

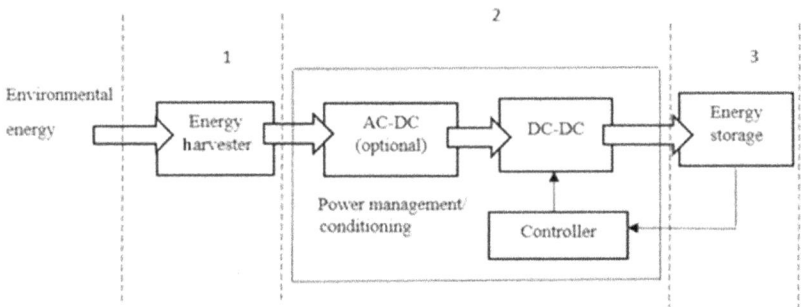

Figure 9.6 Block diagram of EH system.

the sensor nodes without the requirement for battery storage); and (2) external sources (in which it is necessary to store the converted electrical energy before being supplied to the sensor motes). Environmental factors, such as radio frequency (RF)-based EH, solar-based EH, thermal-based EH, and flow-based EH, typically offer free ambient EH sources (e.g., hydro-based EH and wind EH). On the other hand, external EH sources, which can be both mechanical and human-based, are obviously used for EH purposes in such settings [30].

9.9 DIFFERENT TYPES OF ENERGY HARVESTING

Sensor nodes with EH capabilities can continuously operate by collecting ambient energy from nearby sources and storing it in batteries for later use. Typically, various ambient sources have different characteristics and EH systems. For instance, the solar energy that the solar panel collects is both predictable and uncontrollable. The energy picked up by the RF receiver is predictable and controllable. There is no one method that would work for all applications because the energy sources are composed of many types of ambient energy. Many kinds of EH systems are developed to meet a variety of application requirements. As depicted in Figure 9.6, EH systems typically include energy collection components, hardware for power conversion and conditioning, and storage components. The amount of energy that can be captured is determined by the energy collection components. For the sake of directly powering the sensor node or charging the system's batteries, a conversion hardware device and power conditioning system turn the captured energy into electricity. The sensor nodes are powered by the captured energy that is not consumed right away by storage devices [29].

9.9.1 RF-based EH in WSNs

Due to its predictable and controllable characteristics, RF-based EH technology is one of the most widely used EH methods. The sensor nodes in RF-based EH sensor networks gather the energy transmitted by radio waves and transforms them from DC power to self-power. There are various methods (single-stage vs. multistage) for converting RF signals into DC power [29]. The chosen strategy is determined by the intended requirement of the application (i.e., voltage, efficiency, or power). The amount of RF energy that may be gathered depends on a number of factors, including the initial power, antenna gain, distance between the source and the receiver, and the energy conversion efficiency.

These subsystems include (see also Figure 9.7):

1. Receiving antenna – which is used to collect RF energy around the immediate surroundings;

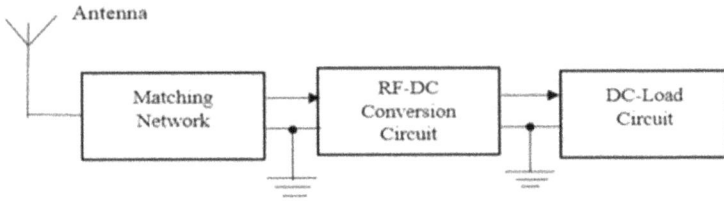

Antenna

Matching
Network

RF-DC
Conversion
Circuit

DC-Load
Circuit

Figure 9.7 RF energy harvesting.

2. Matching network – this technique guarantees the highest RF to DC conversion efficiency;
3. Rectenna – which is used to transform electromagnetic radiation into direct current power [31].

A significant task to investigate is the effectiveness of RF harvesting given its low output power. Additionally, the amount of power gathered may be unexpected due to a number of variables, including the transmission medium, the distance from the transmitting antenna, the directivity of the antenna, and the traffic density. Most studies are done on narrow-band rectenna signals that have a solo frequency and a low DC output power. Voltage multipliers or doublers can be used in a variety of applications to increase the output DC voltage and conversion efficiency. By using ultra-wideband and broadband antennas, the DC output value can be improved.

9.9.2 Solar-based EH in WSNs

Due to its predictable and plentiful qualities in the environment, the solar-based harvesting method is an alternative well-liked EH technique. Thermal conversion and photovoltaic conversion are the two main methods for transforming sunshine into useful energy. Since photovoltaic conversion has a higher power density, it is more widely used. Numerous variables, including sun intensity, cloud cover, heat buildup, and relative humidity, have an impact on solar energy efficacy. Developers must maintain the best possible effectiveness during day time hours because of the limits of solar energy collecting equipment at night in order to assure the feasibility of solar electricity [29]. An example of a solar harvesting unit for industrial WSNs will now be discussed.

A solar energy conversion system that produces electrical energy can be used to power WSN nodes. The block diagram of a typical solar EH system (SEH) is shown in Figure 9.8 [32], which briefs functionality and the performance of solar-powered WSN nodes. The solar panel followed by a DC-DC converter, energy storage unit (battery or super capacitor), followed by a control unit for the DC-DC converter, and finally to the protective circuit for the energy storage unit are the main elements of the SEH.

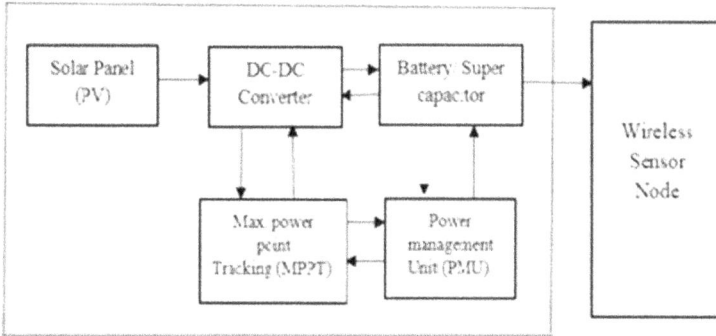

Figure 9.8 Block diagram of solar EH unit.

- Photo-voltaic cell: For, WSN nodes without access to a fixed power, solar energy is a great alternative energy source. Photovoltaic (PV) panels, which are made up of a number of cells of a PV substance, are used to generate solar energy. These PV cells are positioned in the PV generator in parallel and series configurations to produce the necessary output voltage and current.
- DC-DC converter: A DC-DC converter modifies the voltage level of an input DC voltage. This converter is used to match the input requirements of the battery with the output voltage from the solar cells.
- Maximum power point tracking (MPPT) technique: The MPPT method is frequently used in PV applications to capture the most solar energy possible under all sun conditions. The MPPT controller continuously measures the voltage and current flowing from the solar panel in order to determine the duty cycle to be fed to the switching part of the DC-DC converter. The most popular algorithm for the MPPT techniques in all varieties of sun harvesting applications is the Perturb & Observe (P&O) technique [33].

An example circuit for solar EH is shown in Figure 9.9. In this concept, battery charging is controlled by a hardware circuit employing a voltage regulator and switching device. Rechargeable batteries are powered by sunlight. If the battery voltage rises over the regulator's upper limit, which is defined by the output switching diode, battery charging will halt. The switching diode turns on and the battery continues to charge when the voltage of the battery falls below the threshold. High and medium power solar systems frequently use the P&O approach of the MPPT algorithm. However, this necessitates extremely intricate control strategies that are incorporated into the architecture using micro-controllers and digital signal processors (DSPs). This is avoided in this system to keep it straightforward, strong, and long-lasting because it raises the cost and complexity of the system [32].

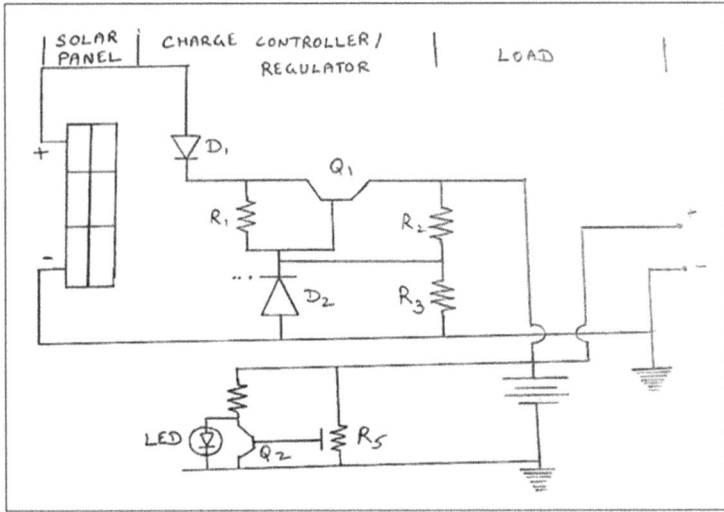

Figure 9.9 A circuit of solar EH.

Due to its availability in the environment, WSNs that employ solar energy for EH are beneficial for environmental surveillance applications.

9.9.3 Wind-based EH in WSNs

The high energy density and environmental abundance of wind-based energy collecting technology make it a good alternative power source. It is captured via a (micro) wind turbine as kinetic energy powering the sensor nodes. The energy supply is always flimsy and unstable because of its unpredictable and uncontrollable characteristics. Numerous studies have been done regarding wind-based EH system design and optimizing network operations. Due to its unpredictability and non-controllability, wind energy is always combined with other energy sources to ensure network operation. The following hybrid EH systems use some of these applications [29].

The block diagram (Figure 9.10) shows several wind energy harvesting systems (WEHSs)-WSN subsystems. The electrical energy used to power the sensor mote is obtained from a micro-wind energy harvester; a Schottky

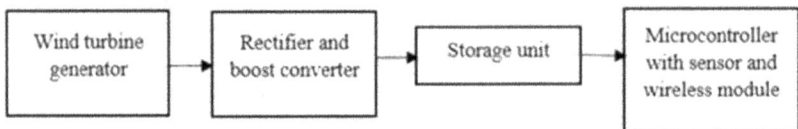

Figure 9.10 Block diagram of wind EH.

diode bridge rectifier is also employed. The rectified electrical signal that was captured is amplified by the boost converter. A different innovative approach is used in the storage unit, using super capacitors as storage devices rather than batteries. The measured signal is transmitted from the structure to the base station by the wireless structural health monitoring system's accelerometer sensor, microprocessor, and XBee transceiver module [34].

9.9.4 Thermal-based EH in WSNs

The idea behind thermal EH is the creation of an electric current whenever there is a difference in temperature between two conducting materials. Both the environmental specifications and the thermoelectric materials are stringent for thermal-based sensors. Environmental research on thermally powered sensor nodes, or thermoelectric generators, already exist. Thermal-based EH amid the air/water contact was used to power sensor nodes using a thermoelectric generator (TEG), which was fastened to a sheep's collar to capture the variation in temperature between the sheep's body and the surrounding environment. Thermal harvesting effectiveness is typically between 5 and 6%, which is poor and poses a significant obstacle to its broad use. Thermal energy harvesters are better suitable for large-scale power generation, such as steam turbines [29].

The following example is a thermal-based EH for temperature monitoring IIOT [35]. Typically, a commercial TEG is a collection of series-connected. As depicted in Figure 9.11, ceramic plates with electrical insulation and heat conductivity are clamped between semiconductor thermocouples. Each thermocouple is made of p-type and n-type semiconductor material and will

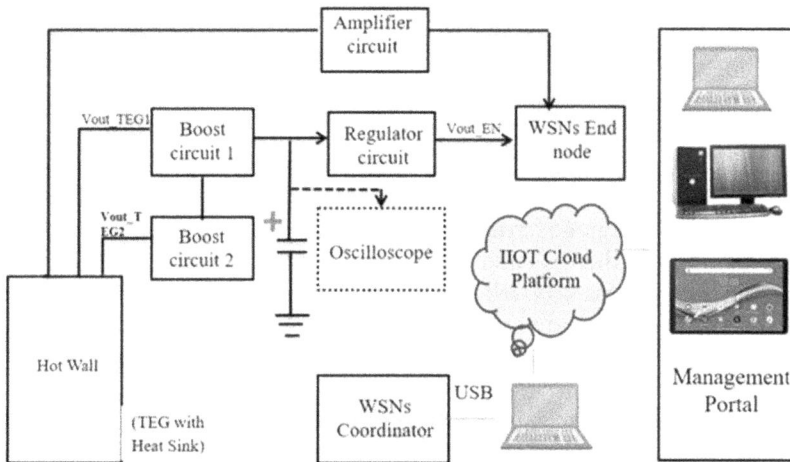

Figure 9.11 Block diagram of thermal-based temperature monitoring in IIOT.

produce electric energy due to the Seebeck effect whenever there is a temperature differential across a TEG; the total output of all the thermocouples will do the same, and microwatts/K of electricity will be delivered in a TEG. A TEG's open circuit voltage (V_G), which may be determined using the formula given below, relates to the difference in temperature across the TEG.

$$V_G = N\alpha_{pn}\Delta T_{TEG} \tag{9.1}$$

where:
   N = the number of semiconductor thermocouples;
   α_{pn} = the Seebeck coefficient of the thermocouple;
   ΔT_{TEG} = the temperature difference across the thermocouples in a TEG.
   α_{pn} and ΔT_{TEG} are defined as:

$$\alpha_{pn} = \alpha_p - \alpha_n \tag{9.2}$$

$$\Delta T_{TEG} = \left(K/K + 2K_{in}\right)\Delta T = \beta\Delta T \tag{9.3}$$

where:
   α_p = Seebeck coefficient of p-type semiconductor;
   α_n = Seebeck coefficient of n-type semiconductor.

1T is the exterior temperature differential between the two ceramic plates, K is the thermal conductivity of the ceramic plate, while K_{in} is the thermocouples' internal thermal conductivity.

9.9.5 Hybrid EH system for WSNs

Combining many energy harvester types in a single sensor node may enhance performance due to the diverse properties of different energy sources. There are a variety of well-liked hybrid energy harvesting devices, such as:

1. Solar panels and a TEG serve as energy converters in the solar/thermal systems in organic fertilizer plants [36].
2. Using solar, thermal, and electromagnetic technologies to make autonomous wireless network nodes more functional [37, 38].
3. Utilizing solar and wind energy to produce energy-autonomous sensing systems [39, 40].

The authors in [41] have given a detailed review of the different types of EH as well as EH for specific applications like self-powered medical wristbands for IoT-based health monitoring of patients and a self-powered campus water supply system.

The authors in [42] have proposed a hybrid EH model which works on a threshold value, that is when the energy level goes down beyond a certain value, EH commences until the node reaches its full charge. If the energy level does not change after a certain amount of time, the node status is broadcast, and it is then ignored from the current route to prevent connection breakdowns. According to the simulation results, AODV/DSDV performs better than LEACH employing H-EHW. Due to end-to-end delay, which is highest for AODV/LEACH, EHW performance worsens. H-EHW controls sufficient energy levels to prevent node failure and link breakage brought on by energy exhaustion.

9.9.6 Human-based EH System for WSNs

There are two sorts of human-based EH devices: activity-based energy gathering and EH based on physiological characteristics. Sensor nodes that harvest energy from human activity, such as movement, finger position changes, body heat, blood flow, and heartbeat, may do so and use that energy to generate the power they require to operate. Recently, a wireless body area network (WBAN) has received a lot of interest. The authors in [43] have discussed the conversion of thermal energy to power in the human body. In this study the efficiency of power transfer from the body part to the application is optimized while human body heat is also scavenged. We contrast two strategies and focus on the critical interaction between the thermal harvester and the power conditioning circuitry. The strategies are: (1) a high output voltage, low thermal resistance lTEG coupled with a high efficiency passively controlled single inductor DC-DC converter; and (2) a high thermal resistance, low electric resistance mTEG paired with a low-input voltage coupled with an inductor-based DC-DC converter.

9.10 OTHER HARVESTING METHODS

The majority of the control systems in intelligent building systems depend on WSNs. In these networks, there are still ways to cut back on the power going to their sensor nodes. These nodes require power in the milliwatt to microwatt range. Typically, primary non-rechargeable batteries are used to power the nodes. It might become impossible to renew or replace these batteries, which would have a negative effect on the environment. This issue might be resolved by energy harvesters (EHs), which could supplement the use of batteries at the sensor nodes while prolonging their lifespans. Mechanical motion, heat, light, radio frequency, and fluid flow have all been taken into account as environmental sources for harvester power extraction. Moreover, the prospective locations and construction methods for power extraction using these harvesters are as follows [44].

9.10.1 Piezo-electric-based EH

Piezoelectric EH is the study of passively extracting energy from ambient dynamic, thermal, and electrical activity using instruments composed of piezoelectric materials. The constitutive equation that connects mechanical deformation to electrical generation governs the direct piezoelectric effect, which in turn controls the piezoelectric energy harvester (PEH):

$$[\delta D] = \left[S^E \, d \, d \in^T \right] [\sigma E] \tag{9.4}$$

PEH equipment comes in a variety of forms, including sheets, cymbal-shaped objects, cylindrical objects, and disks. The deflection level of sheet-shaped devices, which are frequently utilized in strain piezoelectric production, determines its rating. Similar to this, cymbal-shaped piezoelectric transducers produce power as a result of deflection brought on by external loads (such as car and human traffic). It is significant to remember that PEHs should be set up below the flooring's surface. They must function without obstructing human traffic. Some other ways to extract the energy using a PEH system are as follows. The air movements inside the ducts of heating, ventilation, and air conditioning (HVAC) systems can be used as a source of energy by PEHs. Most commercial, industrial, or residential buildings that have a lot of harvesting potential use these ductwork systems. Some studies claim to have replicated ductwork system conditions in wind tunnels. Additionally, pedestrian circulation over PEH-made surfaces can produce electricity for sensors and low-power gadgets like phone chargers.

9.10.2 Pendulum-based EH systems

In order to exploit pendular oscillation mechanisms in power production applications, a great deal of research has been done on them. First, a traditional pendulum mechanism consists of a mass (m) suspended from a pivot on a firm frame that is fastened to a firm base by a string. The pendulum's swing amplitude (θ) and oscillation period (T) are given by:

$$T = 2\pi \sqrt{1/g}; \quad \theta = 1 \tag{9.5}$$

where l is the pendulum's length and g is the acceleration due to gravity.

A diagrammatic representation of a fundamental pendular oscillation mechanism and its key EH components is shown in Figure 9.12. As seen in the illustration, the vibrations felt at this mechanism's base, which in this case is vertical, can cause the pendulum to oscillate (y-axis). The generator will produce power as a result of the pendulum's oscillation's torque over the generator's shaft. To convert the obtained signals to DC power, the power needs to go via a power conditioning stage. This section's main focus is on

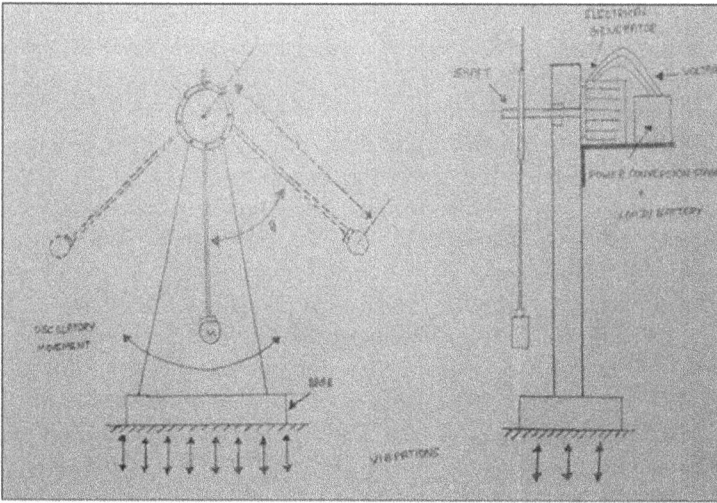

Figure 9.12 Representation of a pendulum.

how this technology is being used at the building level. When installed in big volume reservoirs (such as water tanks, swimming pools, and water cisterns) or mechanical/electrical machinery, a pendular mechanism's movement can be exploited to generate energy.

9.10.3 Electromagnetic Induction-based EH

Faraday's law of electromagnetic induction is the foundation of electromagnetic energy harvesters (EEHs). These gadgets are also referred to as "energy harvesters for magnetic induction". All of the examined literature in this section, including documented realizations of pendulum devices, depicts technology in prototype phases. Systems for reducing vibration are the main topic in this area.

9.10.4 Some other technologies

In this section, **artificial illumination** technologies such as LEDs, fluorescent lighting, and incandescent bulbs are employed for solar cell testing. All the intensities match the normal light bulbs used in workplaces, hotels, restaurants, hospitals, and other types of facilities. WSN nodes could be powered by LEHs with batteries.

Low-power devices, including indoor sensors, can be powered by RF waves, often known as the **RF spectrum**. The input signals for these harvesters must come from the frequency range emitting between 3 and 300 kHz. Base stations that transmit and receive waves are mostly used for communications in the RF band. This spectrum is used to broadcast a variety of services, including TV, AM/FM radio, long term evaluation (LTE), general packet radio service (GPRS), global system for mobiles (GSM), and Wi-Fi. For power transfer, transmission of signals from a base station to a device (such as sensors) is dependent on distance. A power beacon, which shows that these RF harvesters could attain efficiencies of up to 84%, is one method that can be used to lessen the signal intensity loss.

Another application for EHs could be **water flow** in constructing pipelines. The potential energy (high water head), kinetic energy, and water pressure in pipelines might all be used by micro-turbines to generate electricity. The kinetic and pressure energy of the vertical water pipelines has a very high potential for energy collection. The hot/cold water return pipes for substantial air conditioning systems, as well as the tap water supply and drainage systems, are appropriate MEH application sites in buildings. MEHs have a lot of potential applications in wireless sensors, data monitoring systems, and intelligent metering. Because they can recover the energy lost during pressure reductions, some new kinds of control valves for releasing excess pressure in the water pipes have significant promise for EH.

9.11 CONCLUSION

This chapter has provided a brief discussion of M2M to IoT, WSNs, and the usage of these technologies in various fields in the current scenario. The importance of batteries in the sensor nodes has been explained clearly along with the types of batteries used. To extend the lifetime of the battery, researchers have focused on power monitoring and efficient management techniques. The new booming technology in this field is EH to the sensor nodes. We have elaborately discussed the need of EH in the field of WSNs. Various types of harvesting technologies like RF, solar, wind, thermal, human, and hybrid-based EH have been discussed. The last part of the chapter discussed EH on a smaller scale from piezoelectric energy, to pendulum based, artificial illumination, water flow, and RF spectrum.

The potential for enhancing the battery life of sensor node devices is a reality. Many other EH methods may be developed that will eventually replace batteries through additional field study. This last point is crucial because it will lower the danger of contamination if a battery is used longer in the sensor motes or in any other electrical applications provided it is handled and disposed of properly.

REFERENCES

[1] Holler, J., Tsiatsis, V., Mulligan, C., Karnouskos, S., Avesand, S. and Boyle, D., 2014. *Internet of things*. Academic Press.
[2] https://psiborg.in/m2m-communication-and-connected-devices/

[3] Ijemaru, G.K., Ang, K.L.M. and Seng, J.K., 2022. Wireless power transfer and energy harvesting in distributed sensor networks: Survey, opportunities, and challenges. *International Journal of Distributed Sensor Networks*, 18(3), p. 1550 1477211067740.

[4] Devasiya, A., (2021). Technical article, https://control.com/technical-articles/what-is-machine-to-machine-communication-m2m/

[5] Larry-McCreary, (2015). https://www.fierceelectronics.com/person/larry-mccreary

[6] https://www.ionos.com/digitalguide/server/know-how/what-is-machine-to-machine-communication-m2m/, 2020.

[7] Shea, S., (2019). https://www.techtarget.com/iotagenda/definition/machine-to-machine-M2M

[8] Sudarmani, R., Venusamy, K., Sivaraman, S., Jayaraman, P., Suriyan, K. and Alagarsamy, M., (2022). Machine to machine communication enabled internet of things: A review. *International Journal of Reconfigurable and Embedded Systems*, 2089(4864), p. 4864.

[9] Akyildiz, I.F., Su, W., Sankarasubramaniam, Y., Cayirci, E. (2002). Wireless sensor networks: A survey, *Computer Networks*, 38(4), pp. 393–422.

[10] Leung, V. Lecture Slides on "Wireless Sensor Networks", University of British Columbia, Canada.

[11] Aznoli, F. and Navimipour, N.J., (2017). Deployment strategies in the wireless sensor networks: Systematic literature review, classification, and current trends. *Wireless Personal Communications*, 95(2), pp. 819–846.

[12] Bala, T., Bhatia, V., Kumawat, S. and Jaglan, V., (2018). A survey: Issues and challenges in wireless sensor network. *International Journal of Engineering & Technology*, 7(2), pp. 53–55.

[13] Elhoseny, M., Farouk, A., Batle, J., Shehab, A. and Hassanien, A.E., (2020). Secure image processing and transmission schema in cluster-based wireless sensor network. In *Sensor Technology: Concepts, Methodologies, Tools, and Applications* (pp. 698–715). IGI Global.

[14] Wenqi, G.U.O. and Healy, W.M., (2014). Power supply issues in Battery Reliant Wireless Sensor Networks: A review. *International Journal of Intelligent Control and Systems*, 19(1), pp. 15–23.

[15] Culler, D., Estrin, D. and Srivastava, M., (2004). Guest editors' introduction: Overview of sensor networks, *Computer*, 37(8), pp. 41–49.

[16] Reddy, T., (2010). *Linden's battery handbook*, 4th ed., McGraw-Hill Professional.

[17] Young, J. K., (2008). A Practical Guide to Battery Technologies for Wireless Sensor Networking, Available: http://www.sensorsmag.com/networking-communications/batteries/a-practical-guide-battery-technologies-wireless-sensor-netwo-1499

[18] Crossbow-Technology-Inc., (2005). MPR/MIB User's Manual Rev. B Document 7430-0021-06, Available: http://www.xbow.com/Support/Support_pdf_files/MPR-MIB_Series_Users_Manual.pdf

[19] Wikipedia, (2012). Alkaline battery, Available: http://en.wikipedia.org/wiki/Alkaline_battery

[20] Battery Technology, (2012). Battery Technology, Available: http://faculty.bus.olemiss.edu/breithel/b620s02/Humphrey/battery%20research.htm

[21] Rohner, C., Feeney, L.M. and Gunningberg, P., "Evaluating Battery Models in Wireless Sensor Networks", Wired/Wireless Internet Communication, WWIC 2013. Lecture Notes in Computer Science, vol. 7889. Springer, Berlin, Heidelberg, pp. 29–42. https://doi.org/10.1007/978-3-642-38401-1_3.

[22] Kim, T. and Qiao, W., (2011). A hybrid battery model capable of capturing dynamic circuit characteristics and nonlinear capacity effects, *IEEE Transactions on Energy Conversion*, 26(4), pp. 1172–1180.

[23] Manwell, J. and McGowan, J., (1993). "Lead acid battery storage model for hybrid energy system", *Solar Energy*, 50(5), pp. 399–405.

[24] Battery Design LLC: Battery design studio, http://www.batsdesign.com

[25] Bannister, K., Giorgetti, G. and Gupta, S., (2008). "Wireless sensor networking for hot applications: effects of temperature on signal strength, data collection and localization", In *Proceedings of the 5th Workshop on Embedded Networked Sensors (HotEmNets' 08)*, pp. 1–5. Penn State University (psu.edu).

[26] Boano, C.A., Tsiftes, N., Voigt, T., Brown, J. and Roedig, U., (2009). "The impact of temperature on outdoor industrial sensornet applications", *IEEE Transactions on Industrial Informatics*, 6, 451–459. https://doi.org/10.1109/TII.2009.2035111.

[27] Jaguemont, J., Boulon, L., Venet, P., Dubé, Y. and Sari, A., (2015). "Lithium-ion battery aging experiments at subzero temperatures and model development for capacity fade estimation", *IEEE Transactions on Vehicular Technology*, 65, 4328–4343. https://doi.org/10.1109/TVT.2015.2473841.

[28] Chen, M. and Rincon-Mora, G.A., (2006). "Accurate electrical battery model capable of predicting runtime and IV performance", *IEEE Transactions on Energy Conversion*, 21, 504–511. https://doi.org/10.1109/TEC.2006.874229.

[29] Zhang, Y., (2020). Energy Harvesting Technologies in Wireless Sensor Networks. In *Encyclopedia of Wireless Networks* (pp. 414–419). Cham: Springer International Publishing.

[30] Ijemaru, G.K., Ang, K.L.M. and Seng, J.K., 2022. Wireless power transfer and energy harvesting in distributed sensor networks: Survey, opportunities, and challenges. *International Journal of Distributed Sensor Networks*, 18(3), p.15501477211067740.

[31] Sansoy, M., Buttar, A.S. and Goyal, R., (2020, February). Empowering wireless sensor networks with RF energy harvesting. In *2020 7th International Conference on Signal Processing and Integrated Networks (SPIN)* (pp. 273–277). IEEE.

[32] Antony, S.M., Indu, S. and Pandey, R., 2020. An efficient solar energy harvesting system for wireless sensor network nodes. *Journal of Information and Optimization Sciences*, 41(1), pp. 39–50.

[33] Sharma, H., Haque, A. and Jaffery, Z.A., 2018. Modeling and optimisation of a solar energy harvesting system for wireless sensor network nodes. *Journal of sensor and Actuator Networks*, 7(3), p. 40.

[34] Sathiendran, R.K., Sekaran, R.R., Chandar, B. and Prasad, B.A.G., 2014, March). Wind energy harvesting system powered wireless sensor networks for structural health monitoring. *In 2014 International Conference on Circuits, Power and Computing Technologies [ICCPCT-2014]* (pp. 523–526). IEEE.

[35] Hou, L., Tan, S., Zhang, Z. and Bergmann, N.W., (2018). Thermal energy harvesting WSNs node for temperature monitoring in IIoT. *IEEE Access*, 6, pp. 35243–35249.

[36] Chottirapong, K., Manatrinon, S., Dangsakul, P. and Kwankeow, N., (2015). Design of energy harvesting thermoelectric generator with wireless sensors in organic fertilizer plant. In: *2015 IEEE 6th international conference of information and communication technology for embedded systems (IEEE IC-ICTES)*, pp. 1–6.

[37] Virili, M., Georgiadis, A., Collado, A., Niotaki, K., Mezzanotte, P., Roselli, L., Alimenti, F. and Carvalho, N.B., (2015). Performance improvement of rectifiers for wpt exploiting thermal energy harvesting. *Wireless Power Transfer*, 2(1), pp. 22–31.

[38] Virili, M., Georgiadis, A., Mira, F., Collado, A., Alimenti, F., Mezzanotte, P. and Roselli, L., (2016). Eh performance of an hybrid energy harvester for autonomous nodes. In: *2016 IEEE topical conference on wireless sensors and sensor networks (IEEE WiSNet)*, pp. 71–74.

[39] Cammarano, A., Petrioli, C. and Spenza, D. (2016). Online energy harvesting prediction in environmentally powered wireless sensor networks. *IEEE Sensors Journal*, 16(17), pp. 6793–6804.

[40] Spenza, D., Petrioli, C. and Cammarano, A. (2012). Pro-energy: A novel energy prediction model for solar and wind energy-harvesting wireless sensor networks. In: *2012 9th IEEE international conference on mobile ad-hoc and sensor systems (IEEE MASS 2012)*, pp. 75–83.

[41] Bathre, M. and Das, P.K., (2020, July). Hybrid energy harvesting for maximizing lifespan and sustainability of wireless sensor networks: A comprehensive review & proposed systems. In *2020 international conference on computational intelligence for smart power system and sustainable energy (CISPSSE)* (pp. 1–6). IEEE.

[42] Kaur, J. and Bindal, A.K., (2019). Resource aware hybrid energy harvesting for wireless sensor networks. *Journal of Computational and Theoretical Nanoscience*, 16(10), pp. 4117–4124.

[43] Thielen, M., Sigrist, L., Magno, M., Hierold, C. and Benini, L., (2017). Human body heat for powering wearable devices: From thermal energy to application. *Energy Conversion and Management*, 131, pp. 44–54.

[44] Hidalgo-Leon, R., Urquizo, J., Silva, C.E., Silva-Leon, J., Wu, J., Singh, P. and Soriano, G., (2022). Powering nodes of wireless sensor networks with energy harvesters for intelligent buildings: A review. *Energy Reports*, 8, pp. 3809–3826.

[45] Makimaa, Y.P., Sudarmani, R. and Vanithamani, R., (2022). Different energy harvesting, energy saving techniques to improve battery energy of Wireless Sensor Networks – A review. *Stochastic Modeling & Applications*, 26(3), pp. 72–83.

Chapter 10

An efficient lightweight signature verifiable scheme for node authentication using trivariate polynomials over elliptic curve cryptography for decentralized distributed public key infrastructures

Manoj Kumar
Gurukula Kangri Vishwavidyalaya, Haridwar, India

Suryya Farhat
ADGITM, Delhi, India

Kumar Gautam
Gwangju Institute of Science and Technology, Gwangju,
Republic of Korea

CONTENTS

DOI: 10.1201/9781003436461-10

10.1 INTRODUCTION

Due to the rapid growth of internet technology and wireless communications, lightweight cryptography has been gaining popularity over the last few decades. Small computing devices, lightweight devices, or more generally speaking resource-constrained devices are in use, such as radio frequency identification (RFID) tags, contactless smart cards, wireless sensor networks, smartphones, cloud computing, mobile ad hoc networks [1], wireless patient monitoring systems (WPMSs), the Internet of Things (IoT), and wireless ad hoc networks (WANETs). These lightweight devices have a limited power supply and reduced storage and computing capabilities. Therefore, due to these limitations, there is a need for new cryptographic primitives that can satisfactorily be implemented in these ubiquitous small devices.

A WANET is a dynamic type of network [2]. Nowadays it is gaining in popularity due to its property of lacking an infrastructure. It is a self-organized network and consists of dynamic nodes that move independently. As each node has its limited range of signal transfer efficiency, all nodes are capable of interchanging their valuable information only within that range. WANETs use two types of approaches, namely a public key infrastructure (PKI) and identity based information (IBI). In the present work we use the former.

Due to infrastructure deficiency and central authority (CA), secure communication is a challenging task in WANETs. To provide a secure communication infrastructure, one of the most important and widely used infrastructures is public key cryptography (PKC). PKC is a type of management that combines a public key with the corresponding identity of an entity such as a human being or an organization. PKC plays the role of polishing hardware, software, and a procedure to create, control, allocate, utilize, stock, and reverse digital authentication and to manage the encryption of public keys [3–5].

The present work uses the Edwards digital signature algorithm (EdDSA) [6] and trivariate polynomials over elliptic curve cryptography (ECC) to propose an improved lightweight signature verifiable scheme for the authentication of the nodes of a WANET for decentralized distributed PKI. The rest of the chapter is organized as follows. Section 10.2 introduces some basic terminology related to the proposed work. Section 10.3 involves the proposed scheme for node authentication of a WANET for decentralized distributed PKI. Section 10.4 illustrates the proposed scheme with a suitable example. Section 10.5 interprets the encryption/decryption of the proposed scheme. The correctness of the scheme is described in Section 10.6. Finally, the results and a comparison with conventional methods are presented in Section 10.7 followed by the conclusion in the last section.

10.2 PRELIMINARIES

10.2.1 Elliptic curves

Elliptic curves are derived from Weierstrass equations [7], 8] and are defined by:

$$y^2 = x^2 + ax + b$$

where the parameters a and b are taken from a finite field F satisfying the non-singularity condition:

$$4a^3 + 27b^2 \neq 0.$$

10.2.2 Edwards curves

An Edwards curve is a collection of elliptic curves. As its name indicates it was introduced by the great mathematician Harold Edwards [9, 10] in 2007. Mathematically an Edwards curve over a finite field F is defined as

$$E_d = \left\{ (x,y) : x^2 + y^2 = 1 + dx^2 y^2 \right\} \tag{10.1}$$

where $\in F$, $d \neq 0$, 1, and $char(F) \neq 2$.

If $P_1 = (x_1, y_1)$ and $P_2 = (x_2, y_2)$ are two points (which may be equal or distinct) on an Edwards curve E_d, then $P_1 + P_2 = R = (x_3, y_3)$ is defined as:

$$x_3 = \frac{x_1 y_2 + x_2 y_1}{1 + d x_1 x_2 y_1 y_2} \tag{10.2}$$

$$y_3 = \frac{y_1 y_2 - x_1 x_2}{1 - d x_1 x_2 y_1 y_2} \tag{10.3}$$

It is remarkable that the Edwards curve E_d forms an abelian group under the addition operation defined by (10.2) and (10.3). Unlike the elliptic curves, the formulae (10.2) and (10.3) are valid even if $P_1(x_1, y_1) = P_2(x_2, y_2)$.

10.2.3 Twisted Edwards curves

A twisted Edwards curve is the generalization of an Edwards curve and was jointly introduced and studied by Bernstein et al. [11, 12]. A twisted Edwards curve $E_{a,d}$ is a twist of an Edwards curve E_d, and is defined as a set of points (x, y) satisfying the equation

$$E_{a,d} = \left\{ (x,y) : ax^2 + y^2 = 1 + dx^2 y^2 \right\} \tag{10.4}$$

where a,d range over the field F with a characteristic not equal to 2.

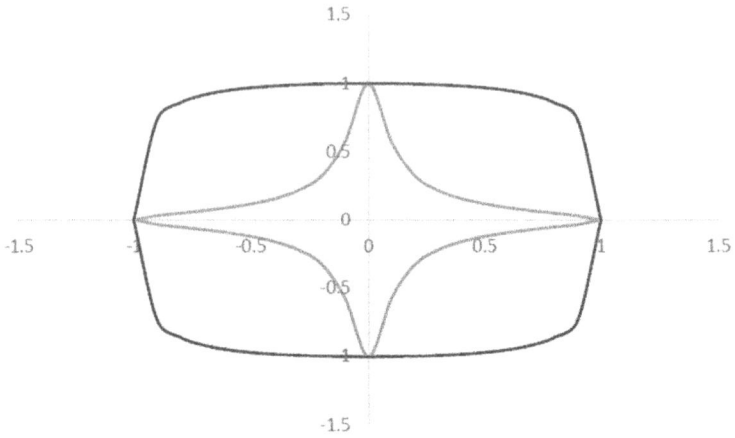

Figure 10.1 Edward curves $x^2 + y^2 = 1 + 200x^2y^2$ (blue), $x^2 + y^2 = 1 - 0.8x^2y^2$ (red).

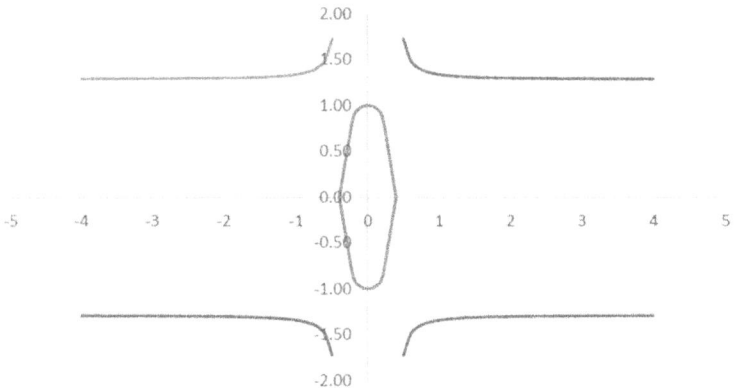

Figure 10.2 Twisted Edwards curve ($T_d : 10x^2 + y^2 = 1 + 6x^2y^2$).

It is remarkable that we use $a = 1$ in twisted Edwards curve $E_{a,d}$.

The addition of two points (which may be equal or distinct) $P_1 = (x_1, y_1)$ and $P_2 = (x_1, y_1)$ on a twisted Edwards curve T_d is a point $R(x_3, y_3) = P_1 + P_2$ on T_d defined by

$$x_3 = \frac{x_1y_2 + x_2y_1}{1 + dx_1x_2y_1y_2} \tag{10.5}$$

$$y_3 = \frac{y_1y_2 - x_1x_2}{1 - dx_1x_2y_1y_2} \tag{10.6}$$

Table 10.1 Brief tabular explanation of EdDSA

Symbol	Explanation
q	Large prime number
F_q	Finite field over q
$E(F_q)$	Elliptic curve over F_q with order $E(F_q) = 2^c l$ where l is a sufficiently large prime and 2^c is a cofactor
B	Base point of $E(F_q)$ and order of B is l
H	Hash function producing $2b - bit$ output with $b > l + q$
k	Private key (b-bit string) chosen randomly
A	Public key with $A = s * B$ where $s = H_0, ...H_{b-1}(k)$
M	Message to be signed
(R, S)	Signature on message M with $R = r * B$ and $S = (r + H(R, A, M)s) \bmod (l)$ where $r = H(H_b, ...H_{b-1}(k), M)$

10.2.4 Edwards digital signature algorithm (EdDSA)

EdDSA was jointly designed by Bernstein et al. [13] and speeds up the conventional DSA. It uses a variant of the Schnorr signature scheme together with the twisted Edwards curve. In brief, EdDSA involves the parameters shown in Table 10.1.

It is remarkable that the above discussed EdDSA is valid only if $2^c SB = 2^c R + 2^c H(R, A, M)A$. A detailed study of the scheme can be found in [13].

10.3 PROPOSED SCHEME

In this section we will propose a lightweight cryptographic scheme for EdDSA using trivariate polynomials of WANET. The proposed scheme consist of the following phases.

Phase 1: Initial Setup
The following symbols are used to initialize the setup phase of the proposed scheme:

$n_1, n_2, n_3, n_4 \rightarrow$ Nodes of WANET
$q \rightarrow$ Very large prime number
$z_q \rightarrow$ Finite field of order q
$E \rightarrow$ An additive group of elliptic curve points over z_q
$g \rightarrow$ Base point of E
$S \rightarrow$ Secret of WANET
$P^k \rightarrow$ Public key of WANET which is computed as $P^k = s * g$
$T \rightarrow$ Threshold value
$H \rightarrow$ Cryptographic hash function from $\{0, 1\}^* \rightarrow z_q$

Phase 2: Share Distribution

This phase is accomplished with the following systematic quotients:

1. Each node n_i, $i = 1, 2, ..., m$, selects a trivariate polynomial, say $F_{x,y,z}^{n_i}(x,y,z)$ over z_q to define an implicit trivariate polynomial as

$$F_{x,y,z}(x,y,z) = \sum_{i=1}^{m} F_{x,y,z}^{n_i}(x,y,z)$$

It is remarkable that according to Shamir secret sharing, secret S of the WANET is given by

$$S = F_{x,y,z}(0,0,0)$$

2. Each node n_i takes hash values $H_{n_i} = H\left(F_{x,y,z}^{n_i}(x,y,z)\right)$ of its trivariate polynomials and securely sends it to all other nodes n_j.
3. On receiving hash values H_{nj} from all other nodes n_j, each node n_i uses its own trivariate polynomial to calculate bivariate polynomials $F_{x,z}^{n_i,j} = F_{x,y,z}^{n_i}(x, H_{n_j}, z)$ and securely sends it to the corresponding node n_j with $j = 1, 2, ..., m$.

    In addition, every node n_i calculates the value $Y^{n_i} = F_{x,y,z}^{n_i}(0,0,0) * g$ and sends it together with the above bivariate polynomials.
4. Every node n_i computes its secret bivariate polynomials as

$$S_{x,z}^{n_i} = \sum_{i=1}^{m} F_{x,z}^{n_i,j}(x,z)$$

Now, a partial share of each node n_i is given by $S_{x,z}^{n_i}(0,0)$.

It is remarkable that the partial shares $S_{x,z}^{n_i}(0,0)$ can also be obtained by substituting the value of x with the hash value H_{ni} in the polynomial

$$f(x) = F_{x,y,z}(x,0,0)$$

Phase 3: Key Generation for Signature

Every node n_i randomly selects a number, say d^{ni} from z_q, as a private key and computes its public key $K_{pub}^{n_i}$ as

$$K_{pub}^{n_i} = d^{n_i} * g$$

Phase 4: Signature Generation

This involves the following systematic quotients:

1. Every node n_i in a WANET uses its share $S_{x,z}^{n_i}(0,0)$ as the secret key for the signature.

2. Each node yields a certificate by combining its own ID and the public key together with the message M to be sent

$$C_{n_i} = \{ 'node\ n_i' \parallel K_{pub}^{n_i} \parallel M \}$$

3. Each node n_i computes its partial signature as $S_p^{n_i} = H(C_{n_i}) * S_{x,z}^{n_i}(0,0)$. Furthermore, all nodes interchange their partial signatures $S_p^{n_i}$ to calculate the fully signed certificate.

4. Now if node n_i wants to calculate its certificate then its signature (R_{ni}, sig_{ni}) is computed as

$$R_{n_i} = H(C_{n_i}) * g$$

$$sig_{n_i} = \{ H(C_{n_i}) + H(R_{n_i}, P^k, C_{n_i}) * F_{x,y,z}(0,0,0) \}$$

Phase 5: Verification of Signature

Signature (R_{ni}, sig_{ni}) of node n_i, generated in a previous phase, will be valid if

$$sig_{n_i} * g = \{ R_{n_i} + H(R_{n_i}, P^k, C_{n_i}) * P^k \}$$

otherwise it is invalid.

10.4 A CONCRETE ILLUSTRATION OF THE PROPOSED SCHEME

10.4.1 Initial setup

Take $q = 101$

Take $T = 2$, where T is the threshold value

$m = 4$

$H = HTR = sha224$

$$E = E(z_{101}) : y^2 = x^3 + 13x + 37 \tag{10.7}$$

It can be easily checked that $g = E(36, 87)$ is the generator of the above elliptic curve E.

Every node n_i, $i = 1, 2, 3, 4$ selects a random symmetric trivariate polynomial in $GF(101)$ as follows:

$$F_{x,y,z}^{m_1} = 5x^2z^2y + 5y^2z^2x + 6xyz + 3x + 3y + 7 \tag{10.8}$$

$$F_{x,y,z}^{n2} = 3x^2z^2y + 3y^2z^2x + 5xyz + 6x + 6y + 5 \tag{10.9}$$

$$F_{x,y,z}^{n3} = 2x^2z^2y + 2y^2z^2x + 3xyz + 5x + 5y + 4 \tag{10.10}$$

$$F_{x,y,z}^{n4} = 4x^2z^2y + 4y^2z^2x + 7xyz + 2x + 2y + 3 \tag{10.11}$$

All the nodes of a WANET can be implicitly expressed by the polynomial:

$$\begin{aligned} F_{x,y,z} &= F_{x,y,z}^{n1} + F_{x,y,z}^{n2} + F_{x,y,z}^{n3} + F_{x,y,z}^{n4} \\ &= 14x^2z^2y + 14y^2z^2x + 21xyz + 16x + 16y + 19 \end{aligned} \tag{10.12}$$

It is remarkable that according to Shamir SST, the secret of a WANET can be taken as $S = F_{x,y,z}(0,0,0) = 19$.

Each node takes hash value H_{ni} of its trivariate polynomial and sends it to all other nodes.

Using Sha224 the hash values of all nodes are given by

$$H_{n_1} = sha224(Node\ n_1) = 94 \tag{10.13}$$

$$H_{n_2} = sha224(Node\ n_2) = 23 \tag{10.14}$$

$$H_{n_3} = sha224(Node\ n_3) = 4 \tag{10.15}$$

$$H_{n_4} = sha224(Node\ n_4) = 74 \tag{10.16}$$

On receiving the hash values H_{n2}, H_{n3}, H_{n4} from the nodes n_2, n_3, n_4 respectively, node n_1 calculates the following bivariate polynomials:

$$F_{x,z}^{11} = 66x^2z^2 + 43z^2x + 59xz + 3x + 87 \tag{10.17}$$

$$F_{x,z}^{12} = 14x^2z^2 + 19z^2x + 37xz + 3x + 76 \tag{10.18}$$

$$F_{x,z}^{13} = 20x^2z^2 + 80z^2x + 24xz + 3x + 19 \tag{10.19}$$

$$F_{x,z}^{14} = 67x^2z^2 + 9z^2x + 40xz + 3x + 43 \tag{10.20}$$

and a special value $Y^{n_1} = 7 * g = 7 * (36, 87) = (33, 12)$.

Similarly, bivariate polynomials and special values Y^{ni} of nodes n_2, n_3, n_4 respectively are given by:

$$F_{x,z}^{21} = 80x^2z^2 + 46z^2x + 66xz + 6x + 64 \tag{10.21}$$

$$F_{x,z}^{22} = 69x^2z^2 + 72z^2x + 14xz + 6x + 42 \tag{10.22}$$

$$F_{x,z}^{23} = 12x^2z^2 + 48z^2x + 20xz + 6x + 29 \tag{10.23}$$

$$F_{x,z}^{24} = 20z^2 + 66z^2x + 67xz + 6x + 45 \tag{10.24}$$

$$Y^{n_2} = 5 * g = 5 * (36,87) = (64,3)$$

$$F_{x,z}^{31} = 87x^2z^2 + 98z^2x + 80xz + 5x + 70 \tag{10.25}$$

$$F_{x,z}^{32} = 46x^2z^2 + 48z^2x + 69xz + 5x + 18 \tag{10.26}$$

$$F_{x,z}^{33} = 8x^2z^2 + 32z^2x + 12xz + 5x + 24 \tag{10.27}$$

$$F_{x,z}^{34} = 47x^2z^2 + 44z^2x + 20xz + 5x + 71 \tag{10.28}$$

$$Y^{n_3} = 4 * g = 4 * (36,87) = (17,11)$$

$$F_{x,z}^{41} = 73x^2z^2 + 95z^2x + 52xz + 2x + 68 \tag{10.29}$$

$$F_{x,z}^{42} = 92x^2z^2 + 96z^2x + 60xz + 2x + 49 \tag{10.30}$$

$$F_{x,z}^{43} = 16x^2z^2 + 64z^2x + 28xz + 2x + 11 \tag{10.31}$$

$$F_{x,z}^{44} = 94x^2z^2 + 88z^2x + 13xz + 2x + 50 \tag{10.32}$$

$$Y^{n_4} = 3 * g = 4 * (36,87) = (29,19)$$

Each node n_1, n_2, n_3, n_4 calculates its secret bivariate polynomial as follows:

$$S_{x,z}^{n_1} = 3x^2z^2 + 80z^2x + 55xz + 16x + 87 \tag{10.33}$$

$$S_{x,z}^{n_2} = 19x^2z^2 + 33z^2x + 79xz + 16x + 84 \tag{10.34}$$

$$S_{x,z}^{n_3} = 56x^2z^2 + 22z^2x + 84xz + 16x + 83 \tag{10.35}$$

$$S_{x,z}^{n_4} = 26x^2z^2 + 5z^2x + 39xz + 16x + 7 \tag{10.36}$$

According to the proposed scheme $P^k = S_k * g = 19 * E(36,87) = E(47,80)$ should be equal to $\sum_{i=1}^{4} Y^{n_i}$.

Now,

$$\sum_{i=1}^{4} Y^{n_i} = (33,12) + E(64,3) + E(17,11) + E(29,19) = E(47,80) = P^k$$

(10.37)

From Equations (10.33), (10.34), (10.35), and (10.36), the partial shares of nodes n_1, n_2, n_3, n_4 are respectively given by

$$S_{0,0}^{n_1} = 87, S_{0,0}^{n_2} = 84, S_{0,0}^{n_3} = 83, S_{0,0}^{n_4} = 7$$

(10.38)

From (10.12) we get that the partial share can also be obtained by putting $x = H_{n1}$:

$$f(x) = F_{x,y,z}(x,0,0) = 16 * x + 19$$

(10.39)

10.4.2 Key generation for signatures

Nodes n_1, n_2, n_3, n_4 randomly select numbers d^{n_1}, d^{n_2}, d^{n_3}, d^{n_4} as their private keys and compute their public keys as follows:

$$K_{pub}^{n_i} = d^{n_i} * g$$

$$node(n_1) = \{d^{n_1} = 11, K_{pub}^{n_1} = 11 * g = (8,59)\}$$

$$node(n_2) = \{d^{n_2} = 17, K_{pub}^{n_2} = 17 * g = (2,77)\}$$

$$node(n_3) = \{d^{n_3} = 29, K_{pub}^{n_3} = 29 * g = (72,4)\}$$

$$node(n_4) = \{d^{n_4} = 31, K_{pub}^{n_4} = 31 * g = (60,24)\}$$

10.4.3 Signature generation

Now every node yields a certificate by combining its own ID and public key together with the message to be sent as follows:

$$C_{n_1} = 'node(n_1)' + (8,59) + 'hello \ word'$$

$$C_{n_2} = 'node(n_2)' + (2,77) + 'hello \ word'$$

$$C_{n_3} = 'node(n_3)' + (72,4) + 'hello \ word'$$

$$C_{n_4} = 'node(n_4)' + (60,24) + 'hello \ word'$$

If node n_1 wants to sign its certificate $(C_{n_1} = 'node1' + K^1_{pub} + 'hello\ word')$ then it requests nodes n_2, n_3, n_4 for their partial signatures which are respectively computed as follows:

$$S_p^{n_2} = H\left(C_{n_1}\right) * S_{0,0}^{n_2} = \left(84,16\right)$$

$$S_p^{n_3} = H\left(C_{n_1}\right) * S_{0,0}^{n_3} = \left(5,96\right)$$

$$S_p^{n_4} = H\left(C_{n_1}\right) * S_{0,0}^{n_4} = \left(52,27\right)$$

where $S_{0,0}^{n_2} = 84, S_{0,0}^{n_3} = 83, S_{0,0}^{n_4} = 7$, and $H(C_{n_1}) = sha224(C_{n_1}) = 38$.

All nodes interchange their partial signatures $S_p^{n_i}$ to calculate a fully signed certificate.

Now signature $(R_{n_1}, sign_{n_1})$ of node n_1 is computed as:

$$R_{n_1} = H\left(C_{n_1}\right) * g = \left(88,87\right)$$

and

$$sign_{n_1} = \left\{H\left(C_{n_1}\right) + H\left(R_{n_1}, P^k, C_{n_1}\right) * F_{x,y,z}\left(0,0,0\right)\right\} = 45$$

where $g = (36, 87)$, $F_{x, y, z}(0, 0, 0) = 19$, and $P^k = 19 * (36, 87) = (47, 80)$.

10.4.4 Verification of signature

Since

$$sign_{n_1} * g = R_{n_1} + H\left(R_{n_1}, P^k, C_{n_1}\right) P^k$$

$$= sign_{n_i} * g = 45 * g = \left(61,67\right)$$

and

$$= R_{n_1} + H\left(R_{n_1}, P^k, C_{n_1}\right) P^k$$

$$= H\left(C_{n_1}\right) * g + H\left(R_{n_1}, P^k, C_{n_i}\right) F_{x,y,z}\left(0,0,0\right) * g$$

$$= \left(H\left(C_{n_i}\right) + H\left(R_{n_i}, P^k, C_{n_i}\right) * F_{x,y,z}\left(0,0,0\right)\right) * g$$

$$= sign_{n_i} * g = 45 * g = 45 * \left(36,87\right) = \left(61,67\right)$$

it verifies the validity of the signature.

10.5 ENCRYPTION/DECRYPTION BY THE PROPOSED SCHEME

If node n_i wants to send a message M to node n_j, then nodes n_i and n_j generate their pairs of private and public keys as $(d^n{}_i, e^n{}_i)$ and $(d^n{}_j, e^n{}_j)$ respectively, where public key $e^n{}_i$ is the inverse of private key $d^n{}_i$. Now message M can be encrypted into ciphertext C as:

$$C = e^{n_j}(x_M, y_M) = (x_C, y_C)$$

where $(x_M, y_M) = M$ is an elliptic curve point. Finally, the ciphertext message can be decrypted as:

$$M = d^{n_j}(x_C, y_C)$$

For example suppose node n_2 wants to send message M to node n_4, node n_1, and node n_3. The node n_i generates their private and public keys on a twisted Edwards curve as a pair:

$$node\ n_1 = (d^{m_1}, e^{m_1}) = (46, 11)$$

$$node\ n_2 = (d^{n_2}, e^{n_2}) = (6, 17)$$

$$node\ n_3 = (d^{n_3}, e^{n_3}) = (7, 29)$$

$$node\ n_4 = (d^{n_4}, e^{n_4}) = (88, 31)$$

where public key e^{n_i} is the inverse of private key d^{n_i}.

If node n_2 wants to send a message $M = 33$ to node n_4, node n_2, it uses the method discussed in [14] to represent M as an elliptic curve point $(x_M, y_M) = M^*$ as:

$$x_M = 11 \text{ and } y_M = 22$$

Now node n_2 uses e^4, the public key of node n_4, to encrypt the message as:

$$C = (x_C, y_C) = e^4(x_M, y_M) = 88(11, 22) = (47, 11)$$

After receiving the encrypted message C from node n_2, node n_4 uses its secret key d^4 to decrypt the message C as:

$$M^* = d^4(x_C, y_C) = 31(47, 11) = (11, 22)$$

which implies that $H(M) = 33$ because $H(M) = x_M + y_M$.

10.6 CORRECTNESS OF THE PROPOSED SCHEME

The correctness of the proposed scheme depends upon the following theorems.

Theorem 10.1

If $F_{x,y,z}^{n_i}$ is a trivariate polynomial of node n_i $i = 1, 2, \ldots m$, then public key $P^k = S * g$ of the WANET will also be equal to the expression ΣY^{n_i}, where $Y^{n_i} = F_{x,y,z}^{n_i}(0,0,0) * g$, g is the base point and the secret S is defined as the constant term of the implicit trivariate polynomial

$$F_{x,y,z} = \Sigma F_{x,y,z}^{n_i}$$

i.e., $S = F_{x,y,z}(0,0,0)$.

Proof: We have

$$S = F_{x,y,z}(0,0,0) \tag{10.40}$$

$$P^k = S * g \tag{10.41}$$

$$Y^{n_i} = F_{x,y,z}^{n_i}(0,0,0) * g \tag{10.42}$$

$$F_{x,y,z}(x,y,z) = \Sigma F_{x,y,z}^{n_i}(x,y,z) \tag{10.43}$$

Operating g on both sides of (10.43) after putting $x = 0 = y = z$ and using (10.40), (10.41), and (10.42) we get the required result.

Theorem 10.2

If $S_{x,z}^{n_i}(0,0)$ and H_{ni} respectively denote the partial share and hash value of the nodes n_i then

$$S_{x,z}^{n_i}(0,0) = F_{x,y,z}(H_{n_i},0,0)$$

where $F_{x, y, z}(x, y, z)$ (as defined in Theorem 10.1) is the implicit polynomial defining all nodes of the WANET.

Proof: we have

$$S_{x,z}^{n_i} = \sum_{i=1}^{m} F_{x,z}^{n_{i,j}}(x,z) \tag{10.44}$$

$$F_{x,z}^{n_i}(x,z) = F_{x,y,z}(x, H_{n_i}, z) \tag{10.45}$$

$$F_{x,z}^{n_i}(x,y,z) = \sum_{i=1}^{m} F_{x,z}^{n_i,j}(x,y,z) \tag{10.46}$$

But $F_{x,y,z}$ is symmetric in x, y so we can write from Equation (10.45)

$$F_{x,z}^{n_i}(x,z) = F_{x,y,z}(H_{n_i}, x, z) \tag{10.47}$$

Now from Equation (10.46) we get

$$F_{x,z}^{n_i}(y,x,z) = \sum_{i=1}^{m} F_{x,z}^{n_i,j}(y,x,z) \tag{10.48}$$

Putting $y = H_{ni}$ in Equation (10.48) we get

$$F_{x,z}^{n_i}(H_{n_i} x, z) = F_{x,y,z}(H_{n_i}, x, z) \tag{10.49}$$

Putting $x = z = 0$ in Equations (10.44), (10.47), and (10.49) we get

$$F_{x,z}^{n_i}(0,0) = F_{x,y,z}(H_{n_i}, 0, 0) \tag{10.50}$$

$$F_{x,y,z}^{n_i}(H_{n_i}, 0, 0) = \sum_{i=1}^{m} F_{x,y,z}^{n_i,j}(H_{n_i}, 0, 0) \tag{10.51}$$

$$S_{x,z}^{n_i}(0,0) = \sum_{i=1}^{m} F_{x,z}^{n_i,j}(0,0) \tag{10.52}$$

Combining Equations (10.50), (10.51), and (10.52) we get

$$F_{x,y,z}(H_{n_i}, 0, 0) = \sum_{i=1}^{m} F_{x,z}^{n_i,j}(0,0) = S_{x,z}^{n_i}(0,0)$$

Hence it is verified.

10.7 RESULTS AND COMPARISON

An elliptic curve digital signal algorithm (ECDSA) is a DSA implementation with an elliptic curve. It can provide 128 bits of security with only 256-bit keys, to give 128 bits of security, while RSA and DSA need key lengths of

Table 10.2 Comparison of running times of RSA, ECDSA, and
Edwards signature algorithms (milliseconds)

	RSA	ECDSA	EdDSA
Execution time	4.69	0.74	0.62

3,072 bits to give the same security. On the other hand, ECDSA uses equal-izer randomness as DSA, so the sole benefit is high speed and key length rather than security.

RSA has received some prominence as a result of the required high speeds of elliptic curves and the associated security threats. EdDSA uses a distinct elliptic curve family known as the twisted Edwards curve to address the same discrete logarithm problem as DSA or ECDSA. While it has a modest speed advantage over ECDSA, its fame stems from the increase in the secu-rity. EdDSA creates a nonce positively as a hash, formatting it as collision resistant, rather than relying on a random integer. Now, taking a move back, the usage of elliptic curves does not ensure a certain level of security. Not all elliptic curves are created equal, and only a few have passed rigorous testing. Fortunately, the PKI sector has gradually adopted Curve25519, particularly for EdDSA. Ed25519 is the public-key signature algorithm as a result of this. Public keys, like ECDSA, are twice as long as the necessary bit security. n terms of key length, EdDSA delivers the maximum level of security. It also alleviates the insecurities present in ECDSA. RSA is widely supported by secure socket shell (SSH) clients, while EdDSA is faster and offers the same level of security with much fewer keys.

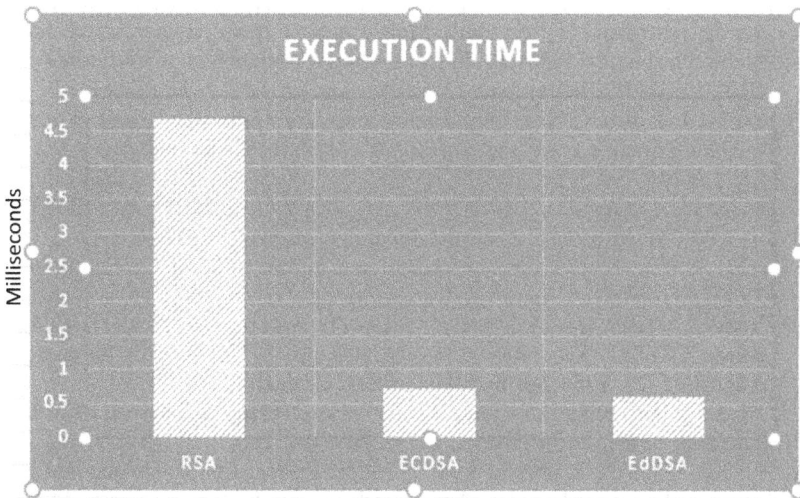

Figure 10.3 Execution time of RSA, ECDSA, and proposed EdDSA scheme.

10.8 CONCLUSION

In the present work, we have introduced a new lightweight signature scheme using EdDSA and trivariate polynomials over ECC for decentralized distributed PKI. The use of twisted Edwards elliptic curves helps to accelerate the execution speed of conventional digital signature schemes (BLS, RSA, ECDSA). It is well known that WANETs behave like resource constraints devices, that is they have less computing power and conventional cryptographic schemes cannot be implemented on these WANETs. Therefore, the proposed scheme is more beneficial than the existing ones. Furthermore, the use of twisted Edwards encryption and decryption make our scheme more lightweight. Our scheme may also be considered suitable for sensor based and wireless networks where the links are liable to be tapped by transmitting the information in pieces that makes the task of an attacker inapprehensible.

REFERENCES

1. V. Daza, J. Herranz, P. Morillo and C. Rafols, Cryptographic techniques for mobile ad-hoc networks, *Computer Networks*, 51(18), 2007, 4938–4950.
2. F. Anjum and P. Mouchtaris, *Security for wireless ad hoc networks*, Wiley-Blackwell, 2007.
3. X. Yao, X. Han and X. Du, "A light-weight certificate-less public key cryptography scheme based on ECC," *2014 23rd International Conference on Computer Communication and Networks (ICCCN)*, Shanghai, 2014, pp. 1–8.
4. M. Kumar, S. Farhat and A. Kumar, "Lightweight and authenticated key exchange based on self Linear pairings," in *Materials Today: Proceedings*, December 2020.
5. Abdul Basit, N. Chaitanya Kumar, V. Ch Venkaiah, Salman Abdul Moiz, Appala Naidu Tentu, and Wilson Naik, Multi-stage Multi-secret sharing scheme for hierarchical access structure. In *Computing, Communication and Automation (ICCCA), 2017 International Conference on*, pp. 557–563. IEEE, 2017.
6. Daniel J. Bernstein, Niels Duif, Tanja Lange, Peter Schwabe, and Bo-Yin Yang, High-speed high-security signatures (PDF), *Journal of Cryptographic Engineering* 2(2), 2012, 77–89. doi:10.1007/s13389-012-0027-1. S2CID 945254. https://ed25519.cr.yp.to/ed25519-20110926.pdf; https://doi.org/10.1007%2Fs13389-012-0027-1; https://api.semanticscholar.org/CorpusID:945254.
7. Neal Koblitz, Elliptic curve cryptosystems. *Mathematics of Computation*, 48, 1987, 203–209.
8. Victor S. Miller. Use of elliptic curves in cryptography. In Hugh C. Williams, editor, *CRYPTO'85, volume 218 of LNCS*, pp. 417–426, Santa Barbara, CA, August 18–22, 1986. Springer, Heidelberg, Germany.
9. Daniel J. Bernstein; Tanja Lange; Peter Birkner, Christiane Peters, *ECM using Edwards curves*. http://eprint.iacr.org/2008/016.pdf.
10. H. M. Edwards, A normal form of elliptic curve, *Bulletin of the American Mathematical Society* 44(3), 2007, 393–423.

11. D. J. Bernstein, P. Birkner, M. Joye, T. Lange, and C. Peters. Twisted Edwards curves. In S. Vaudenay, editor, *Progress in Cryptology – AFRICACRYPT 2008*, volume 5023 of Lecture Notes in Computer Science, pp. 389–405. Springer, 2008.

12. D.J. Bernstein and T. Lange, *Faster addition and doubling on elliptic curves*, Lecture Notes in Computer Science, Progress in Cryptology-Africa Crypt, Springer, 4833, 29–50, 2007.

13. S. Josefsson, I. Liusvaara, *Edwards-Curve Digital Signature Algorithm (EdDSA)*. Internet Engineering Task Force, January 2017. doi:10.17487/RFC8032. ISSN 2070-1721. RFC 8032. Retrieved 2017-07-31. https://tools.ietf.org/html/rfc8032; doi:10.17487%2FRFC8032; https://www.worldcat.org/issn/2070-1721; https://tools.ietf.org/html/rfc8032.

14. M. Boudabra and A. Nitaj. A new public key cryptosystem based on Edwards curves. *Journal of Applied Mathematics and Computing* 61(1–2), 2019, 1–20. doi:10.1007/s12190-019-01257-y.

Chapter 11

Comparative analysis of Rabbit Message Queue in the cloud

The Internet of Things

J. Ebenesar Anna Bagyam, D. Vivesini and K. Preethi Sowndharya

Avinashilingam Institute for Home Science and Higher Education
for Women, Coimbatore, India

CONTENTS

11.1 INTRODUCTION

Cloud computing is a range of services conveyed over the internet or "the cloud". Cloud computing has four types: public cloud, private cloud, hybrid cloud, and community cloud. While public clouds like Amazon's EC2, Windows Azure, and Google's computer engine are prominent examples, the concept of cloud computing extends beyond any particular provider or data center. Familiar messaging-queue programs in cloud computing are Rabbit MQ, Active MQ, Zero MQ, and IBM MQ.

Rabbit MQ reports for devices in the cloud computing environment. It is a well-received, open-source message broker held up commercially by necessity and which is used habitually.

MQ Telemetry Transport (MQTT) has been developed for compact, low-bandwidth devices that can deliver messages via networks. Its simplicity

DOI: 10.1201/9781003436461-11

(only five application programming interfaces; APIs) and minimal wire footprint are well known. Rabbit MQ has been deliberately used to select information sequences of events that have developed over the last 25 years.

Vilaplana et al. (2014) obtained a model based on queueing theory related to cloud computing. The system with an infinite server of dynamic behavior using the cloud reduced the queue length with an increase in production utilization, according to Anupama and Keerthi (2014). Rao et al. (2017) analyzed the single and multi-server systems using cloud data centers based on the queuing system.

Evangelin and Vidhya (2015) analyzed the multi-server model with the first in first out (FIFO) queue discipline, which resulted in reduced service waiting time. The multi-server for cloud computing model was analyzed. Guo et al. (2014) proposed and analyzed work based on simulation with some strategies.

Nazarov et al. (2020) considered the multi-server model solved the queuing system in a retrial in a fixed time using a diffusion limit. The steady-state probability distribution for the busy servers and customers in orbit was estimated, and the simulation method was used for the calculation.

Kumar et al. (2021) examined the M/M/1/N queuing model in a load balancing server of cloud computing system analysis. Adhikari et al. (2021) studied the M/M/c queuing model and reduced the overall waiting time, and mainly focused on the server's utilization factor. Pham et al. (2015) analyzed the cloud computing service for intelligent parking systems, developed a network architecture based on IoT technology, and decreased the waiting time by using simulation methods. Wang and Su (2018) developed two-stage multi-skill customer data centers using a simulation approach in a mixed queuing system. Alotaibi et al. (2021) analyzed multi-classes depending on their service requirements and evaluated a multi-class queuing system based on an analytical model. Li et al. (2015) observed the performance of cloud based message queueing systems (CMQSs) with distributed computing and calculated the waiting times for many messages in the queue, system utilization, and the total delivery time of the message using the Arena tool.

Jafarnejad Ghomi et al. (2019) analyzed cloud computing using the queuing system in modeling techniques. Rao et al. (2018) observed the single and multi-servers in cloud information centers based on queueing theory. Kim et al. (2012) have observed a batch arrival process in a multi-server retrial queueing system and time distribution with a phase type.

Lisovskaya et al. (2022) considered two orbits in the multi-server queue, and performance measures were calculated. Vaquero et al. (2008) analyzed the complete definition of the cloud based on the cloud computing system.

Sencer and Basarir Ozel (2013) analyzed workforce management in a call center based on the simulation model. Zainurin and Kamardan (2021) simulated queueing situations based on theme parks, and performance measures were calculated. Haron and Kamardan (2021) analyzed a busy restaurant based on queueing simulation using Arena software. Safuri and Kamardan

(2022) analyzed the queueing simulation of a supermarket using Arena software; they calculated the customer average waiting time, the number of customers waiting in the queue, and the utilization of the system. Ibrahim et al. (2022) analyzed the weighbridge service process by implementing Rockwell Arena software in the simulation model. Analysis of the software found that it identified barriers to the process, resource allocation adjustments for each weighbridge, and reduced the waiting time for all weighbridges.

Kanoun et al. (2022) proposed different treatments for diabetic retinopathy disease using the simulation approach using Arena software. Mashhood and Ali (2022) developed a demonstrative model of a flexible manufacturing system (FMS) using Arena simulation and the effect of studying operational flexibility on the performance of FMS. Banez (2022) analyzed license insurance in the Land Transportation Office (LTO) during culmination days and culmination hours, the waiting time for complaints made by Arena simulation software, and resource capacity utilization was included in their parameters.

Sinha et al. (2022) analyzed railroad bogies and tapered roller bearings, and optimized the overhauling process using Arena simulation software. The Arena tool helped to develop the model to find potential changes that could lead to a reduction in the waiting time. Ibrahim and Engin (2022) examined the possibility of minimizing operational risks that can arise when Arena simulation software fault reports are resolved. A comparison of the current value stream mapping (VSM) model with the proposed VSM model was made by analyzing the simulation results.

Singh (2021) analyzed the basics of the delivery system to communicate data streams between on-road V2X infrastructure; collected data were delivered using Rabbit MQ with AMQP for communication and with the concurrent versions system (CVS). Catovic et al. (2022) observed microservice development using the Rabbit MQ message broker. An approach using microservice architecture helped to develop the application as a set of small services.

Sowjanya et al. (2011) analyzed the cloud computing applied in queueing theory to decrease waiting time. Bharkad and Durge (2014) analyzed the infinite servers queueing model and calculated the selected measures, such as time spent in system utilization and the response time.

Asria et al. (2022) analyzed the innovative village applications and implementation of the micro-service architecture. Using the Rabbit MQ message broker to implement asynchronous communication, each service's client application response time may be accelerated. The functional and non-functional analysis process implements micro-service analysis and coding.

Ahmad et al. (2022) analyzed the blockchain-based architecture using a message schedule with two access level filters for an approaching message, which was critical and non-critical.

Phung-Duc and Kawanishi (2019) considered the applications in datacenters with an ON-OFF policy using a retrial queue with a multi-server queue and analyzed by a level-dependent death process and quasi-birth. Peniak et al. (2022) analyzed the MQ telemetry transport (MQTT) broker (Rabbit MQ) based on the physical sensor connected to the virtual sensor in the producer and consumer.

The proposed system has chosen Rabbit Messaging Queue as an application, which has been solved theoretically using the supplementary variable technique and experimentally using the Arena simulation method. The comparative statement describes the best method for finding the queue timing.

11.2 RABBIT MESSAGING QUEUE

Rabbit MQ is flimsy and trouble-free when utilizing on premises and in the cloud. It holds up multiple messaging protocols. Rabbit MQ can be utilized to deliver in combined configurations to meet large-scale, high-availability requirements. It is a new economic computing model in which details and computing resources can gain access to a web browser used by customers. The creator sends a message that forms a queue from where the user consumes the data in order. Each communication will be taken in only once and forms a non-parallel queue, in the sense that creators and users will not interconnect unnecessarily and simultaneously in the message queue. The information is kept in reserve for streaming up and getting back into the queue. This prototype is known as point-to-point reporting.

Rabbit MQ has four types of exchange: direct, topic, fan-out, and headers. Every exchange type routes the piece of information individually to a framework and unbreakable set-up customers. Each of the two creates its exchange or uses the pre-arranged default exchange. In the Verizon/Yang-validator project, principles were used to provide a technique for helping equipment vendors to use open config YANG JSON for an existing platform that will later also be open config.

The Yang validator also utilizes nameko and Rabbit MQ to come up with a microservice to verify that JSON messages are, in fact, open config compliant. The messages that are added to the messaging queue schedule often do so in the form of a queue, meaning that each user and each message must wait until the earlier users are attended to. The messaging queue is suitable for different types of applications. It is also an important "multi-point broadcast" application, and you can send messages to multiple target sites (destination list). Rabbit MQ is a messaging broker – an intercessor for messaging. Many applications have been utilized as a joint base to eject and pick up messages, and the message is a secure place to be until pickup.

A message queue is a form of non-parallel service-to-service transmission used in serverless and microservice architecture where pieces of information are kept in reserve, which may queue up on time until they are served out.

Figure 11.1 shows the structural definition of the proposed system containing two different mandatory service providers. The described virtual

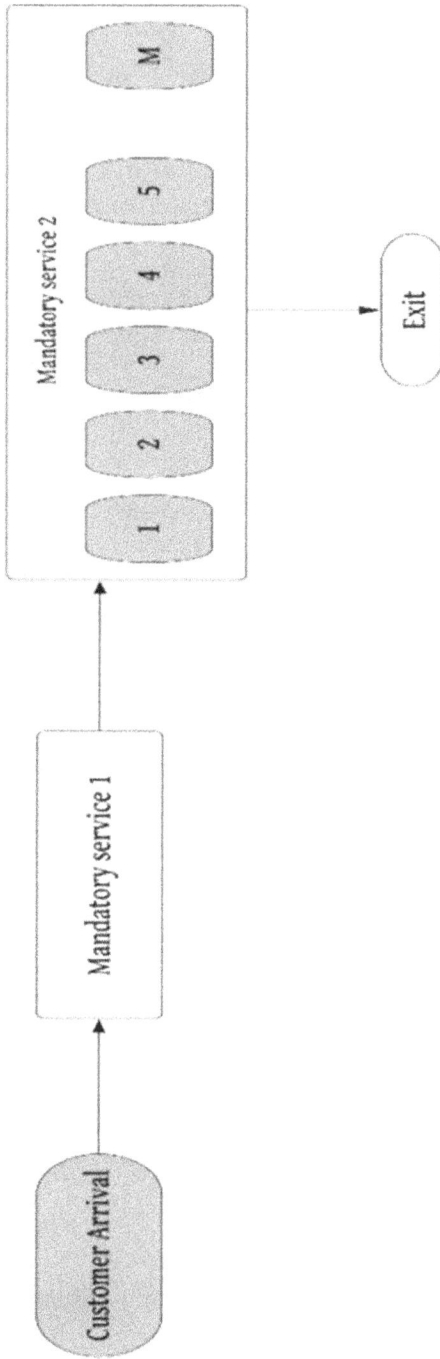

Figure 11.1 Structural outline of Rabbit MQ.

queue was solved and analyzed in two ways: (i) as a theoretical method and (ii) as a simulation analysis.

11.3 QUEUEING DESCRIPTION OF RABBIT MQ

Consider the mandatory two-phase service retrial queuing model. Customers arrive at the system one-by-one according to the Poisson process with an arrival rate α. Blocked customers enter the retrial group termed "orbit". The conditional completion rate, Laplace–Stieltje's transform $A^*(\alpha)$, the distribution function $A(s)$, the density function $a(s)$, and the retrial time are used to generate a general distribution for the retrial time:

$$\tau(x) = \frac{a(x)}{1 - A(x)}$$

The arriving customer takes service immediately when the server is in an ideal state. The very first mandatory service will serve the entire arriving customer, and it is generally distributed with distribution function $B_1(x)$, density function $b_1(x)$, Laplace transform $B_1^*(x)$, and the corresponding first two moments are b_1 and b_2. As soon as the first service is completed, the customer will automatically perform the compulsory second service (which contains the "m" multi-phase) with probability σ_m. After the second mandatory service, the customer will leave the system. The compulsory second service time follows an arbitrary distribution with distribution function $B_m(x)$, density function $b_m(x)$, Laplace transform $B_m^*(x)$, and the first two moments are c_1 and c_2. The conditional completion rates of the first and second mandatory services are:

$$\varepsilon_1 = \frac{b_1(x)}{1 - B_1(x)} \qquad \varepsilon_m = \frac{b_m(x)}{1 - B_m(x)}$$

11.3.1 Steady-state distributions

The system of equations that governs the model under a steady state, by the supplementary variable method, are:

$$\alpha R_0^F = \int_0^\infty \sum_{m=1}^{k} R_{m,0}^Q(x)\varepsilon_m(x)dx \tag{11.1}$$

$$\frac{dR_n^F(x)}{dx} = -(\alpha + \tau(x))R_n^F(x) \quad n \geq 1 \tag{11.2}$$

$$\frac{dR_0^E(x)}{dx} = -\left(\alpha + \varepsilon_1(x)\right)R_0^E(x) \tag{11.3}$$

$$\frac{dR_n^E(x)}{dx} = -\left(\alpha + \varepsilon_1(x)\right)R_n^E(x) + \alpha R_{n-1}^E(x), \quad n \geq 1 \tag{11.4}$$

$$\frac{dR_{m,0}^Q(x)}{dx} = -\left(\alpha + \varepsilon_m(x)\right)R_{m,0}^Q(x) \quad m = 1,2,\ldots,k \tag{11.5}$$

$$\frac{dR_{m,n}^Q(x)}{dx} = -\left(\alpha + \varepsilon_m(x)\right)R_{m,n}^Q(x) + \alpha R_{m,n-1}^Q(x), n \geq 1, \quad m = 1,2,\ldots,k \tag{11.6}$$

The steady-state boundary conditions are:

$$R_0^F(0) = \int_0^\infty \sum_{m=1}^k R_{m,n}^Q(x)\varepsilon_m(x)dx, n \geq 1 \tag{11.7}$$

$$R_0^E(0) = \alpha R_0^F + \int_0^\infty R_1^F(x)\tau(x)dx \quad m = 1,2,\ldots,k \tag{11.8}$$

$$R_n^E(0) = \alpha \int_0^\infty R_n^F(x) + \int_0^\infty R_{n+1}^F(x)\tau(x)dx, n \geq 1 \tag{11.9}$$

$$R_{m,n}^Q(0) = \sigma_m \int_0^\infty R_n^E(x)\varepsilon_1(x)dx \quad n \geq 0 \quad m = 1,2,\ldots,k \tag{11.10}$$

The normalizing condition is:

$$R_0^F + \sum_{n=1}^\infty \int_0^\infty R_n^F(x)dx + \sum_{n=0}^\infty \int_0^\infty R_n^E(x)dx + \sum_{n=0}^\infty \int_0^\infty R_{m,n}^Q(x)dx = 1 \tag{11.11}$$

11.3.2 Steady state solution

The probability generating function (PGF) method is used for different server states to solve the prescribed retrial queueing model. The PGF corresponds to the server's idle state, the server providing the first mandatory service, and the server providing the second multi-phase mandatory service, respectively, as follows:

$$R^F(x,z) = \sum_{n=1}^{\infty} R_n^F(x)z^n$$

$$R^E(x,z) = \sum_{n=0}^{\infty} R_n^E(x)z^n$$

$$R_m^Q(x,z) = \sum_{n=0}^{\infty} R_{m,n}^Q(x)z^n$$

Solving the partial differential equations from Equation (11.2) to (11.10) yields:

$$R^F(x,z) = R^F(0,z)e^{-\alpha x}(1 - A(x)) \tag{11.12}$$

$$R^E(x,z) = R^E(0,z)e^{-\alpha(1-z)x}(1 - B_1(x)) \tag{11.13}$$

$$R_m^Q(x,z) = R_m^Q(0,z)e^{-\alpha(1-z)x}(1 - B_m(x)) \tag{11.14}$$

$$R^F(0,z) = \sum_{m=1}^{k} R_m^Q(0,z)B_m^*(\alpha(1-z)) - \alpha R_0^F \tag{11.15}$$

$$R^E(0,z) = \frac{R^F(0,z)}{z}\left[z + A^*(\alpha)[1-z]\right] + \alpha R_0^F \tag{11.16}$$

$$R_m^Q(0,z) = \sum_{m=1}^{k} \sigma_m R^E(0,z)B_1^*(\alpha(1-z)) \tag{11.17}$$

Substituting Equation (11.15) in (11.16), we get:

$$R^E(0,z) = \frac{\sum_{m=1}^{k} R_m^Q(0,z)B_m^*(\alpha(1-z)) - \alpha R_0^F}{z}\left[z + A^*(\alpha)(1-z)\right] + \alpha R_0^F \tag{11.18}$$

Using the expression $R_m^Q(0,z)$ given in Equation (11.17) into (11.18) and simplifying, we get:

$$R^E(0,z) = \frac{\alpha R_0^F[(A^*(\alpha)(1-z)]}{D(z)} \tag{11.19}$$

where

$$D(z) = \sum_{m=1}^{k} \sigma_m B_1^* \left(\alpha \left(1 - z \right) \right) B_m^* \left(\alpha \left(1 - z \right) \right) \left[z + A^* \left(\alpha \right) \left(1 - z \right) \right] - z$$

Substituting Equation (11.19) in (11.17), we get:

$$R_m^Q (0, z) = \frac{\sum_{m=1}^{k} \sigma_m \alpha R_0^F A^* \left(\alpha \right) \left(1 - z \right) B_1^* \left(\alpha \left(1 - z \right) \right)}{D(z)} \tag{11.20}$$

Substituting $R_m^Q (0, z)$, given in Equation (11.20), in Equation (11.15), we obtain:

$$R^F (0, z) = \frac{\alpha R_0^F}{D(z)} \left[z - \sum_{m=1}^{k} \sigma_m B_1^* \left(\alpha \left(1 - z \right) \right) B_m^* \left(\alpha \left(1 - z \right) \right) z \right] \tag{11.21}$$

Substituting the expression $R^F(0, z)$ in Equation (11.12), we get:

$$R^F (x, z) = \frac{\alpha R_0^F}{D(z)} \left[z - \sum_{m=1}^{k} z \sigma_m B_1^* \left(\alpha \left(1 - z \right) \right) B_m^* \left(\alpha \left(1 - z \right) \right) \right] e^{-\alpha x} \left(1 - A(x) \right)$$

Using the relevant expressions, Equations (11.13) and (11.14) become:

$$R^E (x, z) = \frac{\alpha R_0^F [\left(A^* \left(\alpha \right) \right) \left(1 - z \right)]}{D(z)} e^{-\alpha (1-z)x} \left[1 - B_1 (x) \right]$$

$$R_m^Q (x, z) = \frac{\sum_{m=1}^{k} \sigma_m \alpha R_0^F A^* \left(\alpha \right) \left(1 - z \right) B_1^* \left(\alpha \left(1 - z \right) \right)}{D(z)} e^{-\alpha (1-z)x} \left[1 - B_m (x) \right]$$

11.3.3 Steady state analysis

The PGF of the orbit size when the server is in idle is

$$R^F (z) = \int_0^\infty R^F (x, z) dx.$$

$$= \frac{R_0^F}{D(z)} \left[z - \sum_{m=1}^{k} \sigma_m B_1^* \left(\alpha \left(1 - z \right) \right) B_m^* \left(\alpha \left(1 - z \right) \right) z \right] \left[1 - A^* \left(\alpha \right) \right]$$

The PGF of the orbit size when the server is busy is

$$R^E(z) = \int_0^\infty R^E(x,z)\,dx.$$

$$= \frac{R_0^F\left[\left(A^*(\alpha)\right)\right]}{D(z)}\left[1 - B_1^*\left(\alpha(1-z)\right)\right]$$

The PGF of the orbit size when the server is in the second service is

$$R_m^Q(z) = \int_0^\infty R_m^Q(x,z)\,dx.$$

$$= \frac{\sum_{m=1}^k \sigma_m R_0^F A^*(\alpha) B_1^*\left(\alpha(1-z)\right)}{D(z)}\left[1 - B_m(x)\right]$$

Using the normalizing condition

$$R_0^F + R^F(1) + R^E(1) + \sum_{m=1}^k R_M^Q(1) = 1$$

we get:

$$R_0^F = \frac{D'(1)}{1 - \sum_{m=1}^k \sigma_m\left(1 - A^*(\alpha)\right) - A^*(\alpha)\alpha b_1 - \sum_{m=1}^k \sigma_m A^*(\alpha)\alpha b_2}$$

11.4 PERFORMANCE MEASURES

This section presents the derivation of performance measures of a two-phase mandatory service retrial queue.

The steady state probability that the server is idle during the retrial time is

$$R^F(1) = \frac{R_0^F\left[1 - \sum_{m=1}^k \sigma_m \alpha b_2 - \sum_{m=1}^k \sigma_m \alpha b_1 - \sum_{m=1}^k \sigma_m\right]\left[1 - A^*(\alpha)\right]}{D'(1)}$$

The steady state probability that the server is busy in the mandatory first phase of service is

$$R^E(1) = \frac{R_0^F A^*(\alpha)\alpha b_1}{D'(1)}$$

The steady state probability that the server is busy in the mandatory second phase of service is

$$R_m^Q(1) = \frac{R_0^F \sum_{m=1}^{k} \sigma_m A^*(\alpha)\alpha b_2}{D'(1)}$$

The PGF of the number of customers in the queue is given by

$$P_q(z) = R_0^F + R^F(z) + R^E(z) + R_m^Q(z) =$$
$$\frac{R_0^F}{D(z)}\left[A^*(\alpha)[1-z] - A^*(\alpha)\left(B_1^*(\alpha(1-z))\right) + \sum_{m=1}^{k}\sigma_m A^*(\alpha)B_1^*(\alpha(1-z))\right]$$

$$(11.22)$$

$$N(z) = R_0^F\left[A^*(\alpha)[1-z] - A^*(\alpha)\left(B_1^*(\alpha(1-z))\right) + \sum_{m=1}^{k}\sigma_m A^*(\alpha)B_1^*(\alpha(1-z))\right]$$

The mean number of customers in the queue is

$$L_q = \frac{D'N'' - N''D''}{2D'^2}$$

where

$$D' = \sum_{m=1}^{k}\sigma_m(\alpha b_1 + \alpha b_m)(1 - A^*(\alpha)) - 1$$

$$D'' = \sum_{m=1}^{k}\sigma_m(\alpha^2 c_2 + \alpha^2 c_1)(1 - A^*(\alpha))$$

$$N' = R_0^F\left[-A^*(\alpha) + A^*(\alpha)\alpha b_1 + \sum_{m=1}^{k}\sigma_m A^*(\alpha)\alpha b_2\right]$$

$$N'' = R_0^F \left[A^*(\alpha)\alpha^2 c_1 + \sum_{m=1}^{k} \sigma_m A^*(\alpha)\alpha^2 c_1 \right]$$

The PGF of the number of customers in the system is given by

$$P_s(z) = R_0^F + R^F(z) + zR^E(z) + zR_m^Q(z)$$

$$= \frac{R_0^F}{D(z)} \left[\begin{array}{l} \left(\sum_{m=1}^{k} \sigma_m B_1^*(\alpha(1-z)) B_m^*(\alpha(1-z)) A^*(\alpha) - zA^*(\alpha) B_1^*(\alpha(1-z)) \right) + \\ \left(\sum_{m=1}^{k} \sigma_m A^*(\alpha) B_1^*(\alpha(1-z)) \left[1 - B_m^*(\alpha(1-z)) \right] \right) \end{array} \right]$$

$$(11.23)$$

$$T(z) = R_0^F \left[\begin{array}{l} \left(\sum_{m=1}^{k} \sigma_m B_1^*(\alpha(1-z)) B_m^*(\alpha(1-z)) A^*(\alpha) - zA^*(\alpha) B_1^*(\alpha(1-z)) \right) + \\ \left(\sum_{m=1}^{k} \sigma_m A^*(\alpha) B_1^*(\alpha(1-z)) \left[1 - B_m^*(\alpha(1-z)) \right] \right) \end{array} \right]$$

The mean number of customers in the system is

$$L_s = \frac{D'T'' - T'D''}{2D'^2}$$

where

$$T' = R_0^F \left[\sum_{m=1}^{k} \sigma_m A^*(\alpha) \left[\alpha^2 c_1 + \alpha b_1 \right] - A^*(\alpha)\alpha b_1 - A^*(\alpha) - \sum_{m=1}^{k} \sigma_m A^*(\alpha)\alpha^2 c_1 \right]$$

$$T'' = R_0^F \left[\begin{array}{l} \sum_{m=1}^{k} \left(\sigma_m A^*(\alpha) \right) \left[\alpha^2 c_2 + \alpha^2 c_1 \right] - A^*(\alpha)\alpha^2 c_1 - 2A^*(\alpha)\alpha b_1 \\ \sigma_m A^*(\alpha)\alpha b_2 - \sum_{m=1}^{k} \sigma_m A^*(\alpha)\alpha^2 c_2 - \\ - \sum_{m=1}^{K} \left(\sum_{m=1}^{k} \sigma_m A^*(\alpha)\alpha b_1 \right) \end{array} \right]$$

The busy period is one of the significant performance measures in the retrial situation. The system's busy period B defines the period that starts at an epoch when an arriving customer finds an empty system and ends at the next departure epoch when the system is empty again:

$$
E(B) = \frac{1}{\alpha}\left(R_0^{F-1} - 1\right)
$$

$$
= \frac{\left[\begin{array}{l} 1 - \sum_{m=1}^{k}\left(1 - A^*(\alpha)\right)\left[1 - B_m^{*'}\left(\alpha\left(1 - z\right)\right) + B_1^{*'}\left(\alpha\left(1 - z\right)\right)\right] \\ - A^*(\alpha)\left[B_1^{*'}\left(\alpha\left(1 - z\right)\right) + \sum_{m=1}^{k}\sigma_m B_m^{*'}\left(\alpha\left(1 - z\right)\right)\right] + 1 \end{array}\right]}{\alpha\sum_{m=1}^{k}\sigma_m\left(B_m^{*'}\left(\alpha\left(1 - z\right)\right) + B_1^{*'}\left(\alpha\left(1 - z\right)\right)\right)\left(1 - A^*(\alpha)\right) - 1}
$$

11.5 NUMERICAL ILLUSTRATION

In this section, performance measures are calculated and analyzed in both computational and simulation analysis. Arena simulation software is used for simulation analysis, and MATLAB for computational analysis. Comparative analysis was also carried out.

11.5.1 Computational analysis

Numerical analysis was implemented for suitable values of the system parameters. The system parameters are arrival rate (α), retrial rate (τ), first phase mandatory service rate (ε_1), and the probability that the customer is opting for a multi-phase mandatory service (σ_m) on the following measures:

- L_s is the expected number of customers in the system;
- $E(B)$ is the expected number of busy periods.

Table 11.1 shows that the first mandatory service rate increases, delivering a message during a busy period decreases, and the length of service rate increases. It is observed in Table 11.2 that an increase in the first mandatory service and the probability of opting for a second multi-phase mandatory service results in an increase in the busy period and the length of the system.

We set the various parameters for Figure 11.2 as $\alpha = 29$, $\varepsilon_1 = 20$, $\varepsilon_2 = 28$, $\tau = 14$, and $\sigma_1 = 0.9$.

We set the parameters for Figure 11.3 as $\alpha = 40$, $\varepsilon_1 = 36$, $\varepsilon_2 = 30$, $\tau = 14$, and $\sigma_2 = 0.5$.

We set the parameters for Figure 11.4 as $\alpha = 40$, $\varepsilon_1 = 36$, $\varepsilon_2 = 10$, $\tau = 14$, and $\sigma_1 = 0.8$.

Table 11.1 Impact of τ and ε_1 on E(B) and L_s

	$\varepsilon_1 = 29$		$\varepsilon_1 = 36$		$\varepsilon_1 = 39$	
τ	E(B)	L_s	E(B)	L_s	E(B)	L_s
3	0.0479	0.3091	0.0284	0.2358	0.02008	0.2154
4	0.0477	0.4123	0.0283	0.3156	0.02006	0.2887
5	0.0476	0.5156	0.0283	0.3960	0.02000	0.3628
6	0.0476	0.6190	0.0282	0.4771	0.02000	0.4377
7	0.0473	0.7225	0.0282	0.5588	0.02000	0.5133
8	0.0472	0.8260	0.0281	0.6410	0.01990	0.5898
9	0.0470	0.9297	0.0281	0.7239	0.01990	0.6670
10	0.0469	1.0334	0.0280	0.8074	0.01990	0.7450
11	0.0468	1.1372	0.0280	0.8915	0.01990	0.8237
12	0.0466	1.2412	0.0279	0.9763	0.01990	0.9836

Table 11.2 Impact of ε_1 and σ_1 on E(B) and L_s

	$\sigma_1 = 0.4$		$\sigma_1 = 0.18$		$\sigma_1 = 0.23$	
ε_1	E(B)	Ls	E(B)	Ls	E(B)	Ls
2.1	0.1233	30.095	0.0824	32.7294	0.644	32.2185
2.2	0.1256	30.4023	0.0875	33.1572	0.0751	32.5436
2.3	0.1281	30.7400	0.0928	33.5769	0.0865	32.8291
2.4	0.1307	31.0825	0.0984	33.9849	0.0983	33.0613
2.5	0.1334	31.4295	0.1043	34.3769	0.1108	33.2215
2.6	0.1362	31.7811	0.1104	34.7475	0.1237	33.2843
2.7	0.1391	32.1369	0.1168	35.0899	0.1372	34.2141
2.8	0.1421	32.4968	0.1234	35.3958	0.1511	34.9600
2.9	0.1450	32.8607	0.1303	35.6546	0.1655	35.4467
3	0.1481	33.2283	0.1373	35.8529	0.1802	35.5586

We set the parameters for Figure 11.5 as $\alpha = 28$, $\varepsilon_1 = 16$, $\varepsilon_2 = 10$, $\tau = 14$, and $\sigma_1 = 0.1$.

We set the parameters for Figure 11.6 as $\alpha = 25$, $\varepsilon_1 = 16$, $\varepsilon_2 = 12$, $\tau = 9$, and $\sigma_1 = 0.1$.

Figure 11.2 verifies that when the first phase of service rate increases, the busy period of service also increases; when the retrial rate increases, the busy period of service rate increases. Figure 11.3 verifies that when the retrial rate and first phase of service rate increases, the busy period of service increases and the length of the system size decreases. Figure 11.4 verifies that as the retrial rate and second phase of the service rate increases, the length of the system size increases. Figure 11.5 verifies that when the arrival rate and first phase of the service rate increases, the busy period of service decreases.

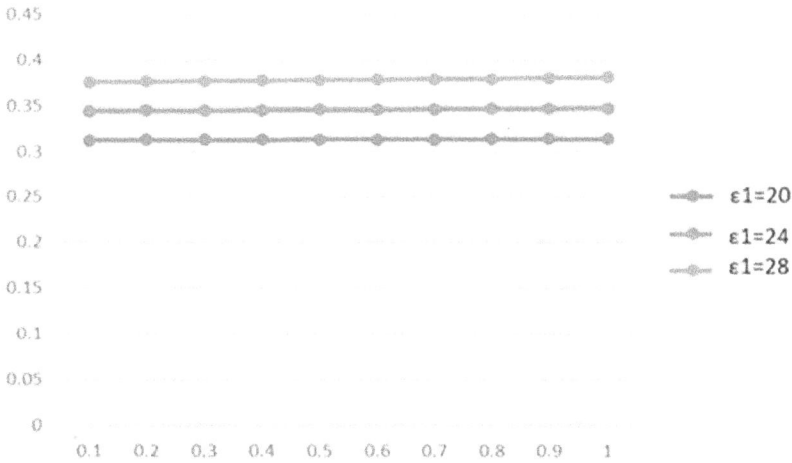

Figure 11.2 Effect of τ and ε_1 on E(B).

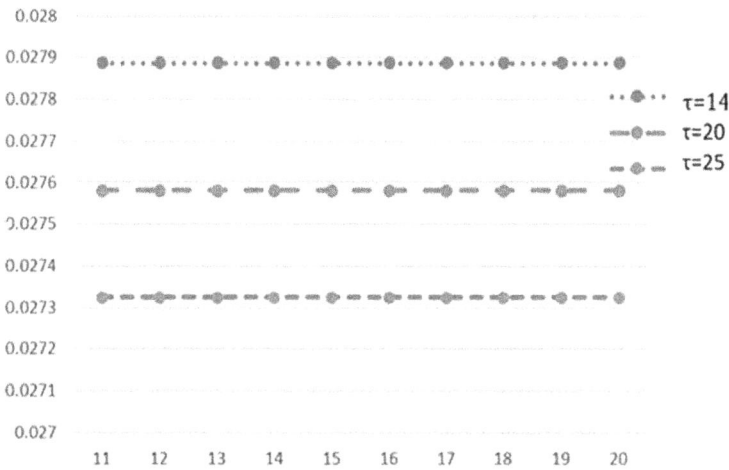

Figure 11.3 Effect of ε_2 and τ on E(B).

Figure 11.6 verifies that when the first phase of the service rate and retrial rate increases, the length of the system size increases.

11.5.2 Arena simulation technique

Arena is a discrete event-based simulation software. It acts as a tool for risk assessments by providing predictive modeling to users. It saves and optimizes resources and time. The software has numerous industrial applications and

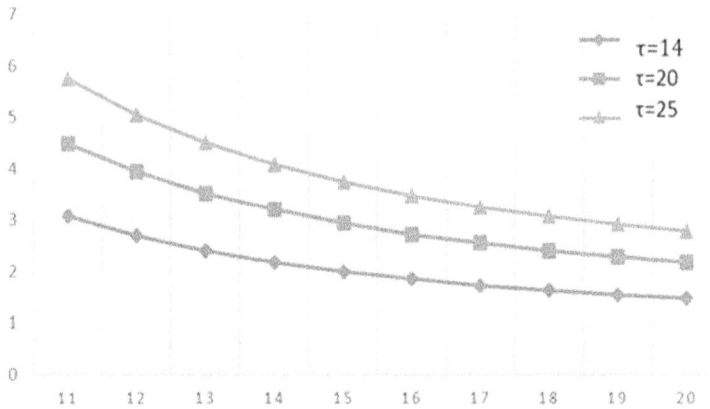

Figure 11.4 Effect of ε_2 and τ on L_s.

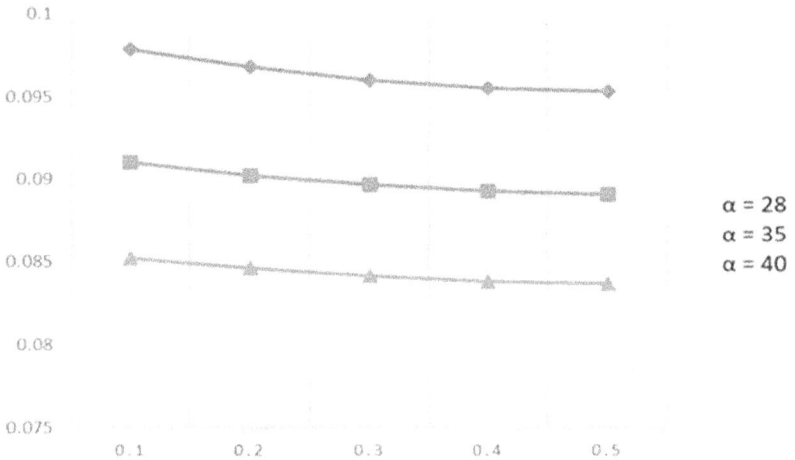

Figure 11.5 Effect of ε_1 and α on E(B).

focuses mainly on three components: input data analysis, process analytics, and data interfacing.

Based on the illustrated flow chart in Figure 11.1, the proposed Rabbit Messaging Queue can be visualized using the various modules of Arena simulation. By use of the drawing board, Rabbit MQ is constructed. The first and foremost step is to adapt the create module, which defines the creation of the entity. Here the entity is a message received from customers, and the arriving messages are considered random (exponential) arrivals.

In the next module used in this model process, the messages should be organized in an assembly process. The first move in the process is to name a

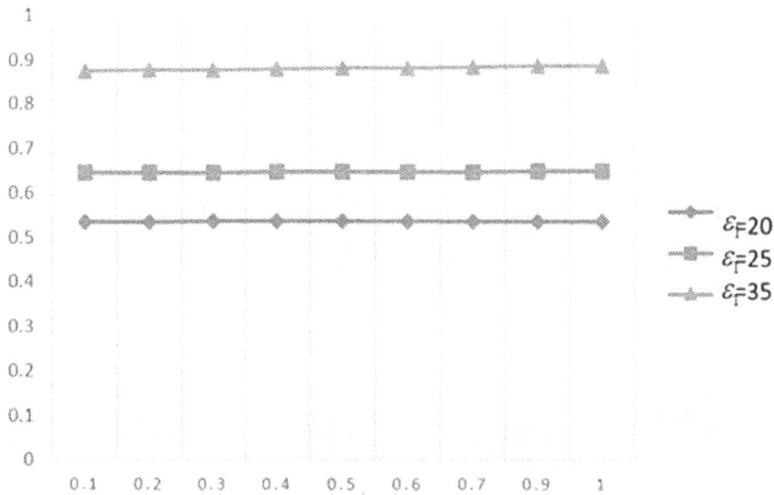

Figure 11.6 Effect of τ and ε_1 on L_s.

process module. Currently, the process uses a standard type, and the logical action is to seize the delay release. In most cases, the seize delay release defines the resources, which is defining the entity (message). Finally, the module is completed by visualizing the message queue. The resource starts to work on a message; the rest of the message will wait in the queue.

The decide module is used to make any decisions in the system. Here the type of decision is n-way by the condition because the system has multi-servers in the second essential part of the service. Mentioning the type of decision is much more important than building a model, which is just the disposal of messages. After service completion, the messages will exit the virtual queue.

Rabbit Messaging Queue has been constructed using the Arena simula-tion model because it is a flexible tool for any application and evaluates performance. In this experiment, the system's performance, such as staff utilization rate, service levels, and average waiting time, has been reported. To estimate the performance of the process, Rabbit MQ has the architecture shown in Figure 11.7. The results of a single replication of the Rabbit MQ model for an average waiting time are displayed in Table 11.3.

Table 11.3 illustrates the average waiting time of virtual message queues. This was derived by assuming the arrival of messages at different times. The average of people using the message applications in the early morning is assumed to be 300 messages. In the morning, only a few people are using the message application, counted as 150 messages. The assumed message counts for the afternoon and evening are 200 and 250, respectively. In the late evening, the message count is considered to be 400 because a significant

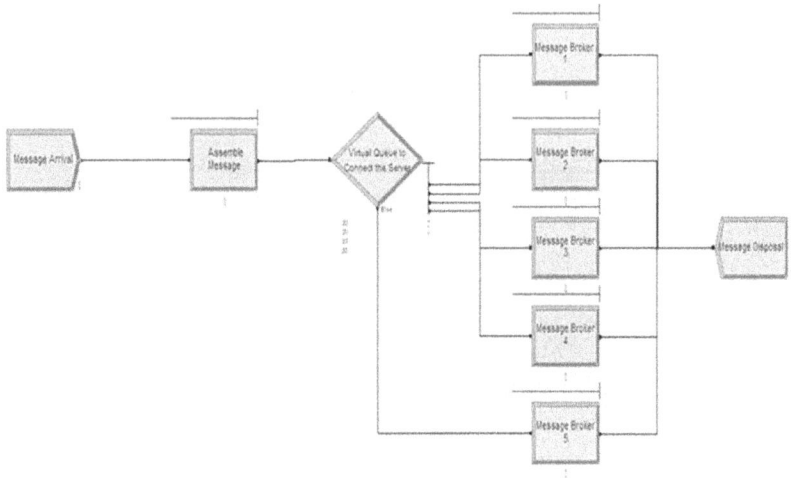

Figure 11.7 Modules of Rabbit Messaging Queue using Arena.

Table 11.3 Waiting time of various queues using simulation technique (hours)

Time period of messages	First essential service queue 1	Second essential service queue 1	Second essential service queue 2	Second essential service queue 3	Second essential service queue 4	Second essential service queue 5
Early morning (6 AM–8 AM)	0.5425	0.5758	1.1041	0.8226	0.7681	0.6599
Morning (9 AM–11 AM)	0.168	0.0385	0.1815	0.2208	0.1759	0.1064
Afternoon (12 PM–2 PM)	0.2036	0.1253	0.2234	0.5799	0.3618	0.2176
Evening (7 PM–9 PM)	0.3076	0.1450	0.5797	0.5412	0.5125	0.2398
Late evening (9 PM–11 PM)	0.6796	1.6054	1.8378	1.0443	1.1973	1.0020

number of people were using the messaging platform at that time. This is the reason why people face delays in sending and receiving messages. The late evening is the peak time for messaging platforms, when a virtual queue is formed. The corresponding average number of busy entities, the average number of scheduled messages, the utilization rate, and the total number of seized messages are given in Table 11.4. In the search for the longest waiting time, system utilization, and seized messages, Figures 11.8 and 11.9 display the monitored results.

Table 11.4 Rate of system utilization

Time period of messages	Average number of busy entities (messages)	Average number of scheduled messages	Average scheduled utilization	Total number of seized messages
Early morning (6 AM–8 AM)	0.1863	11.2708	1	0.1642
Morning (9 AM–11 AM)	0.0984	5.3672	1	0.0911
Afternoon (12 PM–2 PM)	0.0980	5.9019	1	0.0909
Evening (7 PM–9 PM)	0.1277	7.6603	1	0.1158
Late evening (9 PM–11 PM)	0.1960	0.1960	1	0.1960

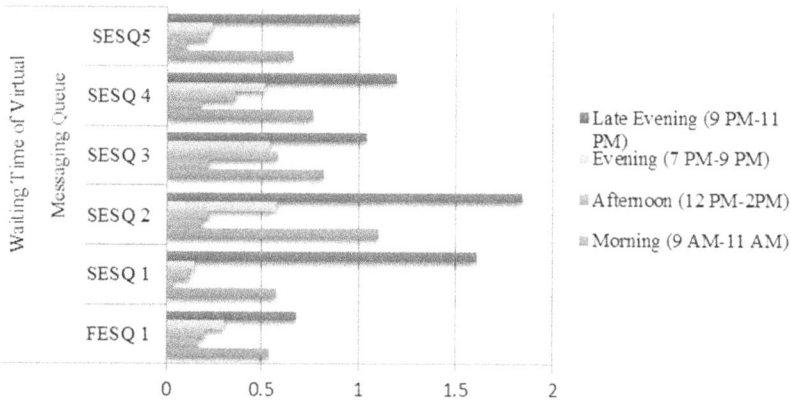

Figure 11.8 Average waiting times (Message) of various queues at different times.

11.6 COMPARATIVE ANALYSIS

By comparing the results of the mathematical analysis using MATLAB and simulation analysis using Arena software, the utilization of the messaging platform is to engage with messages (customers) throughout the hours of the day. Meanwhile, the messages overflow in particular hours of late evening, which form the virtual queue. The queue observes some busy periods for receiving and sending messages. When comparing the busy period results of MATLAB and Arena, there is a slight difference of 0.0482 raised among the analyses which is observed between the maximum utilized time period of the first essential service and the second essential service as shown in Tables 11.2 and 11.4. The obtained value of MATLAB finds a better result for the messaging platform.

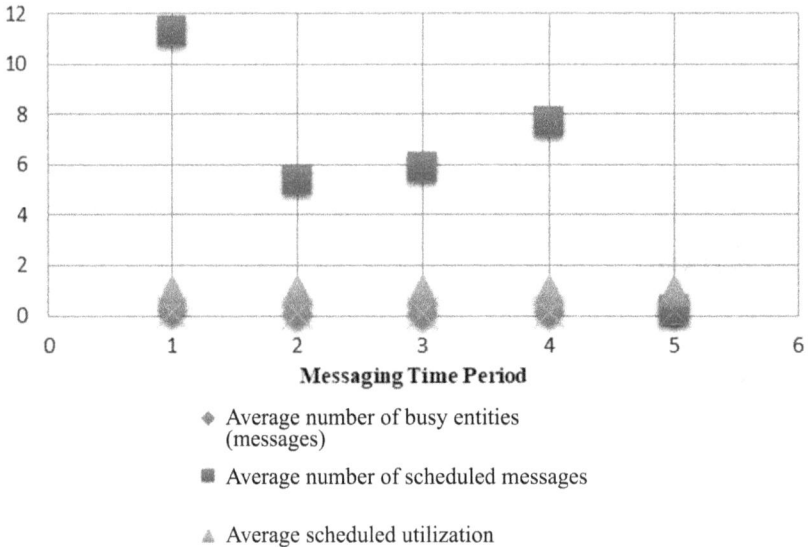

Figure 11.9 Average system utilization with time periods.

11.7 CONCLUSION

In this chapter, we have discussed the essential parameters of Rabbit MQ in the cloud of the Internet of Things. A single server with a two-phase mandatory service model was analyzed. The system performance measures were calculated. The behavior of Rabbit MQ was studied using numerical analysis. Using the Arena tool, we have found the waiting time and usage of the system. In this model, the fanout exchange key was used to deliver the message to the customers. Regarding future analysis, the process of a direct exchange key in Rabbit Messaging Queue is also used to deliver the messages, like the fanout exchange key, which can formulate the mathematical queuing model.

REFERENCES

Adhikari, S., Hutaihit, M. A., Chakraborty, M., Mahmood, S. D., Durakovic, B., Pal, S., & Obaid, A. J. (2021). Analysis of average waiting time and server utilization factor using queueing theory in cloud computing environment. *International Journal of Nonlinear Analysis and Applications*, *12*(Special Issue), 1259–1267. http://dx.doi.org/10.22075/ijnaa.2021.5636

Ahmad, I., Abdullah, S., Bukhsh, M., Ahmed, A., Arshad, H., & Khan, T. F. (2022). Message scheduling in Blockchain based IoT environment with additional fog broker layer. *IEEE Access*, *10*, 97165–97182. https://ieeexplore.ieee.org/abstract/document/9885201

Alotaibi, F. M., Ullah, I., & Ahmad, S. (2021). Modeling and performance evaluation of multi-class queuing system with QoS and priority constraints. *Electronics*,
10(4), 500. https://doi.org/10.3390/electronics10040500

Anupama, A., & Keerthi, G. S. (2014). Using Queuing theory the performance measures of cloud with infinite servers. *International Journal of Computer Science
and Engineering Technology (IJCSET)*, *5*(1). http://ijcset.com/docs/IJCSET14-05-
01-026.pdf

Asria, S. A., Astawaa, I. N. G. A., Sunayaa, I. G. A. M., Nugroho, I. M. R. A., &
Setiawanb, W. (2022). Implementation of asynchronous microservices architecture on smart village application. *12*(3), 1236–1243. https://www.researchgate.
net/profile/I-Nyoman-Gede-Astawa-2/publication/361620172

Banez, J. L. M. (2022). Improvement of drivers' license issuance by reducing waiting
time using system simulation, *1*(1), 19–35. https://bulsu-ovprde.com/wp-content/
uploads/2022/10/Jeremy-Laurence-M.-Banez.pdf

Bharkad, N. N., & Durge, M. H. (2014). The application of queue theory in cloud
computing to reduce the waiting time. *International Journal of Engineering
Research and Applications*, *4*(10), 64–69. https://www.semanticscholar.org/paper

Catovic, A., Buzadija, N., & Lemes, S. (2022). Microservice development using
RabbitMQ message broker. *Science, Engineering and Technology*, *2*(1), 30–37.
https://doi.org/10.54327/set2022/v2.i1.19

Evangelin, K. R., & Vidhya, V. (2015). Performance measures of queuing models using
cloud computing. *Asian Journal of Engineering and Applied Technology*, *4*(1),
8–11. https://trp.org.in/wp-content/uploads/2015/10/AJEAT-Vol.4-No.1-Jan-Jun-
2015-pp.8-11.pdf

Guo, L., Yan, T., Zhao, S., & Jiang, C. (2014). Dynamic performance optimization for cloud computing using M/M/m queueing system. *Journal of Applied
Mathematics*, *2014*.1–8. https://doi.org/10.1155/2014/756592

Haron, N. A. A., & Kamardan, M. G. (2021). Queuing system of a busy restaurant
using simulation software. *Enhanced Knowledge in Sciences and Technology*, *1*(2),
66–71. https://publisher.uthm.edu.my/periodicals/index.php/ekst/article/view/2099

Ibrahim, Y. E. L., & Engin, O. (2022). Process improvement for error records in a
financial institution with the analysis of simulation on using value stream mapping.
Journal of Advanced Research in Natural and Applied Sciences, *8*(2), 171–187.
https://doi.org/10.28979/jarnas.907042

Ibrahim, M. F., Putri, M. M., Sari, D. N., & Utama, D. M. (2022). Industrial
area weighbridge simulation model considering vehicle capacity and destination using arena software. In *AIP Conference Proceedings*, *2453*(1). https://doi.
org/10.1063/5.0094266

Jafarnejad Ghomi, E., Rahmani, A. M., & Qader, N. N. (2019). Applying queue
theory for modeling of cloud computing: A systematic review. *Concurrency and
Computation: Practice and Experience*, *31*(17), e5186. https://doi.org/10.1002/
cpe.5186

Kanoun, S., Jerbi, B., & Kamoun, H. (2022). Discrete event simulation to evaluate different treatments of diabetic retinopathy disease. *American Journal of
Operations Research*, *12*(6), 250–260. https://doi.org/10.4236/ajor.2022.126014

Kim, C. S., Mushko, V. V., & Dudin, A. N. (2012). Computation of the steady
state distribution for multi-server retrial queues with phase type service process. *Annals of Operations Research*, *201*(1), 307–323. https://doi.org/10.1007/
s10479-012-1254-7

Kumar, R., Soodan, B. S., Kuaban, G. S., Czekalski, P., & Sharma, S. (2021). Performance analysis of a cloud computing system using queuing model with correlated task reneging. *Journal of Physics: Conference Series*, *2091*(1), 012003. https://doi.org/10.1088/1742-6596/2091/1/012003

Li, J., Cui, Y., & Ma, Y. (2015). Modeling message queueing services with reliability guarantee in cloud computing environment using colored petri nets. *Mathematical Problems in Engineering*, (1–20). https://doi.org/10.1155/2015/383846

Lisovskaya, E., Fedorova, E., Salimzyanov, R., & Moiseeva, S. (2022). Resource retrial queue with two orbits and negative customers. *Mathematics*, *10*(3), 321. https://doi.org/10.3390/Math10030321

Mashhood, H., & Ali, M. (2022). A simulation study to evaluate the performance of FMS using routing flexibility. *International Journal of Simulation and Process Modelling*, *18*(3), 222–236. https://doi.org/10.1504/IJSPM.2022.126894

Nazarov, A., Moiseev, A., Phung-Duc, T., & Paul, S. (2020). Diffusion limit of multi-server retrial queue with setup time. *Mathematics*, *8*(12), 2232. https://doi.org/10.3390/math8122232

Peniak, P., Rástočný, K., Kanáliková, A., & Bubeníková, E. (2022). Simulation of virtual redundant sensor models for safety-related applications. *Sensors*, *22*(3), 778. https://doi.org/10.3390/s22030778

Pham, T. N., Tsai, M. F., Nguyen, D. B., Dow, C. R., & Deng, D. J. (2015). A cloud-based smart-parking system based on Internet-of-Things technologies. *IEEE Access*, *3*, 1581–1591. https://doi.org/10.1109/ACCESS.2015.2477299.

Phung-Duc, T., & Kawanishi, K. I. (2019). Multiserver retrial queue with setup time and its application to data centers. *Journal of Industrial & Management Optimization*, *15*(1), 15. https://doi.org/10.3934/jimo.2018030

Rao, N. T., Srinivas, P., Rajkumar, C., Bhattacharyya, D., & Kim, H. J. (2017). Studies on performance analysis of cloud computing based data centers with queuing models using MATLAB. *International Journal of Grid and Distributed Computing*, *10*(6), 55–69. http://dx.doi.org/10.14257/ijgdc.2017.10.6.05.

Rao, N. T., Srinivas, P., Sudha, K., Bhattacharyya, D., & Kim, T. H. (2018). Performance of M/M/1 and M/D/1 queuing models on data centers with cloud computing technology using MATLAB. *International Journal of Grid and Distributed Computing*, *11*(3), 11. http://dx.doi.org/10.14257/ijgdc.2018.11.302

Safuri, A. A. M., & Kamardan, M. G. (2022). Queue management and application in supermarket. *Enhanced Knowledge in Sciences and Technology*, *2*(1), 352–360. https://publisher.uthm.edu.my/periodicals/index.php/ekst/article/view/5281

Sencer, A., & Basarir Ozel, B. (2013). A simulation-based decision support system for workforce management in call centers. *Simulation*, *89*(4), 481–497. https://doi.org/10.1177/0037549712470169

Singh, P. (2021). Middleware support for communication in connected vehicles. https://www.scss.tcd.ie/publications/theses/diss/2021/TCD-SCSS-DISSERTATION-2021-064.pdf

Sinha, P., Mund, P., Mishra, R., Deshmukh, J., & Mungilwar, S. (2022). Cartridge tapered roller bearing of rail bogies overhauling process optimization by using ARENA simulation. In *AIP Conference Proceedings*, *2653*(1). https://doi.org/10.1063/5.0111212

Sowjanya, T. S., Praveen, D., Satish, K., & Rahiman, A. (2011). The queueing theory in cloud computing to reduce the waiting time. *International Journal of Computer Science Engineering & Technology, 1*(3). https://www.semanticscholar. org/paper/The-Queueing-Theory-in-Cloud-Computing-to-Reduce-Sowjanya/ a40cdccb20acb4e28a4c7263e7d416b9aebeba14

Vaquero, L. M., Rodero-Merino, L., Caceres, J., & Lindner, M. (2008). A break in the clouds: towards a cloud definition. *ACM Sigcomm Computer Communication Review, 39*(1), 50–55. https://doi.org/10.1145/1496091.149610

Vilaplana, J., Solsona, F., Teixidó, I., Mateo, J., Abella, F., & Rius, J. (2014). A queuing theory model for cloud computing. *The Journal of Supercomputing, 69*(1), 492–507. https://doi.org/10.1007/S11227-014-1177-Y

Wang, C. H., & Su, P. (2018). A simulation study of workforce management for a two-stage multi-skill customer service center. *International Journal of Operations Research, 15*(1), 15–28. https://www.researchgate.net/publication/325397378

Zainurin, N. S. A. M., & Kamardan, M. G. (2021). Analysis of queuing at theme park using arena simulation software. *Enhanced Knowledge in Sciences and Technology, 1*(2), 152–159. https://publisher.uthm.edu.my/periodicals/index.php/ ekst/article/view/2121

Chapter 12

Novel drug delivery systems

D. Jeslin

Bharath Institute of Higher Education & Research, Chennai, India

Muralikrishnan Dhanasekaran

Auburn University, Auburn, AL, USA

P. Panneerselvam

Bharath Institute of Higher Education & Research, Chennai, India

CONTENTS

12.1 INTRODUCTION

Controlled drug delivery systems can include the maintenance of drug levels within a desired range, the need for fewer administrations, optimal use of the drug in question, and increased patient compliance [1]. While these advantages can be significant, the potential disadvantages cannot be ignored, like the possible toxicity or non-biocompatibility of the materials used, undesirable by-products of degradation, any surgery required to implant or remove the system, the chance of patient discomfort from the delivery device, and the higher cost of controlled-release systems compared with traditional pharmaceutical formulations. The ideal drug delivery system should be inert, biocompatible, mechanically strong, comfortable for the patient, capable of achieving high drug loading, safe from accidental release, simple to administer and remove, and easy to fabricate and sterilize. The goal of many of the original controlled-release systems was to achieve a delivery profile that would yield a high blood level of the drug over a long period of time. With traditional drug delivery systems, the drug level in the blood rises after each administration of the drug, and then decreases until the next administration. The key point with traditional drug administration is that the blood level of the agent should remain between a maximum value, which may represent a toxic level, and a minimum value, below which the drug is no longer effective.

12.2 BACKGROUND

The science of controlled release first originated from the development of oral sustained release products in the 1940s and early 1950s [2]. First of all, the controlled release of marine antifoulants (in the 1950s) and the controlled release of fertilizers (1970s) were formulated which had only a single application in soul science. The development of pharmacology and pharmacokinetics demonstrated the importance of the drug release rate in determining the therapeutic effectiveness of the therapy. This is the reason behind the development of controlled release drugs. Modified release dosage forms are entirely new. The first time the scientist Rhozes formulated mucilage coated pills was in about AD 900. This technique was widely adopted in the 10th century by European countries in the form of gold, silver, and pearl coated tablets; this coating modifies the drug release rates. Advancement in coating technology included sugar and enteric coating on pills and tablets in the late 1800s. Further coating developed to the enteric coating of tablets followed by the incorporation of a second drug in the sugar coating layer, which happened in 1938.

However, the first patent for an oral sustained release preparation went in the favour of Zbigniew Lipowski; his preparation contained small coated beads that released the drug slowly and constantly. This idea was later developed by Blythe who launched the first marketed sustained release product in 1952. Over the past 30 years, as the sophistication involved in the marketing of new drugs increased and various advantages were recognized in controlled release drug delivery systems (CRDDSs), greater attention was paid to this field. Today the oral controlled drug delivery system has becomes a major drug delivery system as drugs have a high water solubility and short biological half-life. Other than oral, various routes like transdermal, ocular, vaginal, and parenteral are used for the controlled release of various drugs.

The history of controlled release technology is divided into three periods: 1950 to 1970 saw sustained drug release; 1970 to 1990 concerned the needs of controlled drug delivery; post-1990 was the modern era of controlled release technology [3].

12.3 TERMINOLOGY OR DEFINITIONS OF CONTROLLED RELEASE DOSAGE FORMS

The United States Pharmacopoeia (USP) defines the modified-release (MR) dosage form as "the one for which the drug release time course and/or location are chosen to accomplish therapeutic or convenience objectives not offered by conventional dosage forms such as solutions, ointments, or promptly dissolving dosage forms" [4]. One class of MR dosage form is an extended-release (ER) form which is defined as the one that allows at least a two-fold reduction in dosing frequency or significant increase in patient compliance or therapeutic performance when compared with that presented as a conventional dosage form (a solution or a prompt drug-releasing dosage form). The terms "controlled release (CR)", "prolonged release", "sustained or slow release (SR)", and "long-acting (LA)" have been used synonymously with "extended release".

Controlled drug delivery is one which delivers the drug at a predetermined rate, locally or systemically, for a specified period of time.

SR allows delivery of a specific drug at a programmed rate that leads to drug delivery for a prolonged period of time [5]. Prolonged-release products release the active ingredients slowly and work for a longer time.

Prolonged release or SR systems, which only prolong the therapeutic blood or tissue levels of the drug for an extended period of time, cannot be considered as CR systems by this definition. They are distinguished from rate controlled drug delivery systems, which are able to specify the release rate and duration *in vivo* precisely, on the basis of simple *in vitro* tests.

The difference between CR and SR is controlled drug delivery which delivers the drug at a predetermined rate for a specified period of time. Controlled release is a perfect zero-order release over time, irrespective of concentration. An SR dosage form is defined as the type of dosage form in which a portion (i.e., initial dose) of the drug is released immediately, in order to achieve the desired therapeutic response more promptly; the remaining (maintenance dose) is then released slowly by achieving a therapeutic level which is prolonged, but not maintained constantly. SR implies the slow release of the drug over a certain time period. It may or may not be CR.

Drug targeting, on the other hand, can be considered as a form of CR in that it exercises spatial control of drug release within the body.

12.4 RATIONALE

The basic rationale of a CRDDS is to optimize the biopharmaceutical, pharmacokinetical, and pharmacodynamical properties of a drug in such a way that its utility is maximized through the reduction in side effects and the cure or control of the disease condition in the shortest possible time by using the smallest quantity of drug, administered by the most suitable route [6].

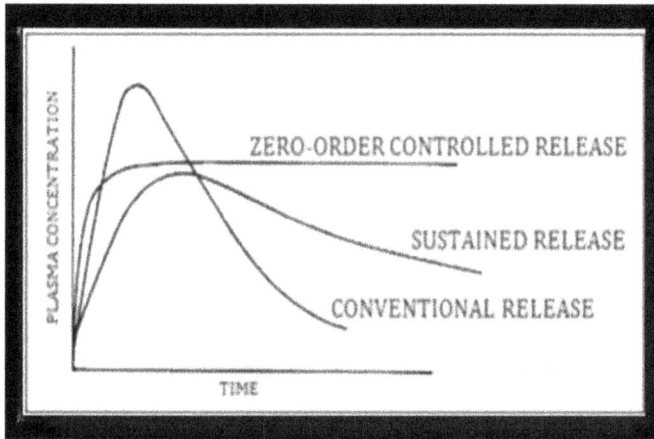

Figure 12.1 Plasma drug concentration: time profile.

The immediate release drug delivery system lacks some features like dose maintenance, CR rate, and site targeting. An ideal drug delivery system should deliver the drug at a rate dictated by the needs of the body over a specified period of treatment (Figure 12.1).

12.5 ADVANTAGES OF CONTROLLED RELEASE DOSAGE FORMS

12.5.1 Clinical advantages

- Reduction in frequency of drug administration;
- Improved patient compliance;
- Reduction in drug level fluctuation in blood;
- Reduction in total drug usage when compared with conventional therapy;
- Reduction in drug accumulation with chronic therapy;
- Reduction in drug toxicity (local/systemic);
- Stabilization of medical condition (because of more uniform drug levels);
- Improvement in bioavailability of some drugs because of spatial control;
- Economical to the health care providers and the patient.

12.5.2 Commercial/industrial advantages

- Illustration of innovative/technological leadership;
- Product life-cycle extension;
- Product differentiation;
- Market expansion;
- Patent extension.

12.5.3 Disadvantages

- Delay in onset of drug action;
- Possibility of dose dumping in the case of a poor formulation strategy;
- Increased potential for first pass metabolism;
- Greater dependence on gastro-intestinal residence time of dosage form;
- Possibility of less accurate dose adjustment in some cases;
- Cost per unit dose is higher when compared with conventional doses;
- Not all drugs are suitable for formulating into ER dosage form.

12.6 SELECTION OF DRUG CANDIDATES OR CHARACTERISTICS THAT MAY MAKE A DRUG UNSUITABLE FOR CONTROL RELEASE DOSAGE FORM

The selection of a drug for formulation into ER dosage form is the key step. The following candidates are generally not suitable for ER dosage forms:

- Short elimination half-life;
- Long elimination half-life;
- Narrow therapeutic index;
- Poor absorption;
- Active absorption;
- Low or slow absorption;
- Extensive first pass effect [7].

12.7 PARAMETERS FOR DRUG SELECTION

- Parameter: preferred value;
- Molecular weight/size: <1000;
- Solubility: >0.1 µg/mL for pH 1 to pH 7.8 Pka;
- Non-ionized moiety: >0.1% at pH 1 to pH 7.8;
- Apparent partition co-efficient: high;
- Absorption mechanism: diffusion;
- General absorbability: from all GI segments;
- Release: should not be influenced by pH and enzymes.

12.8 BIOPHARMACEUTICS AND PHARMACOKINETIC ASPECTS IN THE DESIGN OF CONTROLLED RELEASE PER ORAL DRUG DELIVERY SYSTEMS

CRDDSs are dosage forms from which the drug is released at a predetermined rate and which is based on a desired therapeutic concentration and the drug's pharmacokinetic characteristics [8].

12.8.1 Biological half-life (t½)

The shorter the t½ of a drug the larger will be the fluctuations between the minimum steady state concentration and the maximum steady state concentration upon repeated dosing. Thus the drug product needs to be administered more frequently.

12.8.2 Minimum effective concentration (MEC)

If an MEC is required, either frequent dosing of a conventional drug product is necessary or a CR preparation may be chosen.

12.8.3 Dose size and extent of duration

The longer the extent of duration the larger the total dose per unit delivery needs to be. Hence there is a limitation to the amount of drug that can be practically incorporated into such a system.

12.8.4 Relatively long t½ or fluctuation desired at steady state

It is the belief of some that neither an SR nor a CRDDS is needed or useful for drugs having a t½ of 12 hours or more [9]. This is not so because there are two cases for which a 12 or 24-hour CRDDS seems to be indicated: A drug having a t½ between 12 and 72 hours may be designed for a CRDDS requiring application every two to three days. The decline of the blood level time curve after release of the drug from the system will depend on the drug's t½. Naturally, the fluctuation between the therapeutic concentration (Css) max and Css min may accordingly be relatively large; in other words it adds SR to the slow elimination process. For some drugs having a t½ between 20 and 100 hours, and which are intended for long term use, one may desire small fluctuations between peaks and troughs at steady states, either to achieve a certain therapeutic effect or because the therapeutic range is narrow.

12.9 DESIRED BIOPHARMACEUTICAL CHARACTERISTICS OF A DRUG TO QUALIFY FOR CDDS

12.9.1 Molecular weight or size

Small molecules may pass through the pores of a membrane by convective transport [10]. This applies to both the drug release from the dosage form and the transport across a biological membrane. For biological membranes the limit may be a molecular weight of 150 and 400 respectively for spherical molecules and chain-like compounds.

12.9.2 Solubility

For all mechanisms of absorption the drug must be present at the site of absorption in the form of a solution. During the preformulation study it is necessary to determine the solubility of the drug at various pH values. If the solubility is less than 0.1 µg/mL (in an acidic medium) one may expect variable and reduced bioavailability. If the solubility is less than 0.01 µg/mL absorption and availability most likely become limited. Hence the driving force for diffusion may be inadequate. It seems that drugs are well absorbed by passive diffusion from the small intestine upon oral administration if at least 0.1 to 1.0% is in a non-ionized form.

12.9.3 Apparent partition coefficient (APC)

Drugs being absorbed by passive diffusion must have a certain minimal APC. The higher the APC in an n-octanol/buffer system the higher is the flux across a membrane for many drugs. The APC should be determined for the entire pH range in the GI tract. The APC must also be applied for the partition of the drug between the CRDDS and the biological fluid.

12.9.4 General absorption mechanism

For a drug to be a variable candidate for an oral CRDDS, its absorption mechanism must be by diffusion throughout the entire GI tract [11]. The term "diffusion" here refers to the dual pathway of absorption either by partitioning into the lipid membrane (across the cells) or by passing through water filled channels (between the cells). It is also important that absorption occurs from all segments of the GI tract which may depend on the drug's pKa, the pH in the segment, the binding of the drug to mucus, the blood flow rate, and so on. The absorption process seems to be highly dependent on the hydrodynamics in the GI lumen. Even though the first order and square root of the time release can result in highly effective drug delivery systems it is widely believed that the ultimate goal is a zero-order release profile.

Zero-order release *in vitro* will produce zero-order release *in vivo* only if: (1) the entire GI tract behaves as a one-compartment model, that is, the various segments throughout the GI tract are homogeneous with respect to absorption; and (2) the drug release rate is the rate limiting step in the absorption process.

With first-order release, on the other hand, smaller and smaller amounts are released per unit of time with increasing time. Assuming that the rate of absorption gets slower beyond the small intestine due to increased viscosity, decreased mixing, and decreased intestinal surface area, less drug is absorbed.

In any case, the drug release from the CRDDS should not be influenced by pH changes within the GI tract, such as by enzymes present in the lumen or peristalsis. For all practical terms, the one-compartment open model is quite suitable for designing a CRDDS for most drugs.

12.9.5 Elimination half-life of pharmacokinetic parameters

Drugs having a t½ of 8 hours are ideally suited to CRDDSs [12]. If t½ is less than 1 hour the dose size required to be incorporated for a 12 or 24 hour duration dosage form may be too large. If t½ is very long there is usually no need for a CRDDS, unless it is simply intended for a reduction in the fluctuation of steady state blood levels.

12.9.6 Total clearance (CL)

CL is a measure of the volume of distribution cleared by drug per unit time. It is the key parameter in estimating the required dose rate for a CRDDS and predicting the steady state concentration.

12.9.7 Terminal disposition rate constant (Ke or λz)

The terminal disposition rate constant or elimination rate constant can be obtained from t½ and is required to predict a blood level time profile.

12.9.8 Apparent volume of distribution (Vz)

Vz is the hypothetical volume a drug would occupy if it were dissolved at the same concentration as that found in blood [13]. It is the proportionality constant relating the amount of drug in the body to the measured concentration in the blood. Among the trio, CL, Vz, and t½, the former two parameters are the independent variables and the last one is the dependent variable.
   The Vz or CL is required to predict the concentration time profile.

12.9.9 Absolute bioavailability (F)

F is the percentage of drug taken up into systemic circulation upon extra-vascular administration. For drugs to be suitable for CRDDSs one wants an F value close to 100%.

12.9.10 Intrinsic absorption rate constant (Ka)

The Ka of the drug administered orally in the form of a solution should be high, generally by an order of magnitude higher than the desired release rate constant of the drug from the dosage form, in order to ensure that the release process is the rate controlling step.

12.9.11 Therapeutic concentration (Css)

The therapeutic concentrations are the desired or target steady state peak concentrations (Css max), the desired or target steady state minimum concentrations (Css min), and the mean steady state concentration (Css avg) [14]. The difference between Css max and Css min is the fluctuation. The smaller the desired fluctuation the greater must be the precision of the dosage form performance. The lower the Css, the smaller the Vz; the longer the t½, the higher the F and the less amount of drug is required to be incorporated into a CRDDS.

12.10 APPROACHES TO DESIGN CONTROLLED RELEASE FORMULATIONS

1. Dissolution CR:
   - Encapsulation dissolution control;
   - Seed or granule coated;
   - Micro-encapsulation;
   - Matrix dissolution control.
2. Diffusion CR:
   - Reservoir type devices;
   - Matrix type devices.
3. Diffusion and dissolution controlled systems.
4. Ion exchange resins.
5. Osmotically controlled release.

12.11 MECHANISTIC ASPECTS FOR ORAL CONTROLLED RELEASE DRUG DELIVERY FORMULATION

12.11.1 Dissolution controlled release

Dissolution is defined as solid substance solubilized in a given solvent [15]. It is a rate determining step when liquid is diffused from solid. Several theories explain dissolution: diffusion layer theory, surface renewal theory, and limited salvation theory.

$$dc/dt = kD \cdot A (Cs - C)$$

$$dc/dt = D/h \, A.(Cs - C)$$

where dc/dt = dissolution rate, k = dissolution rate constant (first order), D = diffusion coefficient/diffusivity, Cs = saturation/maximum drug solubility, C = concentration of drug in bulk solution, Cs–C = concentration gradient, and h = thickness of diffusion layer.

Two common formulation system rely on dissolution to determine the release rate of drugs: (i) the encapsulated dissolution system and (ii) the matrix dissolution system.

12.11.2 Encapsulated dissolution system

This is also known as a coating dissolution controlled system [16]. The dissolution rate of the coat depends upon the stability and thickness of the coating. It masks colour, odour, and taste and minimizes GI irritation. CR products, by decreasing the dissolution rate of drugs which are highly water soluble, can be formulated by preparing appropriate salts or derivatives, by coating the drug with a slowly dissolving material, or by incorporating the drug into a slowly dissolving carrier. Examples are ornadespansules and chlortrimeto repetabs.

12.11.3 Matrix dissolution system

This is also known as a monolithic dissolution controlled system [17]. This dissolution is controlled by altering the porosity of the tablet, decreasing its wetability, or dissolving at a slower rate. It follows first-order drug release, which can be determined by the dissolution rate of the polymer. Examples are demeanedextencaps and dimetappextentabs.

12.11.4 Diffusion controlled system

This is a major process for absorption in which no energy is required. In this drug molecules diffuse from a region of higher concentration to lower concentration until equilibrium is attained and it is directly proportional to the concentration gradient across the membrane. In this system the release rate is determined by its diffusion through a water-insoluble polymer. There are two types of diffusion devices:

- The reservoir diffusion system;
- The matrix diffusion system.

12.11.5 Reservoir diffusion system

This is also called a laminated matrix device [18]. It is a hollow system containing an inner core surrounded by a water insoluble membrane and a polymer which can be applied by coating or microencapsulation. The rate controlling mechanism is that the drug will partition into a membrane and exchange with the fluid surrounding the drug by diffusion. Commonly used polymers are HPC, ethylcellulose, and polyvinylacetate. Examples are Nico-400 and Nitro-Bid (Figure 12.2).

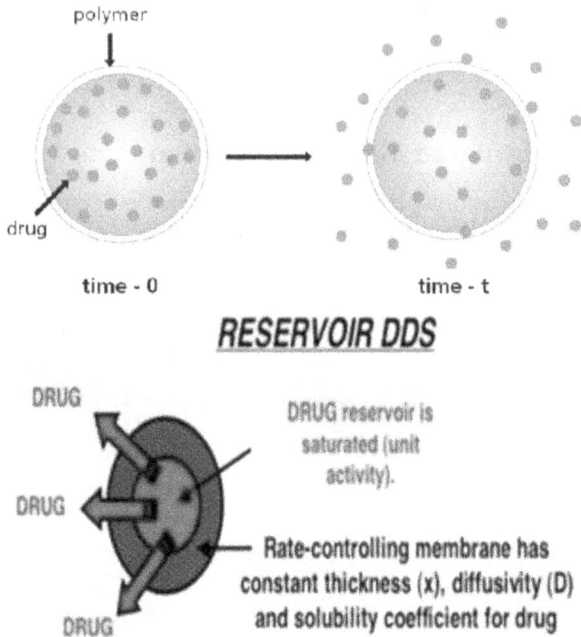

Figure 12.2 Reservoir diffusion system.

Rate controlling steps: Polymeric contenting coating, thickness of coating, hardness of microcapsule.

12.11.6 Matrix dissolution system

1. Rigid matrix diffusion: materials used are insoluble plastics such as PVP and fatty acids.
2. Swellable matrix diffusion: also called glassy hydro gels and popular for sustaining the release of highly water soluble drugs. Materials used are hydrophilic gums [19]. Examples are natural guar gum and Tragacanth.

Semisynthetic-HPMC, CMC, Xanthum gum. Synthetic-Polyacrilamides. Examples: GlucotrolXL, ProcardiaXL

The Higuchi equation describes the drug release from this system [20]:

$$Q = \left[D\varepsilon / T (2A - \varepsilon Cs.t) \right] 1/2$$

where Q = amount of drug release per unit surface area at time t, D = diffusion coefficient of drug in the release medium, ε = porosity of the matrix,

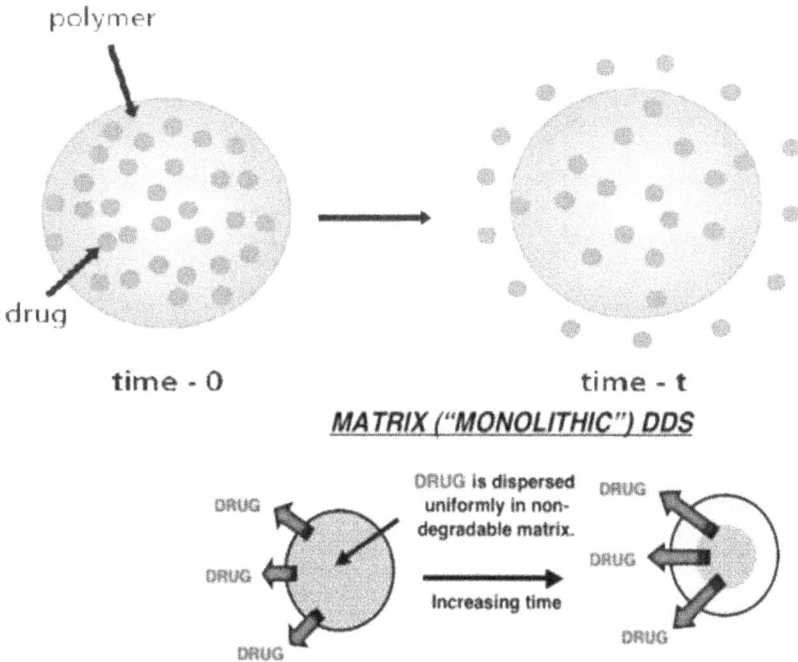

Figure 12.3 Matrix dissolution system.

Cs = solubility of drug in release medium, T = tortuosity of matrix, and A = concentration of drug present in matrix per unit volume (Figure 12.3).

Rate controlling step: Diffusion of dissolved drug in matrix.

12.11.7 Dissolution and diffusion controlled release system

This drug is encased in a partially soluble membrane and pores are created due to dissolution of parts of the membrane [14]. This permits entry of an aqueous medium into the core and the drug is dissolved or diffused out of the system. Example: Ethylcellulose and PVP mixture dissolves in water and creates spores of insoluble ethylcellulose

12.11.8 Ion exchange resin controlled release system

Ion exchange resins are cross-linked water insoluble polymers carrying ionizable functional groups. These resins are used for taste masking and CR systems. The formulations are developed by embedding the drug molecules in the ion-exchange resin matrix; this core is then coated with a semi-permeable

Table 12.1 Marketed drug products with their mechanism based classification

Serial number	Technology		Brand name	Drug	Manufacturer
1	Diffusion controlled system		Welbutrin XL	Bupropion	GlaxoSmith Kline
2	Matrix system tablet		Ambien CR	Zolpidem tartarate	Sanofi-Aventis
3	Method using ion exchange resin		Tussionex pennkinetics ER Suspension	hydrocodon Polistirex and chlorpheneramine polistirex	UCB Inc.
4	Methods using osmotic pressure	Elementary osmotic pump	Efidac 24@	Chlorpheneramine maleate	Novartis
		Push-pull osmotic system	Glucotrol XL@	Glipizide	Pfizer Inc.
5	pH independent formulation		Inderal@ LA	Propranolol HCL	Wyeth Inc.
6	Altered density formulation		Modapar	Levodopa and benserazide	Roche Products, USA

coating material such as ethyl cellulose. This system reduces the degradation of a drug in the gastro-intestinal tract (GIT). The most widely used and safe ion-exchange resin is divinyl benzene sulphonate. In tablet formulations ion-exchange resins have been used as a disintegrant (Table 12.1).

12.11.8.1 Principle

This is based on the preparation of totally insoluble ionic materials:

- Resins are insoluble in acidic and alkaline media.
- They contain ionizable groups which can be exchanged for drug molecules.
- Ipn exchange resins (Ers) are capable of exchanging positively or negatively charged drug molecules to form insoluble poly salt resinates.
- Types: there are two types of IER:
  - Cationic exchange resins: $RSO_3^- H^+$ Resin functional groups;
  - Anionic exchange resins: RNH_3^+OH.

12.11.8.2 Mechanism of action

IERs combine with drugs to form insoluble ion complexes:

$$R - SO_3H^+ + H_2N - A \rightleftarrows R - SO_3 - NH_3^+ - A$$

$$R - SO_3^+OH^- + HOOC - B \rightleftarrows RNH_3^+ - OOC - B + H_2O$$

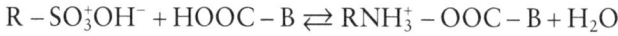

where A-NH$_2$ is a basic drug and B-COOH is an acidic drug.

These resinates are administered orally
↓
Two hours in the stomach in contact with acidic fluid at pH 1.2
↓
Intestinal fluid, remain in contact with slightly basic pH for 6 hrs
↓
Drug can be slowly liberated by exchange with ions present in GIT

In the stomach

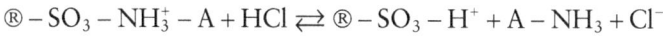

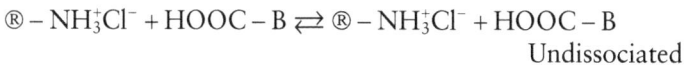

$$® - SO_3 - NH_3^+ - A + HCl \rightleftarrows ® - SO_3 - H^+ + A - NH_3 + Cl^-$$

$$® - NH_3^+Cl^- + HOOC - B \rightleftarrows ® - NH_3^+Cl^- + HOOC - B$$
$$\text{Undissociated}$$

Thus carboxylic acid will be poorly dissociated in the stomach and thus absorbed.

In the intestine

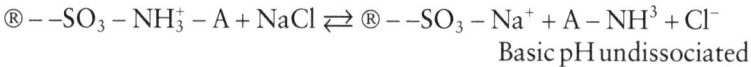

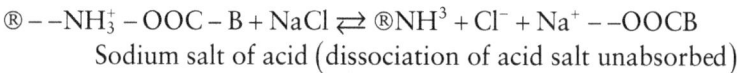

$$® - -SO_3 - NH_3^+ - A + NaCl \rightleftarrows ® - -SO_3 - Na^+ + A - NH^3 + Cl^-$$
$$\text{Basic pH undissociated}$$

$$® - -NH_3^+ - OOC - B + NaCl \rightleftarrows ®NH^3 + Cl^- + Na^+ - -OOCB$$
$$\text{Sodium salt of acid (dissociation of acid salt unabsorbed)}$$

Amine salt will be poorly dissociated in the intestine and thus absorbed.

12.12 CLASSIFICATION OF CONTROLLED RELEASE SYSTEM

The CR system is divided into the following major classes based on the release pattern:

1. Rate pre-programmed drug delivery system;
2. Activated modulated drug delivery system;
3. Feedback regulated drug delivery system;
4. Site targeting drug delivery system.

12.12.1 Rate pre-programmed drug delivery system

In this, the release of drug molecules from the delivery system is pre-planned with a particular flow rate profile of the medicine [6]. The system controls the molecular diffusion of drug molecules in or across the barrier medium within or surrounding the delivery system.

12.12.2 Polymer membrane permeation controlled system

In this system, the drug is completely or partially encapsulated in a drug reservoir cubicle whose drug-releasing surface is covered by a flow rate controlling polymeric membrane. In the drug reservoir, the drug can be solid or a dispersion of solid drug particles or a concentrated drug solution in a liquid or in a solid type dispersion medium. The polymeric membrane may be made up of the fabricated form of a homogeneous or heterogeneous non-porous or partially microporous or semi-permeable membrane.

12.12.3 Polymer matrix diffusion controlled system

In this drug, the reservoir is prepared by the homogeneously dispersing drug particles in the rate controlling hydrophilic or lipophilic polymer matrix. The resultant medicated polymer matrix provides the medicated disc with a defined surface area and controlled thickness.

12.12.4 Micro-reservoir partition controlled system

The drug reservoirs are a suspension of solid particles in the aqueous solution of the water-miscible polymer. A micro-dispersion partition controlled system is prepared by applying high dispersion techniques. In short a reservoir and matrix dispersion forms a micro-reservoir.

12.13 FACTORS INFLUENCING THE DESIGN AND ACT OF CONTROLLED RELEASE PRODUCTS

12.13.1 Physiological properties

12.13.1.1 Aqueous solubility

Most of the active pharmaceutical moieties (APIs) are weakly acidic or basic in nature and affect the water solubility of APIs [9]. Weak water soluble drugs are difficult for designing controlled release formulations. High aqueous solubility drugs show a burst release followed by a rapid increment in plasma drug concentration. These types of drugs are a good candidate for CRDDS. The pH dependent solubility also creates a problem in formulating

CRDDS. Biopharmaceutical classification system (BCS) class III and IV drugs are not suitable candidates for this type of formulation.

12.13.1.2 Partition coefficient (P value)

The P value denotes the fraction of the drug in an oil and aqueous phase and that is a significant factor affecting the passive diffusion of the drug across the biological membrane. The drugs having a high or low P value are not suitable for CR, though they should be appropriate for dissolving in both phases.

12.13.1.3 Drug pKa

pKa is the factor that determines the ionization of a drug at the physiological pH in GIT. Generally, these highly ionized drugs are poor candidates for CRDDS. The absorption of the unionized drug occurs rapidly as compared to ionized drugs from the biological membranes. The pKa range for an acidic drug that ionizes depends on the pH being 3.0–7.5 and for a basic drug it lies between 7 and 11.

12.13.1.4 Drug stability

Drugs that have a stable acid/base, enzymatic degradation, and other gastric fluids are good candidates for CRDDS. If a drug degrades in the stomach and small intestine, it is not suitable for CR formulations because it will decrease in the bioavailability of the concerned drug.

12.13.1.5 Molecular size and weight

The molecular size and weight are two important factors which affect the molecular diffusibility across a biological membrane. If the molecular size is less than 400D it is easily diffused but when greater than 400D drug diffusion is a problem.

12.13.1.6 Protein binding

The drug–protein complex acts as a reservoir in plasma for the drug. Drugs showing high plasma protein binding are not good candidates for CRDDS because protein binding increases the biological half-life. So there is no need to sustain the drug release.

12.13.2 Biological factors
12.13.2.1 Absorption

Uniformity in the rate and extent of absorption is an important factor in formulating the CRDDS. However, the rate-limiting step is the drug to be released from the dosage form. The absorption rate should be more rapid than the release rate to prevent dose dumping. The various factors like aqueous solubility, log P, and acid hydrolysis affect the absorption of drugs.

12.13.2.2 Biological half-life (t½)

In general a drug that has a short half-life requires frequent dosing and is a suitable candidate for a CR system. A drug with a long half-life requires dosing after a long time interval. Ideally, the drugs having a t½ of two to three hours are a suitable candidate for CRDDS. Drugs with a t½ of more than seven to eight hours are not used for a CR system.

12.13.2.3 Dose size

A CRDDS is formulated to eliminate repetitive dosing, so it must contain a larger dose than a conventional dosage form [12]. But the dose used in a conventional dosage form gives an indication of the dose to be used in a CRDDS. The volume of the sustained dose should be as large as it comes under the acceptance criteria.

12.13.2.4 Therapeutic window

The drugs with a narrow therapeutic index are not suitable for a CRDDS. If the delivery system failed to control release, it would cause dose dumping and ultimately toxicity.

12.13.2.5 Absorption window

The drugs which show absorption from the specific segment in GIT are a poor candidate for a CRDDS. Drugs which are absorbed throughout the GIT are good candidates for CR (Figure 12.4).

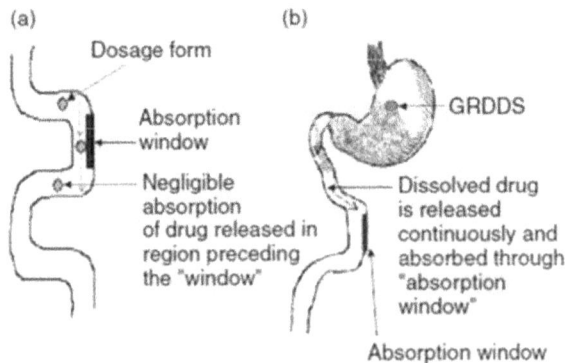

Figure 12.4 Absorption window.

Table 12.2 Pharmacokinetic parameters for drug selection

Parameter	Comment
Biological or elimination half-life	Should be between two and six hours
Elimination rate constant (KE)	Required for design
Total clearance (CLT)	Dose independent
Intrinsic absorption rate	Should be greater than the release rate
Apparent volume of distribution (Vd)	Vd affects the required amount of the drug
Absolute bioavailability	Should be 75% or more
Steady-state concentration (Css)	Lower Css and smaller Vd
Toxic concentration	The therapeutic window should be broader

12.13.2.6 Patient physiology

The physiological condition of the patient, such as the gastric emptying rate, residential time, and GI diseases, influences the release of the drug from the dosage form directly or indirectly.

Pharmacokinetic parameters considered during drug election are listed in Table 12.2.

12.14 POLYMERS USED IN CONTROL DRUG DELIVERY SYSTEMS

Polymers are becoming increasingly important in the field of drug delivery [11]. The pharmaceutical applications of polymers range from their use as binders in tablets to viscosity and flow controlling agents in liquids, suspensions, and emulsions. Polymers can be used as film coatings to disguise the unpleasant taste of a drug, to enhance drug stability, and to modify drug release characteristics. This review focuses on the significance of pharmaceutical polymers for controlled drug delivery applications. Sixty million patients benefit from advanced drug delivery systems today, receiving safer and more effective doses of the medicines they need to fight a variety of human ailments, including cancer. Controlled drug delivery (CDD) occurs when a polymer, whether natural or synthetic, is judiciously combined with a drug or other active agent in such a way that the active agent is released from the material in a predesigned manner. The release of the active agent may be constant over a long period, it may be cyclic over a long period, or it may be triggered by the environment or other external events. In any case, the purpose behind controlling the drug delivery is to achieve more effective therapies while eliminating the potential for both under and overdosing.

12.14.1 Polymers as biomaterials for delivery systems

A range of materials have been employed to control the release of drugs and other active agents [15]. The earliest of these polymers were originally intended for other, non-biological uses, and were selected because of their desirable physical properties; for example:

- Poly (urethanes) for elasticity;
- Poly (siloxanes) or silicones for insulating ability;
- Poly (methyl methacrylate) for physical strength and transparency;
- Poly (vinyl alcohol) for hydrophilicity and strength;
- Poly (ethylene) for toughness and lack of swelling;
- Poly (vinyl pyrrolidone) for suspension capabilities.

To be successfully used in controlled drug delivery formulations, a material must be chemically inert and free of leachable impurities. It must also have an appropriate physical structure, with minimal undesired ageing, and be readily processable. Some of the materials that are currently being used for controlled drug delivery include:

- Poly(2-hydroxyethylmethacrylate);
- Poly (N-vinylpyrrolidone);
- Poly (methylmethacrylate);
- Poly (vinylalcohol);
- Poly (acrylic acid);
- Polyacrylamide;
- Poly (ethylene-co-vinylacetate);
- Poly (ethyleneglycol);
- Poly (methacrylic acid).

Polymers include:

- Insoluble, inert polyethylene, polyvinylchloride, methylacrilate, ethylcellulose;
- Insoluble, erodible carnauba wax, stearylalcohol, castor wax;
- Hydrophilic–methylcellulose, hydroxyl ethyl cellulose, sodium carboxy methyl cellulose, sodium alginate.

In a matrix system the drug is dispersed as solid particles within a porous matrix formed of a water insoluble polymer, such as polyvinylchloride.

Initially, drug particles located at the surface of the release unit will be dissolved and the drug released rapidly. Thereafter, drug particulates, at a successively increasing distance from the surface of the release unit, will be dissolved and released by diffusion in the pores to the exterior of the release unit. The main formulation factor by which the release rate from the matrix system can be controlled are the amount of the drug in the matrix, the porosity of the release unit, and the solubility of the drug.

However, in recent years additional polymers designed primarily for medical application have entered the arena of CR. Many of these materials are designed to degrade within the body, including:

- Polylactides (PLA);
- Polyglycolides (PGA);

- Poly (lactide-co-glycolides) (PLGA);
- Polyanhydrides;
- Polyorthoesters.

Originally, polylactides and polyglycolides were used as absorbable suture material, and it was a natural step to work with these polymers in controlled drug delivery systems [16]. The greatest advantage of these degradable polymers is that they are broken down into biologically acceptable molecules that are metabolized and removed from the body via normal metabolic pathways. However, biodegradable materials do produce degradation by-products that must be tolerated with little or no adverse reactions within the biological environment.

These degradation products, both desirable and potentially non-desirable, must be tested thoroughly, since there are a number of factors that will affect the biodegradation of the original materials. The various important factors indicating the breadth of structural, chemical, and processing properties that can affect biodegradable drug delivery systems are:

- Chemical structure;
- Chemical composition;
- Distribution of repeat units in multimers;
- Presence of ionic groups;
- Presence of unexpected units or chain defects;
- Configuration structure;
- Molecular weight;
- Molecular-weight distribution;
- Morphology (amorphous/semi-crystalline, microstructures, residual stresses);
- Presence of low-molecular-weight compounds;
- Processing conditions;
- Annealing;
- Sterilization process;
- Storage history;
- Shape;
- Site of implantation;
- Adsorbed and absorbed compounds (water, lipids, ions, etc.);
- Physico-chemical factors (ion exchange, ionic strength, pH);
- Physical factors (shape and size changes, variations of diffusion coefficients).

REFERENCES

1. John C, Morten C. *The science of dosage form design, Aulton: Modified release per oral dosage forms.* 2nd ed. Churchill Livingstone, 2002;290–300.
2. Brahmankar DM, Jaiswal SB. *Biopharmaceutics and pharmacokinetics: Pharmacokinetics.* 2nd ed. Delhi: Vallabh Prakashan, 2009;399–401.

3. Allen LV, Popvich GN, Ansel HC. *Ansel's pharmaceutical dosage form and drug delivery system* Baltimore, MD: Lippincott Williams and Wilkins, 8th ed., 2004;260–263.
4. Lee VHL. *Controlled drug delivery fundamentals and applications: Influence of drug properties on design.* 2nd ed. New York: Marcel Dekker, 1987;16–25.
5. Kushal M, Monali M, Durgavati M, Mittal P, Umesh S, Pragna S. Oral controlled release drug delivery system: An overview. *Int. Res. J. Pharm.* 2013;4(3):70–76.
6. Vyas SP, Khar RK. *Controlled drug delivery: Concepts and advances.* 1st ed. Vallabh Prakashan, 2002;156–189.
7. Lachaman L, Liberman HA, Kanig JL. *The theory and practice of industrial pharmacy.* 3rd ed. Bombay: Varghese Publishing House, 1987.
8. Patrick JS. *Martin's physical pharmacy and pharmaceutical sciences.* 3rd ed. Bombay: Varghese Publishing House, 1991;512–519.
9. Kar RK, Mohapatra S, Barik BB. Design and characterization of controlled release matrix tablets of Zidovudin. *Asian J. Pharm. Clin. Res.* 2009;2:54–6.
10. Jain NK. *Controlled and novel drug delivery.* CBS Publisher and Distribution, 1997;1–25.
11. Robinson JR, Lee VH. *Controlled drug delivery.* 2nd ed. Marcel Dekker, 1987;4–15.
12. Venkataraman DSN, Chester A, Kliener L. An overview of controlled release system. *Handbook of pharmaceutical controlled release technology.* Marcel Dekker, 2000;1–30.
13. Mamidala R, Ramana V, Lingam M, Gannu R, Rao MY. Review article factors influencing the design and performance of oral sustained/controlled release dosage form. *Int. J. Pharm. Sci. Nanotechnol.* 2009;2:583.
14. Kamboj S, Gupta GD. Matrix Tablets: An important tool for oral controlled release dosage form. *Pharmainfo. Net.* 2009;7:1–9.
15. Gupta S, Singh RP, Sharma R, Kalyanwat R, Lokwani P. Osmotic pumps: A review. *Int. J. Compr. Pharm.* 2011;6:1–8.
16. Wise DL. *Handbook of pharmaceutical controlled release technology.* New York: Marcel Dekker; 2002;432–460.
17. Gibaldi M. *Biopharmaceutics and clinical pharmacokinetics.* 3rd ed. Philadelphia: Lea & Febiger; 1984.
18. Bechgaard H, Nielson GH. Controlled release multiple units and single unit dosage. *Drug Dev. Ind. Pharm.* 1978;4:53–67. doi:10.3109/03639047809055639
19. Tripathi KD. *Essentials of Medical pharmacology.* 5th ed. New Delhi: Jaypee Brothers Medical Publishers (P) Ltd; 2003.
20. Chien YW. Novel drug delivery system. Volume 50. (n.d.).

Chapter 13

Integration of Blockchain for IoT communication

Jyoti Dargan and Neha Gupta
Sushant University, Gurugram, India

CONTENTS

DOI: 10.1201/9781003436461-13

13.1 INTRODUCTION

The whole world has experienced the novel virus COVID-19 threat which mandated social distancing and an online work culture. This scenario has resulted in the exponential growth of a number of connected devices. As already predicted by the Deloitte report (Deloitte 2020), "The IoT cloud platform market is expected to grow from US$ 6.4 billion in 2020 to USD$ 11.5 billion by 2025". The number of IoT devices is expected to reach 21.5 billion in 2025 from 9.9 billion in 2020 (Statistic.com n.d.). This brings about great concern related to data collection, storage, privacy, and reuse. The IoT has penetrated into everyday fields like lifestyle, health and most specialised applications, for example, financial transactions and nuclear plant automation devices. This has created the necessity of using a technology which will provide relief from all these problems and security threats.

The IoT assimilates various devices' sensing, connecting, analysing, and computing abilities over resource constrained environments. These devices usually rely on client server architecture or Edge or Fog computing techniques for fast and real-time data processing or controlling actuators.

Blockchain technology, which was firstly utilised by Satoshi Nakamoto (Nakamoto 2008) for electronic financial transactions, has shown its ample capability for creating tamper proof, incorruptible records or for transactions in a distributed mode without any central authority. With the integration of Blockchain technology with the IoT ecosystem, security and data integrity issues are mostly readdressed. The key features which enable the use of Blockchain for the redressal of data security issues are distributed ledger technology, immutability, privacy and anonymity.

This chapter elucidates two influential technologies (i.e., Blockchain technology and the IoT) and the significance of their integration for applications in various areas like smart cities, smart homes, electric vehicle management, supply chain management, agriculture, e-governance, and distributed energy management.

13.2 THE INTERNET OF THINGS

The Internet of things (IoT) is a system of interrelated computing devices, mechanical digital machines, objects, animals or people that are provided with unique identifiers and the ability to transfer data over a network without requiring human-to-human or human-to-computer interaction.

(Oracle.com 2020)

The IoT has created a vision of a connected world with all things, living and non-living, communicating with each other through small wearable devices

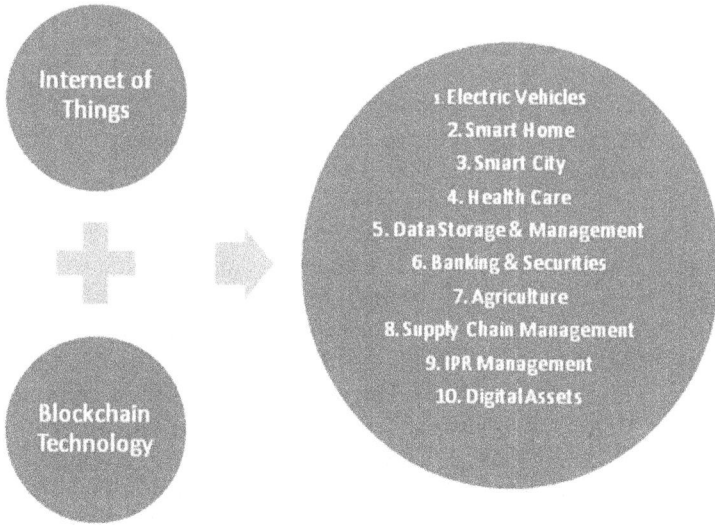

Figure 13.1 Applications of Blockchain based IoT systems.

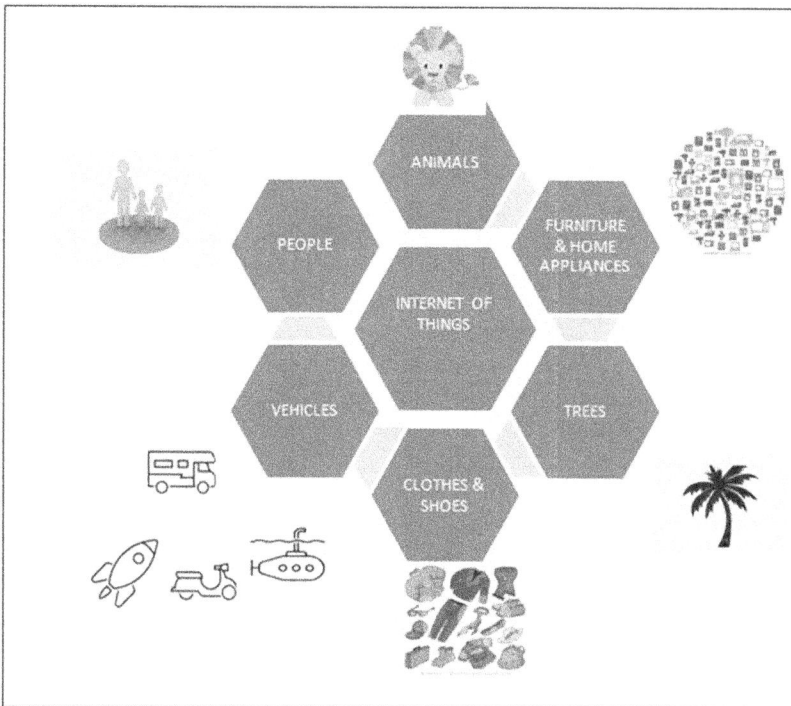

Figure 13.2 The Internet of Things.

which sense or collect data through sensors, store the data efficiently and securely, communicate or transfer the data through the internet, analyse the data intelligently with processing techniques, and utilise the results when responding to or controlling some precision machines through actuators (Mohd Aman et al. 2020). This has improved the quality of and streamlined our life by its entering into various fields like smart homes, smart devices, smart cities, smart grids, healthcare systems, lifestyle management, industry automation, traffic management, RFID (Radio Frequency Identification), transportation, the hospitality industry, logistics, agriculture, and supply chain management. Their application is never ending and represents the real world in digital form. During the time of COVID-19the IoT helped health agencies to gather and process real-time data remotely without physical intervention. It helped in mapping epidemic occurrence, spreading patterns, and intensity. For real-time applications and a time constrained environment the IoT uses Fog, Edge, or cloud computing. Use of artificial intelligence, machine learning techniques, and cloud computing in IoT architecture has played a pivotal role in spreading the wings of IoT into every field of life.

13.2.1 IoT architecture

An IoT architecture (Borgia 2014) refers to the overall design and structure of an IoT system. It typically includes several key components:

1. IoT devices: These are the physical devices that make up the IoT system, such as sensors, cameras, and actuators. They collect and transmit data to the IoT platform.
2. IoT gateway: This is a device or software that acts as an intermediary between IoT devices and an IoT platform. It is responsible for collecting data from the devices, performing data filtering, and forwarding the data to the IoT platform.
3. IoT platform: This is the backbone of the IoT system. It is responsible for managing, storing, and analysing the data collected from the IoT devices. It can also be used to control the devices and platform tasks such as sending notifications or triggering actions based on the data.
4. Communication protocols: These are the rules and standards that govern the communication between the IoT devices and the IoT platform. They can be based on technologies such as Bluetooth, Wi-Fi, Zigbee, and cellular networks.
5. Cloud services: Cloud-based services such as storage, analytics, and machine learning are often used to provide scalability, flexibility, and security for IoT data and the IoT platform.
6. Application layer: This is the topmost layer of the IoT architecture; it is where the end-user interacts with the IoT system, through web or mobile applications.

The IoT works in three layers of network architecture (Fakhri and Mutijarsa 2019):

1. Perception layer: This consists of sensors and actuators for sensing and gathering information and performing functions, e.g., smart phones are equipped with many applications through smart sensors for keeping a record of your health, movement, location, etc.
2. Network layer: Information is transferred using various communication protocols from the sending post to receiving port. Various protocols like Wi-Fi, GSM, Bluetooth, and RFID are used for communication of information.
3. Application layer: This layer relates to the application-specific service deliverance using software and user interfaces e.g., mobile apps developed to record behavioural patterns, exercise routines, physical movement, and location details.

For a better understanding of the functioning of the IoT the three-layer structure may be replaced by a five-layer structure (Sethi and Sarangi 2017) with three more layers in addition to the perception and application layers (Ngu et al. 2017). Here the network layer is divided into:

1. Transport layer: This transfers the data to a middleware layer, i.e., the processing layer.
2. Processing layer: This stores as well as processes the data received from the transport layer.
3. Business layer: This is the top-most layer above the application layer which manages the whole IoT network related to a particular system or business.

13.2.2 IoT communication technologies

IoT communication refers to the way that IoT devices communicate with each other and with the wider internet. There are several different communication technologies and protocols that are used in IoT systems. The most common communication technologies and protocols are as given below, but new technologies continue to be developed and adopted as the technology evolves:

1. Wi-Fi: This is a popular communication technology for IoT devices that are located within the range of a Wi-Fi network. It offers high bandwidth and low latency, making it well suited for applications such as streaming video or controlling smart home devices.
2. Bluetooth: This is a low-power, short-range, wireless technology that is commonly used for connecting IoT devices to smart phones and other devices.

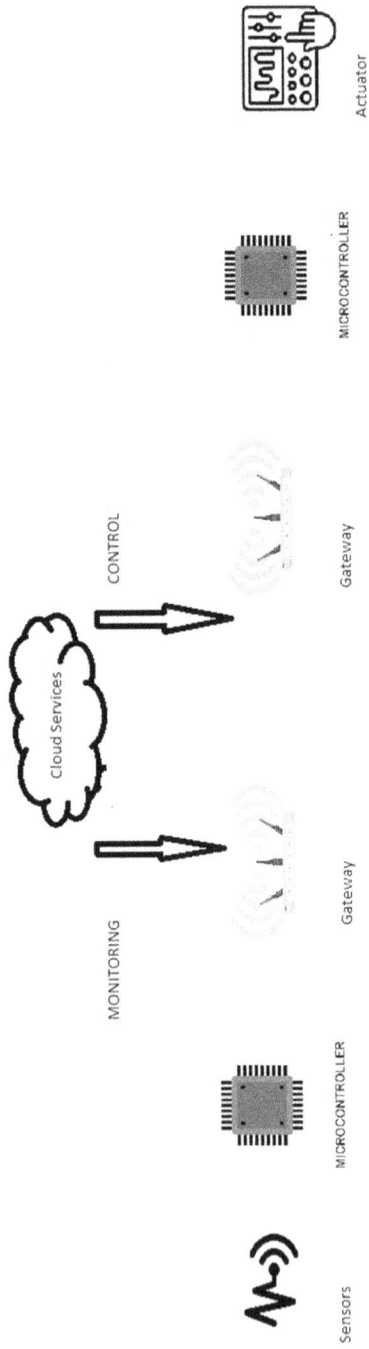

Figure 13.3 General block diagram of the IoT.

3. Zigbee: This is a low power, low data rate, wireless technology that is well suited for IoT devices that need to operate for long periods of time on batteries.
4. LoRaWAN: This is a long-range, low-power, wireless technology that is well suited for IoT devices that are located in remote or hard-to-reach areas.
5. Cellular networks: IoT devices can also use cellular networks (2G, 3G, GSM, 4G,5G) to communicate with the internet. This is particularly useful for devices that are mobile or located in areas without Wi-Fi or other wireless networks.
6. Satellite: For devices located in remote areas, satellite communication may be necessary to connect to the internet.

Most of the IoT data is used in preventive maintenance and predictive assessment of critical infrastructure used in industry, healthcare, agriculture, and so on. Hence the reliability of the data is the pivotal challenge. Under the given circumstances, it is not wise to rely upon the centralised data validation system where any company can modify or fake the data (Reyna et al. 2018).

13.3 BLOCKCHAIN

Although Blockchain was first used in the time stamping of documents so that they cannot be backdated (Abideen n.d.), the power of Blockchain was recognised and implemented by Satoshi Nakamoto in 2008 in the cryptocurrency Bitcoin (Nakamoto 2008). Bitcoin is a secure encrypted e-cash transaction platform with peer-to-peer data validation called "consensus". With the advent of new technologies, the core functionalities of Blockchain have been enhanced. Now Blockchain is becoming the foundational technology for most of the upcoming infrastructural and business models. Blockchain is a distributed ledger of digital transactions where each transaction is added as a block to the chain after due validation by the blocks earlier in the chain. Each block is connected to the previous block of the chain through the hash of the previous block. The transactions are stored in all the blocks of the Blockchain after a consensual validation process called "mining". Hence there is no need for a central validation agency since each block carries a record of transactions (Drescher 2017). Three components essential to define the Blockchain are data structure, algorithm, and cryptographic or security techniques. Blockchain easily overcomes security threats through identification, authorisation, and authentication.

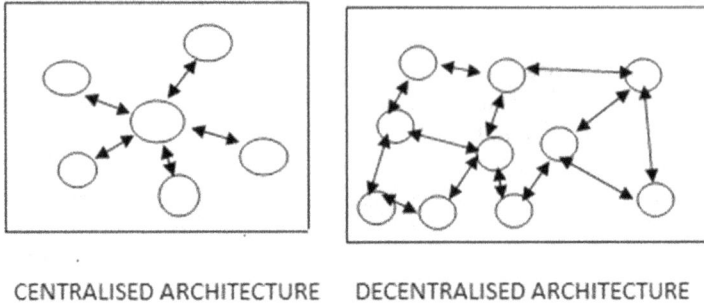

CENTRALISED ARCHITECTURE   DECENTRALISED ARCHITECTURE

Figure 13.4 Difference between centralised and distributed architecture.

13.3.1 Blockchain architecture: The Blockchain utilises the distributed or decentralised architecture

The benefits of distributed architecture are improved reliability, reduced latency, high computational abilities, and reduced infrastructural cost due to the absence of a central server. Blockchain (Dargan et al. 2021) maintains the integrity and reliability of a system. In a peer-to-peer or distributed system no central or middle nodes are involved in the transaction, but the direct transaction is done between the contractual nodes in much less time and at much less cost. The validity of the transaction is done by every node in the Blockchain, hence the reliability of the data is maintained. Every new transaction is added to every node as append only data ledger format. Each node or block in the chain has the details of the data stored in the Blockchain. Since each node shares the same data, trust and integrity are a major component of a Blockchain. In Blockchain every node can use the verified data available to them but no node can modify or tamper with the data. In case of technical failure, the data can easily be retrieved from the neighbouring nodes and there is no threat of system failure.

13.3.2 Blockchain data structure

The data structure of a Blockchain is composed of a series of blocks, where each block contains a number of transaction records. These blocks are linked together in a chain-like structure, hence the name "Blockchain". In Blockchain data are stored in each node or block. Each block in the chain contains a reference to the previous block, creating a secure and tamper-proof chain of data.

The basic structure of a block in a Blockchain typically includes the following fields:

- Block header: This contains metadata about the block, such as the block number, time stamp, and reference to the previous block.

- Transactions: This is the data stored in the block, typically a list of transactions that occurred in the Blockchain network.
- Nonce: A nonce is a 32-bit random value that is used in the block header to create a unique hash for the block. This is important for ensuring the integrity of the Blockchain.
- Hash: This is a unique, fixed-size, digital fingerprint that is generated for each block. It is created by applying a cryptographic hash function algorithm, usually SHA 256. Hash is calculated by adding the nonce value to the data of the block, i.e., (nonce + block data). Every node is known or identified by its hash value which is unique to every block and fingerprint of the block. Each block is connected to the previous block through the hash function of that block. If at any stage the data of any block changes or is corrupted, its hash value changes, due to which the connection between the blocks cannot be maintained or the continuity of the Blockchain ceases.
- Merkle tree: This is a specific type of data structure that is used to summarise all the transactions in a block and allows for the efficient and secure verification of the contents of large blocks.

The data structure of the Blockchain can vary, or other fields can be added to a block depending on the specific implementation of the Blockchain and the consensus mechanism used.

A Merkle tree is constructed by repeatedly hashing pairs of transactions until there is only one hash, called the root hash or Merkle root. This root hash is included in the block header and serves as a summary of all the transactions in the blocks.

The advantage of using a Merkle tree is that it allows for the efficient verification of the contents of a block without the need to download and verify the entire block. Instead, a user can download only the small portion of the tree (Merkle proof) that is necessary to verify a specific transaction.

Each leaf node of the tree represents a transaction, and each non-leaf node is the hash of its child nodes; this way it forms a tree structure where

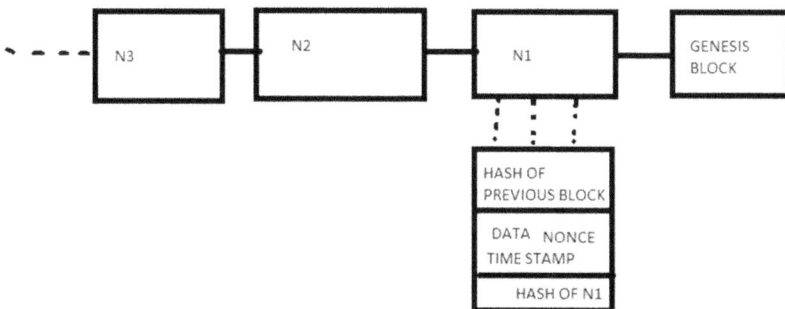

Figure 13.5 Blockchain data structure.

the root node is the Merkle root. The Merkle root is also used to prove the authenticity of a transaction, by providing a proof that a specific transaction is included in the block. It provides a way to prove authenticity and integrity of data in a secure and efficient way.

13.3.3 Block validation or mining

Block validation or mining is the process of finding the hash of the newly added block. Whenever a new block is added by a new transaction or addition of data, all blocks or miners present in the Blockchain start computing the hash of the function with the help of some algorithm (e.g., SHA 256 or SHA 512). The block which finds the hash first, transmits the data alongwith the nonce to all the nodes. The hash is confirmed through the consensus which is the data validation of the newly added block. Data validators are the nodes in the Blockchain which validate the new block and obtain some kind of incentive for the act of validation. This kind of incentive is different in different Blockchains, e.g., in Bitcoin every block validator which adds a new block is rewarded with the cryptocurrency called BTC or Bitcoin. Various methods to find out the block validators are Proof of Work (PoW), Proof of Stake (PoS), Proof of Activity (PoA) and so on.

13.3.4 Blockchain consensus algorithms

Blockchain consensus protocols (Tapscott and Tapscott 2016) are the mechanisms by which different participants in a Blockchain network agree on the state of the ledger. There are several different types of consensus protocols that have been developed for Blockchain networks, each with their own

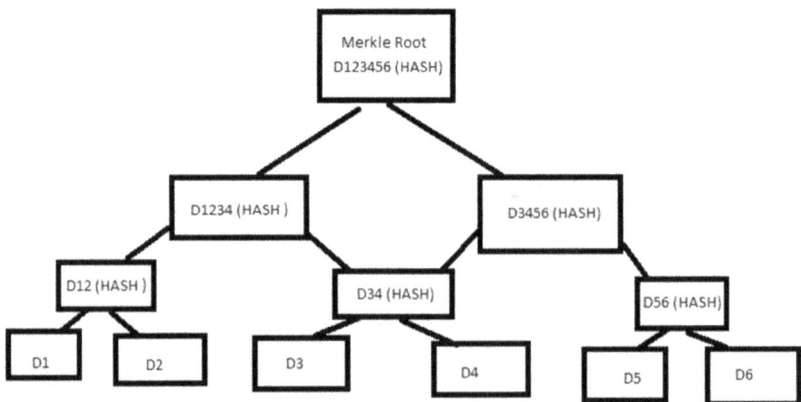

Figure 13.6 Merkle tree architecture.

strengths and weaknesses and suitability for different types of Blockchain networks and use-cases.

1. Proof of Work (PoW): This is the original consensus algorithm used by Bitcoin. It is based on solving complex mathematical problems in order to create new blocks and add them to the chain.
2. Proof of Stake (PoS): This is an algorithm that allows Blockchain participants to validate transactions and create new blocks based on the number of coins they hold or "stake" in the network.
3. Delegated Proof of Stake (DPoS): This is a variation of PoS where token holders can vote for a set of validators to create new blocks and validate transactions.
4. Practical Byzantine Fault Tolerance (PBFT): This is a consensus algorithm that can be used in permissioned Blockchains where all blocks are known and trusted.
5. Proof of Elapsed Time (PoET): This is a consensus algorithm which is designed for permissioned networks and is based on the assumption that a trusted entity, such as a hardware security module (HSM), is available to all participants in the network
6. Proof of Burn (PoB): This is an algorithm which involves destroying or burning the token as a way to prove ownership and secure the network. This is typically done by sending the tokens to a probably unspendable address, such as an address which has a public key but no corresponding private key. The idea is that the act of burning tokens demonstrates a commitment to the network and reduces the overall supply of the tokens, which can increase their value. It is also used in some decentralised autonomous organisations (DAOs) as a form of stake to give more voting rights to token holders who have burned their tokens.
7. Proof of Authority (PoA): This is a consensus algorithm used in some Blockchain networks, particularly those that are permissioned or private. In a PoA network, the nodes that validate transactions and add them to the Blockchain are designated "validators" and they are chosen based on their reputation or identity. These validators are typically known individuals or organisations that have been vetted and approved by the network's governing authority. Because the validators are known, their actions can be audited and held accountable, which can help to improve the security and transparency of the network. In a PoA network, validators are responsible for validating transactions, creating new blocks, and maintaining the network's consensus. Unlike other consensus mechanisms, such as PoW or PoS, PoA does not require any computational work or tokens to be held in order to validate transactions. This can make the network more efficient and faster, but it also means that it is less decentralised than PoW and PoS networks.
8. Proof of Existence (PoE): This allows users to cryptographically prove the existence of a document or other digital file at a specific point of time.

13.3.5 Blockchain platforms

There are several types of Blockchain platforms, including:

1. Public Blockchains: These are the decentralised networks that are open to anyone, allowing for transparent and anonymous transactions. Examples include Bitcoin and Etherium.
2. Private Blockchains: These are the decentralised networks that are permissioned, meaning that only authorised parties are allowed to participate in the network and access the data. Private Blockchains are typically used by organisations for internal-use cases as supply chain management.
3. Consortium Blockchains: These are the hybrids of public and private Blockchains, in which a group of organisations collectively owns and manages the network. Consortium Blockchains are typically used in industries where multiple companies need to share data and work together, such as finance and logistics.
4. Hybrid Blockchains: These are Blockchains that combine features of both public and private Blockchains to offer the best of both worlds. They are private but can be configured to be public as well.
5. Federated Blockchains: These are the Blockchains that are controlled by a group of pre-selected nodes, rather than a decentralised network of users.
6. Directed Acyclic Graph (DAG) Based: These are Blockchains that use a different structure called a DAG instead of a chain of blocks. In DAG, the transactions are linked to each other in a way that forms a graph rather than a chain. Examples include IOTA and Nano.
7. Sidechains: These are separate Blockchains that are attached to a parent Blockchain and can be used to scale the parent and enable new functionality.

The above Blockchains are not mutually exclusive and some platforms may have elements of multiple types. Additionally, new types of Blockchain platforms may emerge in the future as the technology continues to evolve.

There are several types of Blockchain platforms that are in use, each with their own unique features and characteristics. Some of the most popular types include:

1. Bitcoin: This is the first and most well-known Blockchain, created in 2009. It is a decentralised digital currency that allows for peer-to-peer transactions without the need for a central authority.
2. Etherium: Thisis a Blockchain platform that allows for the creation of decentralised applications (dApps) and smart contracts. It also has its own cryptocurrency: Ether (ETH).
3. Ripple: This is a Blockchain platform that is primarily focused on facilitating cross-border payments and currency exchanges. It has its own cryptocurrency: XRP.

4. Hyperledger: This is an open source Blockchain platform that is focussed on enterprise-use cases. It is designed to be highly modular and customisable, allowing businesses to build their own Blockchain solutions.

5. EOS: This is a Blockchain platform that is focused on high performance decentralised applications. It uses a unique consensus algorithm that allows for faster and more efficient transactions than other Blockchain platforms.

6. Corda: This is an open source Blockchain platform that is focussed on financial services and other industries where privacy and security are critical. It uses a unique consensus algorithm that allows only authorised parties to see the transaction details.

7. IOTA: This is a Blockchain platform that uses a different consensus algorithm called Tangle, which is designed to enable fast and fee-free transactions, making it ideal for use in IoT applications.

8. Solana: This is a Blockchain platform that uses a consensus mechanism called Proof of Stake and allows for high-throughput and low-latency transactions.

9. Multichain: This refers to the type of Blockchain system that allows for the creation and use of multiple separate Blockchains within the same network. Each individual Blockchain, also known as a "side-chain", can also have its own set of rules, permissions, and functions; and it can be used for different purposes or by different groups of users. Babelchain or multichain integrates multiple Blockchain networks into a single system. It aims to overcome the limitations of traditional Blockchain networks by allowing different chains to communicate and interact with each other, and share data and assets seamlessly. Babelchain allows different Blockchains to work together, regardless of their underlying technology, enabling a wide range of use cases and business models, One of the main advantages of multichain Blockchain is the ability to create separate, isolated environments for different use cases or groups of users, while still being able to share data and assets between them. This can increase security and privacy as well as enabling new types of business models such as direct-to-customer sales.

10. Hyper Ledger Fabric Blockchain: This is an open-source Blockchain platform that is part of the Hyperledger project, which is a collaborative effort by the Linux Foundation to advance cross-industry Blockchain technologies. Hyperledger Fabric is designed for use in enterprise-grade Blockchain applications and is intended to provide a foundation for building robust and secure Blockchain networks. One of the key features of Hyperledger Fabric is its modular architecture, which allows for flexibility in design and implementation of Blockchain networks. It provides a plug-and-play model for different components such as consensus, membership services, and smart

contracts. Hyperledger Fabric also has a permissioned network model, where the network participants are known and identified entities, and it provides a fine-grained access control mechanism to manage access to the network and its data. This allows for more control over the network and its participants, making it more suitable for use in industries such as finance, healthcare, and supply chain management, where privacy and security are of paramount importance.

13.3.6 Smart contracts

Created by Nick Szabo in 1994, these are the self-executable digital contracts with the terms of the agreement written directly into code. It is a computer program that automatically executes the terms of a contract when certain conditions are met. These contracts are executed without the intervention of the blocks. Solidity language is mostly used to write smart contracts, which are run on an Etherium Virtual Machine (EVM). EVM is a decentralised, consensus-driven computer. Each smart contract is stored as a new block in the Blockchain, which makes them transparent, tamper-proof, and decentralised.

A smart contract can be used to automate a wide range of processes, such as the transfer of assets, the management of supply chains, and the execution of financial transactions and legal agreements. It eliminates the need for intermediaries, such as banks and lawyers, and reduces the risk of fraud and error. Smart contracts are used in various fields like e-payments, cryptocurrency, digital identities, property dealings, e-commerce, digital assets, notary services, tax, compliance, audit, and energy transactions.

In the context of the IoT, smart contracts can be used to automate communication and decision making between IoT devices, further increasing efficiency and security in IoT networks. For example, a smart contract can be used to automatically control the temperature in a smart building based on sensor data, or to automatically initiate a maintenance schedule for a piece of equipment based on usage data.

Smart contracts are still a new technology for which a legal and regulatory framework for their use is still under development. Also, the implementation of smart contracts requires a certain level of technical knowledge and resources.

13.4 INTEGRATION OF THE IoT AND BLOCKCHAIN

IoT devices are under a continuous security threat at the physical layer e.g. data storage, the network layer e.g. insecure communication between devices, gateway/cloud & devices etc. and at the application layer at various levels like device and mobile application etc. A major issue of corrupted data is due to inappropriate authentication and authorisation procedures. MQTT, DDS, Zigbee, and Zwave are the few protocols which support authentication. Physical security is also of utmost concern as the hardware used is simple and vulnerable. Insecure cloud and web interfaces are the threats at the application level. Unreliable communications, hostile environments, inadequate trust management, and inadequate protection of data are the challenges for IoT communication (Mohamad Noor and Hassan 2019). In addition to the above-mentioned security issues, Blockchain technology also efficiently caters for other issues related to Quality of Service (QoS), trust, reliability and immutability, autonomy, secure code deployment, latency, and so on. The decision of adoption of Blockchain integration design depends on the fact that where the IoT interaction takes place. If the interaction is to take place inside the IoT with low latency, then IoT–IoT design is used which is fastest and stores only a part of the IoT data in the Blockchain. In IoT–Blockchain design all interactions are through Blockchain. In this type, all the data are stored in the Blockchain as immutable records. A hybrid approach is used as a middle option between the previous two as it stores a part of the data in the Blockchain and the rest of the data is communicated through IoT devices. This approach efficiently utilises the benefits of Blockchain and saves the bandwidth and data storage capacity in the real-time IoT. Combining Fog or Edge computing with a hybrid design enables the mining process. The availability of various devices like Ethembedded, Raspnode, and Ethraspbian allows the installation of Etherium, Bitcoin, and Litecoin, respectively, on Raspberry Pi or Beaglebone Black. The Wi-Fi router "Antroiter R1-LTC" has mining capabilities and so can be used with IoT devices for mining processes in Blockchain. Fog computing brings the computing near to the IoT device, hence complementing the distributed property of the Blockchain and making the device an end node or a miner. Blockchain can also be provided as a service layer by adding one additional layer to the architecture [Delloitte Report, 2020]. Blockchain agents are involved to provide services to sensors. Agents provide data security and miners verify the transactions and add each validating transaction as a block.

13.5 ADVANTAGES OF THE INTEGRATION OF BLOCKCHAIN TECHNOLOGY FOR IoT COMMUNICATION

There are several advantages to using Blockchain technology in IoT communication, including:

1. Security: Blockchain provides a tamper-proof and decentralised way to store and transmit data which can help to protect IoT devices and networks from hacking and other forms of cyber threats.

2. Privacy: Blockchain can be used to encrypt and secure communication between IoT devices, keeping data private and secure.
3. Trust: Blockchain's decentralised nature allows for the creation of trustless networks in which devices can securely communicate without the need for a central authority.
4. Automation: Smart contracts can be used to automate communication and decision making between IoT devices, further increasing efficiency and security in IoT networks.
5. Scalability: Blockchain technology allows for the creation of a decentralised network of IoT devices, which can help to increase scalability and reduce the reliance on centralised servers.
6. Interoperability: Blockchain can be used to create a common language and set of protocols for IoT devices, allowing them to communicate and interact with each other regardless of their manufacturer or technology.
7. Provenance: Blockchain can be used to identify and authenticate IoT devices, which can help to prevent unauthorised access and improve overall security.

While Blockchain technology offers many potential benefits for IoT communication, it is not without challenges, which are provided in the next section.

13.6 MAJOR CHALLENGES IN THE INTEGRATION OF IoT WITH BLOCKCHAIN

Major challenges faced in the practical implementation of IoT–Blockchain (Mendling et al. 2018) are:

1. Storage capability: IoT devices are massive in number and the data involved with these smart devices may range up to several gigabytes. Using Blockchain with the IoT is therefore a challenge as, in Blockchain, entire data are stored in each block for which each device requires huge data storage capabilities. The application of a hybrid structure of IoT–Blockchain is where only some of the data (refined information) is stored in the Blockchain.
2. Scalability: Due to the large data storage with the IoT device, the number of IoT devices (blocks) in the Blockchain is limited. Also due to the consensus mechanism, the latency of the IoT–Blockchain system is increased.
3. Privacy: In Blockchain every node has access to the data and any kind of privacy of data is out of the question. Although the data are protected from any kind of fraudulent and malicious alterations, they are available to all. Since each block is assigned with a unique ID created

by an asymmetric cryptographic technique (the use of Publi5c Key and Private Key) the anonymity of each block is ensured.

4. Legal Issues: The legality of the Blockchain is also an important factor alongwith IoT laws and policies. At present, there is no dedicated legislation in India that comprehensively regulates blockchain technology. The absence of specific regulations can create uncertainties and challenges in certain areas. whereas the Meity introduced a draft policy for the IoT in 2016 (Meity 2016). The Reserve Bank of India (RBI), the country's central bank, has taken a cautious approach towards cryptocurrencies. In April 2018, the RBI issued a circular that prohibited banks and financial institutions from dealing with cryptocurrencies or providing services to entities dealing with cryptocurrencies. However, this circular was set aside by the Supreme Court of India in March 2020, deeming it unconstitutional. As a result, cryptocurrencies are not illegal in India, but their legal status and regulatory framework remain uncertain. It is also affected by the data privacy laws of the country.

5. Consensus: Most IoT devices are constrained devices, therefore it is not appropriate on their part to participate in the process of consensus. In Babelchain Blockchain, a novel concept of consensus appeared which is called "Proof of Understanding" and which came into the picture for IoT devices. Here, instead of mining the hash function, the device translates from different protocol and neighbouring devices to agree on message meaning [Deloitte 2020].

6. Security: Security issues in IoT are attacks to end devices, communication channels, network protocols, sensory data, software attacks, and denial of service (DoS) attacks, whereas the security issues for Blockchain are attacks on the consensus protocol, eclipse attacks, double spending, the vulnerability of smart contracts, programming frauds, distributed denial of service attacks (DDoSs), leakage of private keys and so on. Therefore the type of Blockchain used must be capable of handling and averting all these attacks on the system.

13.7 USE CASES OF BLOCKCHAIN-BASED IoT APPLICATIONS

Blockchain and the IoT are two technologies that are often discussed in the context of their potential to develop and improve various industries. Together they can be integrated to create a secure and distributed system for collecting, storing, and sharing data. One of the main advantage of integrating Blockchain and IoT is that it allows for secure and tamper-proof tracking of data, as well as secure and transparent sharing of that data. This can be useful in industries where transparency and traceability are important. Self-driven vehicles (SDVs), smart grids, microgrids, automated vehicular systems, smart city traffic surveillance systems, supervisory control and data acquisition systems (SCADAs), industrial automation systems, remote healthcare monitoring and control, e-governance and so on are a few of the examples where IoT devices are used for data acquisition, monitoring, analysis, and control. A few of the applications of Blockchain-based IoTs (BIoTs) are given below.

13.7.1 BIoTs inthehealthcare industry

IoTs are used in the healthcare industries as:

1. Remote patient monitoring: IoT devices, such as wearable sensors and medical devices can be used to collect data on patients' vital signs and other health metrics. This data can be used to monitor patients remotely and to provide early warnings of potential health issues.
2. Electronic health records (EHRs): IoT devices can be used to collect data on patients' health and transmit it to EHRs, making it more complete and upto date. Blockchain-based IoTs can provide tamper-proof, immutable, and authenticated records which can be help to improve the quality of care and reduce the risk of errors.
3. Clinic trials: BIoT devices can be used to collect data on patients participating in clinical trials, which can help to improve the accuracy and efficiency of the trials.
4. Medication management: BIoTs can be used to track the movement of medicines through the supply chain, providing transparency and immutability of data. This can help to prevent medication errors and improve patient safety.
5. Telemedicine: BIoT devices can be used to enable remote consultations between patients and healthcare professionals, which can help to improve access to care, especially in remote or underserved areas.
6. Medical equipment management: IoT devices can be used to monitor the status and usage of medical equipment which can help to reduce down time and increase the efficiency of healthcare facilities.

13.7.2 Application of Blockchain-based IoTs in supply chain management

Here are a few ways in which BIoTs are being used in supply chain management (Bahga and Madisetti 2016) to gain better visibility and control over operations, improve efficiency, responsiveness, and reduce costs and risks:

1. Real time tracking: IoT devices such as RFID tags and GPS sensors (Wang, Han, and Beynon-Davies 2019) can be used to track the movement of goods through the supply chain in real time providing transparency and visibility to the entire process.

2. Predictive maintenance: BIoT devices can be used to monitor the condition and performance of equipment and vehicles, providing real-time data that can be used to predict and prevent equipment failures.

3. Inventory management: BIoT devices can be used to track inventory levels in real time, providing accurate information on stock levels and helping to reduce the risk of stockouts.

4. Quality control: BIoT devices can be used to collect data on the quality of goods, such as temperature, humidity, and vibrations, throughout the supply chain. These data can be utilised to identify and address quality issues before they reach the customer.

5. Logistics optimisation: IoT devices can be used to collect data on the movement of goods and vehicles (Mathijsen and Sadouskaya 2017), which can be used to optimise logistics and reduce transportation costs.

6. Compliance and security: BIoT devices can be used to track the movement of goods and monitor compliance with regulations such as those related to food safety and security.

13.7.3 Application of Blockchain-based IoTs in e-governance

Blockchain-based IoTs are used in e-governance (Ølnes, Ubacht, and Janssen 2017) to improve the delivery of public services and increase transparency, which in turn lead to more efficient and effective governance. A few ways in which BIoTs can be used are:

1. Smart cities: BIoTs can be used to collect data on a wider range of urban services, such as traffic, parking, and air quality. This data can be used to improve the efficiency of the services and liveability of the cities.

2. Public safety: BIoT devices such as CCTV cameras and sensors can be used to monitor public spaces, providing real-time data that can be used to prevent and respond to crime and other public safety issues.

3. Environmental monitoring: BIoT devices can be used to collect data on environmental conditions, such as air and water quality, which can be further used to identify and address environmental issues.

4. Public services: BIoT devices may be used to monitor the delivery of public services, such as the electricity and water supply, and to provide real-time information on the status of these services.

5. Citizen engagement: BIoT devices can be used to provide citizens with real-time information on government services and to facilitate citizen engagement through online platforms and apps.

6. E-voting: BIoT devices can be used to facilitate secure and transparent e-voting (Hjalmarsson et al. 2018), which can help to increase voter turnout and reduce the risk of fraud.

7. Unique identity identifiers: BIoT devices may be used to identify, authenticate, and authorise a user or a device in an identity management system. A BIoT device can be equipped with biometric sensors such as finger print scanners or facial recognition cameras to authenticate users, or be used to securely communicate between devices and an identity management system through secure communication protocols such as HTTPS or VPN. It can help to improve security and protect against unauthorised access to sensitive information.

13.7.4 Application of Blockchain-based IoTs in the real estate industry

By integrating BIoT devices, the real estate industry (Veuger 2018) can improve the efficiency and liveability of properties, reduce costs, and improve the overall real estate experience for both property managers and residents. A few uses are:

1. Smart building management: BIoT devices such as sensors and smart devices can be used to monitor and control building systems such as lighting, heating, and security. This can help in improving energy efficiency and reducing operating costs.
2. Predictive maintenance: BIoT devices may be used to monitor the condition of buildings and equipment, providing real-time data that can be used to predict and prevent equipment failures.
3. Building automation: A wide range of building systems, such as elevators, HVAC systems, and security systems, can be automated using BIoT devices.
4. Smart homes: The use of BIoT devices can create smart homes, which provide residents with the ability to control lighting, heating, and other systems remotely through mobile apps.
5. Real estate marketing: Virtual tours of properties alongwith interactive, data-driven, marketing material can be created with the use of data collected through BIoT devices. Blockchain can be used to provide secure and transparent record-keeping for property ownership, mortgages, and other important real estate documents. This can improve the overall transparency and trust in the real estate market.
6. Smart leasing: BIoT devices can be used to collect data on the usage of properties, which can be used to inform leasing decisions and to improve the management of rental properties.
7. Property management: BIoT devices can be used to monitor the status of properties, such as occupancy rates, energy consumption, and security, which can help to improve the management of property and reduce costs. Some ways this can be done include using smart contracts to automate the process of buying and selling property, using a Blockchain-based land registry to record and track property

ownership, and using digital tokens to represent ownership of fractional shares in a property.

13.7.5 Application of Blockchain-based IoTs in the finance industry

Major applications of BIoT devices in the field of banking and finance are:

1. Fraud detection: BIoT devices may be used for the authorisation or authentication of the data of financial transactions and hence help to identify and prevent fraud.
2. Risk management: Data collected from BIoT devices on a wide range of risks, such as market risks and credit risks, can be used to inform risk management decisions.
3. Smart banking: BIoT devices such as wearables and smart devices can be used to enable secure and convenient banking transactions such as account access and money transfer.
4. Investment management: BIoT devices can be used to collect data on a wide range of investment opportunities, which can be used to inform investment decisions.
5. Digital asset management: BIoT devices can be used to create, issue, and track the ownership of digital assets such as stocks, bonds, and other financial instruments.
6. KYC (Know Your Customer)/AML (Anti Money Laundering): BIoT devices can be used to create a shared digital ledger of customer information, which can be accessed by multiple financial institutions to verify the identity of customers.
7. Cryptocurrencies: Blockchain technology is the backbone of many cryptocurrencies, such as Bitcoin and Ethereum, which can be used as an alternative to traditional currencies.
8. Digital identity: Blockchain can be used to create and store digital identities, which can be used to verify the identity of individuals and organisations in financial transactions.
9. Smart contracts: Blockchain-based smart contracts can be used to automate financial transactions, such as the execution of trades or the release of funds in an escrow.
10. Payment processing: Blockchain can be used to facilitate fast, secure, and low-cost cross-border payments.
11. Insurance: BIoT devices can be used to collect data on a wide range of risks, such as weather risks and vehicle risks, which can be used to inform insurance decisions.
12. Smart contracts: BIoT devices can be used to automate financial transactions, such as stock trades and loan agreements, by using smart contracts, which are self-executing contracts (Huckle et al. 2016) with the terms of the agreement written directly into the code.

13.7.6 Use of Blockchain-based IoTs in agriculture

In agriculture, the integration of Blockchain and the IoT (Lin et al. n.d.) refers to the use of sensors, devices, and technology to collect and analyse data in order to optimise crop production and improve efficiency in farming operations. This can include monitoring soil moisture, temperature, and nutrient levels, as well as tracking weather patterns and crop growth. It can also be used to control irrigation systems and automate tasks such as planting, harvesting, and pest management. IoT devices can collect and transmit data to the Blockchain, which can then be analysed and used to improve crop yields and reduce costs. This not only increases yield and reduces cost but also improves sustainability and reduces environmental impact. It enables the secure and transparent tracking of food from farm to table, allowing consumers to know exactly where their food comes from, how it is grown, and who handled it, that is, from its origin, cultivation, harvesting, packaging to distribution. This allows for improved food safety, traceability, and transparency.

This technology also helps in creating a marketplace for agricultural products that enables farmers to sell directly to consumers and bypass intermediaries. This can help farmers increase their income. Smart contracts can also be used to automate the process of buying and selling goods, creating a more efficient and transparent food supply chain.

13.7.7 Use of Blockchain-based IoT in energy management systems

Blockchain technology can be used in IoT applications to improve the efficiency and security of management systems. Some ways in which this can be done include:

1. Smart metering: Blockchain-based smart meters (Grid et al. 2019) can be used to accurately measure and record the consumption of energy in realtime. These data can be stored on a Blockchain, providing transparency and security for energy consumption data.
2. Peer-to-peer (P2P) energy trading: Blockchain technology can enable P2P energy trading (Li et al. 2018), allowing individuals and organisations to buy and sell excess energy directly with each other, rather than relying on a centralised utility.
3. Renewable energy certificates (RECs): Blockchain-based RECs (Dargan et al. 2021) can be used to track the generation and trade of renewable energy, providing transparency and security for clean energy transactions.
4. Demand response: Blockchain-enabled demand response (Kim and Huh 2018) programs can be used to incentivise energy conservation by allowing customers to be rewarded for reducing their energy consumption during peak demand periods.

5. Microgrids: Blockchain technology can be used to create decentralised energy systems, known as microgrids(Mollah et al. 2021), which can improve the resilience and efficiency of energy distribution.

13.8 CONCLUSION

Blockchain-based IoT communication refers to the use of Blockchain technology to secure and facilitate communication between IoT devices. Most IoT applications handle sensitive and confidential data and are therefore vulnerable to security threats due to data leaks and misuse of data by central authorities. Previous to Blockchain, devices used other methods for security like MQTT (Message Queuing Telemetry Transport) or encrypted passcode technology (RSA (Rivest-Shamir-Adleman algorithm) and DH (Diffie-Hellman)) in the form of a public and private key. Blockchain's decentralised and tamper-proof nature can be used to ensure secure and private communication between IoT devices. It also provides a way to validate and authenticate these devices. The use of Blockchain will solve most of the problems related to authentication, anonymity, and data security by unauthorised sources. Here the control of data is not by a single node but is distributed throughout the Blockchain, and validation is done through consensus by each participating node in the validation process called mining. Additionally, smart contracts can be used to automate communication and decision making between IoT devices, making the communication system more prompt and efficient. IoTs with Blockchain capabilities are providing more data security and integrity through the execution of smart contracts.

REFERENCES

Abideen, Sainul. n.d. "www.cybrosys.com www.blockchainexpert.uk1," 1–97.
Bahga, Arshdeep, and Vijay K. Madisetti. 2016. "Blockchain Platform for Industrial Internet of Things." *Journal of Software Engineering and Applications*. https://doi.org/10.4236/jsea.2016.910036
Borgia, Eleonora. 2014. "The Internet of Things Vision: Key Features, Applications and Open Issues." *Computer Communications* 54: 1–31. https://doi.org/10.1016/j.comcom.2014.09.008
Dargan, J., Gupta, N., and Singh, L.2021. "Blockchain Based Energy Management System: A Proposed Model." In *2021 International Conference on Technological Advancements and Innovations (ICTAI)*, 510–14. Tashkant: IEEE. https://doi.org/10.1109/ICTAI53825.2021.9673233
Deloitte. 2020. "Deloitte IoT Report-2020." https://www2.deloitte.com/content/dam/Deloitte/in/Documents/technology-media-telecommunications/in-tmt-IoT_Theriseoftheconnectedworld-28aug-noexp.pdf
Drescher, Daniel. 2017. *Blockchain Basics: A Non-Technical Introduction in 25 Steps*. https://doi.org/10.1007/978-1-4842-2604-9

Fakhri, Dinan, and Kusprasapta Mutijarsa. 2019. "Secure IoT Communication Using Blockchain Technology." *ISESD 2018 –International Symposium on Electronics and Smart Devices: Smart Devices for Big Data Analytic and Machine Learning.* https://doi.org/10.1109/ISESD.2018.8605485

Grid, Smart, National Smart, Grid Mission, The Nsgm, Subsequently Guidelines, State Level, Project Management, et al. 2019. "Blockchain Applications in Smart Grid-Review and Frameworks." *IEEE Access* 7 (1): 1–5. https://doi.org/10.1109/ACCESS.2019.2920682

Hjalmarsson, Friorik P., Gunnlaugur K. Hreioarsson, Mohammad Hamdaqa, and Gisli Hjalmtysson. 2018. "Blockchain-Based E-Voting System." In *IEEE International Conference on Cloud Computing, CLOUD.* https://doi.org/10.1109/CLOUD.2018.00151

Huckle, Steve, Rituparna Bhattacharya, Martin White, and Natalia Beloff. 2016. "Internet of Things, Blockchain and Shared Economy Applications." *Procedia Computer Science.* https://doi.org/10.1016/j.procs.2016.09.074

Kim, Seong Kyu, and Jun Ho Huh. 2018. "A Study on the Improvement of Smart Grid Security Performance and Blockchain Smart Grid Perspective." *Energies* 11 (8). https://doi.org/10.3390/en11081973

Li, Zhetao, Jiawen Kang, Rong Yu, Dongdong Ye, Qingyong Deng, and Yan Zhang. 2018. "Consortium Blockchain for Secure Energy Trading in Industrial Internet of Things." *IEEE Transactions on Industrial Informatics.* https://doi.org/10.1109/TII.2017.2786307

Lin, Y. P. et al. n.d. "Blockchain: The Evolutionary next Step for ICT e-Agriculture." *Environ. MDPI.* https://doi.org/10.3390/environments4030050

Sadouskaya, Krystsina. 2017. "Adoption of Blockchain Technology in Supply Chain and Logistics" Unpublished Thesis, Bachelor of Business Logistics, South-Eastern Finland University of Applied Sciences, XAMK.

Meity. 2016. "IoT Policy-2016." https://www.meity.gov.in/content/internet-things#:~:text=Department of Electronics and Information, which focuses on following objectives%3A&text=To create an IoT industry,6%25 of global IoT industry, last visited on 20.12.2020 at 1.32 pm.

Mendling, Jan, Ingo Weber, Wil Van Der Aalst, Jan Vom Brocke, Cristina Cabanillas, Florian Daniel, and Søren Debois, et al. 2018. "Blockchains for Business Process Management - Challenges and Opportunities." *ACM Transactions on Management Information Systems.* https://doi.org/10.1145/3183367

Mohamad Noor, Mardiana Binti, and Wan Haslina Hassan. 2019. "Current Research on Internet of Things (IoT) Security: A Survey." *Computer Networks* 148: 283–94. https://doi.org/10.1016/j.comnet.2018.11.025

Mohd Aman, Azana Hafizah, Elaheh Yadegaridehkordi, Zainab Senan Attarbashi, Rosilah Hassan, and Yong Jin Park. 2020. "A Survey on Trend and Classification of Internet of Things Reviews." *IEEE Access* 8: 111763–82. https://doi.org/10.1109/ACCESS.2020.3002932

Mollah, Muhammad Baqer, Jun Zhao, Dusit Niyato, Kwok Yan Lam, Xin Zhang, Amer M.Y.M. Ghias, Leong Hai Koh, and Lei Yang. 2021. "Blockchain for Future Smart Grid: A Comprehensive Survey." *IEEE Internet of Things Journal* 8 (1): 18–43. https://doi.org/10.1109/JIOT.2020.2993601

Nakamoto, Satoshi. 2008. "Bitcoin: A Peer-to-Peer Electronic Cash System." *Whitepaper.* https://bitcoin.org/bitcoin.pdf

Ngu, Anne H., Mario Gutierrez, Vangelis Metsis, Surya Nepal, and Quan Z. Sheng. 2017. "IoT Middleware: A Survey on Issues and Enabling Technologies." *IEEE Internet of Things Journal* 4 (1): 1–20. https://doi.org/10.1109/JIOT.2016.2615180

Ølnes, Svein, Jolien Ubacht, and Marijn Janssen. 2017. "Blockchain in Government: Benefits and Implications of Distributed Ledger Technology for Information Sharing." *Government Information Quarterly*. https://doi.org/10.1016/j.giq.2017.09.007

Oracle.com. 2020. "Definitin of IoT." https://www.oracle.com/internet-of-things/what-is-iot.html

Reyna, Ana, Cristian Martín, Jaime Chen, Enrique Soler, and Manuel Díaz. 2018. "On Blockchain and Its Integration with IoT. Challenges and Opportunities." *Future Generation Computer Systems* 88 (2018): 173–90. https://doi.org/10.1016/j.future.2018.05.046

Sethi, Pallavi, and Smruti R. Sarangi. 2017. "Internet of Things: Architectures, Protocols, and Applications." *Journal of Electrical and Computer Engineering* 2017. https://doi.org/10.1155/2017/9324035

Statistic.com. n.d. "IoT Report-Statistics.Com." https://www.statista.com/statistics/1101442/iot-number-of-connected-devices-worldwide/#:~:text=Internet of Things - active connections worldwide 2015-2025&text=The total installed base of, billion units worldwide by 2025.

Tapscott, Don, and Alex Tapscott. 2016. *Blockchain Revolution*. https://doi.org/10.1515/ngs-2017-0002

Veuger, Jan. 2018. "Trust in a Viable Real Estate Economy with Disruption and Blockchain." *Facilities*. https://doi.org/10.1108/F-11-2017-0106

Wang, Yingli, Jeong Hugh Han, and Paul Beynon-Davies. 2019. "Understanding Blockchain Technology for Future Supply Chains: A Systematic Literature Review and Research Agenda." *Supply Chain Management*. https://doi.org/10.1108/SCM-03-2018-0148

Index

Pages in **bold** refer to tables.

For Product Safety Concerns and Information please contact our EU
representative GPSR@taylorandfrancis.com
Taylor & Francis Verlag GmbH, Kaufingerstraße 24, 80331 München, Germany